Handbook of Intelligent Healthcare Analytics

Scrivener Publishing
100 Cummings Center, Suite 541J
Beverly, MA 01915-6106

Publishers at Scrivener
Martin Scrivener (martin@scrivenerpublishing.com)
Phillip Carmical (pcarmical@scrivenerpublishing.com)

Handbook of Intelligent Healthcare Analytics

Knowledge Engineering with Big Data Analytics

Edited by

A. Jaya

Department of Computer Application, B.S. Abdur Rahman Crescent Institute of Science, Technology, Chennai, India

K. Kalaiselvi

Department of Computer Science, Vels Institute of Science, Technology and Advanced Studies, Chennai, India

Dinesh Goyal

Poornima Institute of Engineering & Technology, Jaipur, India

and

Dhiya AL-Jumeily

Faculty of Engineering and Technology, Liverpool John Moores University, UK

Scrivener Publishing

WILEY

This edition first published 2022 by John Wiley & Sons, Inc., 111 River Street, Hoboken, NJ 07030, USA and Scrivener Publishing LLC, 100 Cummings Center, Suite 541J, Beverly, MA 01915, USA

For more information about Scrivener publications please visit www.scrivenerpublishing.com.

Wiley Global Headquarters
111 River Street, Hoboken, NJ 07030, USA

For details of our global editorial offices, customer services, and more information about Wiley products visit us at www.wiley.com.

Library of Congress Cataloging-in-Publication Data

ISBN 978-1-119-79179-9

Cover image: Pixabay.Com
Cover design by Russell Richardson

Set in size of 11pt and Minion Pro by Manila Typesetting Company, Makati, Philippines

Printed in the USA

10 9 8 7 6 5 4 3 2 1

Contents

Preface

The power of healthcare data analytics is being increasingly used in the industry. With this in mind, we wanted to write a book geared towards those who want to learn more about the techniques used in healthcare analytics for efficient analysis of data. Since data is generally generated in enormous amounts and pumped into data pools, analyzing data patterns can help to ensure a better quality of life for patients. As a result of small amounts of health data from patients suffering from various health issues being collectively pooled, researchers and doctors can find patterns in the statistics, helping them develop new ways of forecasting or diagnosing health issues, and identifying possible ways to improve quality clinical care. Big data analytics supports this research by applying various processes to examine large and varied healthcare data sets. Advanced analytics techniques are used against large data sets to uncover hidden patterns, unknown correlations, market trends, customer preferences, and other useful information. This book covers both the theory and application of the tools, techniques and algorithms for use in big data in healthcare and clinical research. It provides the most recent research findings to derive knowledge using big data analytics, which helps to analyze huge amounts of real-time healthcare data, the analysis of which can provide further insights in terms of procedural, technical, medical, and other types of improvements in healthcare. In addition, this book also explores various sources of personalized healthcare data.

For those who are healthcare researchers, this book reveals the innovative hybrid machine learning and deep learning techniques applied in various healthcare data sets. Since machine learning algorithms play a major role in analyzing the volume, veracity and velocity of big data, the scope of this book focuses on various kinds of machine learning algorithms existing in the areas such as supervised, unsupervised, semi-supervised, and reinforcement learning. It guides readers in implementing the Python environment for machine learning in various application domains. Furthermore, predictive analytics in healthcare is explored, which can help to detect early signs of patient deterioration from the ICU to a general ward, identify at-risk patients in their homes to prevent hospital readmissions, and prevent avoidable downtime of medical equipment.

Also explored in the book are a wide variety of machine learning techniques that can be applied to infer intelligence from the data set and the capabilities of an application. The significance of data sets for various applications is also discussed along with sample case studies. Moreover, the challenges presented by the techniques and budding research avenues necessary to see their further advancement are highlighted.

Patient's healthcare data needs to be protected by organizations in order to prevent data loss through unauthorized access. This data needs to be protected from attacks that can encrypt or destroy data, such as ransomware, as well as those attacks that can modify or corrupt a patient's data. Security is paramount since a lot of devices are connected through the internet of things and serve many healthcare applications, including supporting smart healthcare systems in the management of various diseases such as diabetes, monitoring heart functions, predicting heart failure, etc. Therefore, this book explores the various challenges for smart healthcare, including privacy, confidentiality, authenticity, loss of information, attacks, etc., which create a new burden for providers to maintain compliance with healthcare data security.

In addition to inferring knowledge fusion patterns in healthcare, the book also explores the commercial platforms for healthcare data analytics. The new benefits that healthcare data analytics brings to the table, run analytics and unearth information that could be used in the decision-making of practitioners by providing insights that can be used to make immediate decisions. Also investigated are the new trends and applications of big data analytics for medical science and healthcare. Healthcare professionals, researchers, and practitioners who wish to figure out the core concepts of smart healthcare applications and the innovative methods and technologies used in healthcare will all benefit from this book.

Editors
Dr. A. Jaya
Dr. K. Kalaiselvi*
Dr. Dinesh Goyal
Prof. Dhiya AL-Jumeily
**Corresponding Editor*

1

An Introduction to Knowledge Engineering and Data Analytics

D. Karthika* **and K. Kalaiselvi†**

Department of Computer Applications, Vels Institute of Science, Technology & Advanced Studies (Formerly Vels University), Chennai, Tamil Nadu, India

Abstract

In recent years, the philosophy of Knowledge Engineering has become important. Information engineering is an area of system engineering which meets unclear process demands by emphasizing the development of knowledge in a knowledge-based system and its representation. A broad architecture for knowledge engineering that manages the fragmented modeling and online learning of knowledge from numerous sources of information, non-linear incorporation of fragmented knowledge, and automatic demand-based knowledge navigation. The project aims to provide petabytes in the defined application domains with data and information tools. Knowledge-based engineering (KBE) frameworks are based on the working standards and core features with a special focus on their built-in programming language. This language is the key element of a KBE framework and promotes the development and re-use of the design skills necessary to model complex engineering goods. This facility allows for the automation of the process preparation step of multidisciplinary research (MDA), which is particularly important for this novel. The key types of design rules to be implemented in the implementation of the KBE are listed, and several examples illustrating the significant differences between the KBE and the traditional CAD approaches are presented. This chapter discusses KBE principles and how this technology will facilitate and enable the multidisciplinary optimization (MDO) of the design of complex products. This chapter discusses their reach goes beyond existing CAD structure constraints and other practical parametric and space exploration approaches. There is a discussion of the concept of KBE and its usage in architecture that supports the use of MDO. Finally, this chapter discusses on the key measures and latest trends in the development of KBE.

**Corresponding author*: d.karthi666@gmail.com

†*Corresponding author*: kalairaghu.scs@velsuniv.ac.in

A. Jaya, K. Kalaiselvi, Dinesh Goyal and Dhiya Al-Jumeily (eds.) Handbook of Intelligent Healthcare Analytics: Knowledge Engineering with Big Data Analytics, (1–20) © 2022 Scrivener Publishing LLC

Keywords: Data analytics, knowledge, knowledge engineering, principles, knowledge acquisition

1.1 Introduction

1.1.1 Online Learning and Fragmented Learning Modeling

Applied artificial intelligence (AI) was defined as knowledge engineering [1], with three major scientific questions: knowledge representation, the use of information, and the acquisition of knowledge. In the big data age, the three fundamental problems must evolve with the basic characteristics of the complex and evolving connections between data objects, which are autonomous sources of information. Big data not only rely on domain awareness but also distribute knowledge from numerous information sources. To have knowledge of engineering tools for big data, we need tons of experience. Three primary research issues need to be addressed for the 54-month, RMB 45-million, 15-year Big Data Knowledge Engineering (BigKE) project sponsored by China's Ministry of Science and Technology and several other domestic agencies: 1) online learning and fragmented learning modeling; 2) nonlinear fusion of fragmented information; and 3) multimedia fusion of knowledge. Discussing these topics is the main contribution of this article. With 1), we examine broken information and representation clusters, immersive online content learning with fragmented knowledge, and simulation with spatial and temporal characteristics of evolving knowledge. Question 2) will discuss connections, a modern pattern study, and dynamic integration between skills subsections of fragmented information. The key issues mentioned in Figure 1.1 will be collaborative, context-based computing, information browsing, route discovery, and the enhancement of interactive knowledge adaptation.

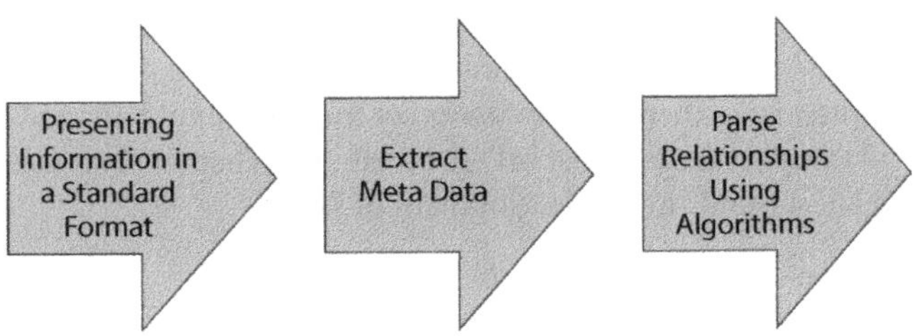

Figure 1.1 Knowledge engineering.

Due to these features of several channels, traditional offline data mining methods cannot stream data since it is important to reform the data. Online learning methods help solve this issue and adapt easily to the drift of the effects of streaming. But typical methods to online learning are explicitly configured for single-source info. Thus, the maintenance of these features concurrently provides great difficulties and opportunities for large-scale data production. Big data starts with global details, tackles dispersed data like data sources and function streams, and integrates diverse understanding from multiple data channels, as well as domain experience in personalized demand-driven knowledge services. In the age of big data, many data sources are usually heterogeneous and independent and require evolving, complex connections among data objects. These qualities are considered by substantial experience. Meanwhile, major suppliers of information provide personalized and in-house demand-driven offerings through the usage of large-scale information technologies [2].

Centered on the characteristics of multiple data sets, the key to a multi-source retrieval of information is fragmented data processing [3]. To create global awareness, local information pieces from individual data points can be merged. Present online learning algorithms often use linear fitting for the retrieval of dispersed knowledge from local data sources [4]. In the case of fragmented knowledge fusion, though, linear fitting is not successful and may even create problems of overfitting. Several studies are ongoing to improve coherence in the processing and interpretation of fragmented knowledge [6], and the advantage of machine learning for large data interpreting is that most samples are efficient, thus eliminating the possibility of over-adjustment at any rate [7]. Big data innovation acquires knowledge mostly from user-produced content, as opposed to traditional information engineering's focused on domain experience, in addition to authoritative sources of knowledge, such as technical knowledge bases. The content created by users provides a new type of database that could be used as a main human information provider as well as to help solve the problem of bottlenecks in traditional knowledge engineering. The information created by the consumer is broad and heterogeneous which leads to storage and indexing complexities [5], and the knowledge base should be able to build and disseminate itself to establish realistic models of data relations. For instance, for a range of reasons, clinical findings in survey samples can be incomplete and unreliable, and preprocessing is needed to improve analytical data [8].

Both skills are essential for the creation of personalized knowledge base tools as the knowledge base should be targeted to the needs of individual users. Huge information reinforces distributed expertise to develop any ability. Big data architecture also requires a customer interface to overcome user-specific problems. With the advent of science and innovations, in the fast-changing knowledge world of today, the nature of global economic growth has changed by introducing more communication models, shorter product life cycles, and a modern product production rate. Knowledge engineering is an AI field that implements systems based on information. Such structures provide computer applications with a broad variety of knowledge, policy, and reasoning mechanisms that provide answers to real-world issues. Difficulties dominated the early years of computer technology. Also, knowledge engineers find that it is a very long and expensive undertaking to obtain appropriate quality knowledge to construct a reliable and usable system. The construction of an expert system was identified as a knowledge learning bottleneck. This helped to gain skills and has been a big area of research in the field of information technology.

The purpose of gathering information is to create strategies and tools that make it as simple and effective as possible to gather and verify a professional's expertise. Experts tend to be critical and busy individuals, and the techniques followed would also minimize the time expended on knowledge collection sessions by each expert. The key form of the knowledge-based approach is an expert procedure, which is intended to mimic an expert's thinking processes. Typical examples of specialist systems include bacterial disease control, mining advice, and electronic circuit design assessment. It currently refers to the planning, administration, and construction of a system centered on expertise. It operates in a broad variety of aspects of computer technology, including data baselines, data collection, advanced networks, decision-making processes, and geographic knowledge systems. This is a big part of software computing. Wisdom engineering also falls into connection with mathematical reasoning as well as a strong concern with cognitive science and social-cognitive engineering, where intelligence is generated by socio-cognitive aggregates (mainly human beings) and structured according to the way humans thought and logic operates. Since then, information engineering has been an essential technology for knowledge incorporation. In the end, the exponentially increasing World Wide Web generates a growing market for improved usage of knowledge and technological advancement.

1.2 Knowledge and Knowledge Engineering

1.2.1 Knowledge

Knowledge is characterized as abilities, (i) which the person gains in terms of practice or learning; theoretical or practical knowledge of a subject; (ii) what is known inside or as a whole; facts and information; or (iii) knowing or acquainted with the experience of life or situation. This knowledge is defined accordingly. The retrieval of information requires complex cognitive processes including memory, understanding, connectivity, association, and reasoning. Knowledge of a subject and its capacity for usage for a specific purpose are also used to suggest trustworthy comprehension. Information may be divided into two forms of knowledge: tacit and clear. Tacit knowledge is the awareness that people have and yet cannot get. Tacit knowledge is more relevant since it gives people, locations, feelings, and memories a framework. Efficient tacit information transfer typically requires intensive intimate correspondence and trust. The tacit understanding is not easily shared. Still, consciousness comprises patterns and culture that we still cannot comprehend. On the other hand, information, which is easy to articulate, is called explicit knowledge. Coding or codification is the tool used to translate tacit facts into specific details. The awareness expressed, codified, and stored in such media is explicitly facts. Explicit information. The most common simple knowledge sources are guides, manuals, and protocols. Audio-visual awareness may also be an example of overt intelligence, which is based on the externalization of human skills, motivations, and knowledge.

1.2.2 Knowledge Engineering

Edward Feigenbaum and Pamela McCorduck created Knowledge Engineering in 1983: To address difficult issues that typically need a great deal of human experience in the fields of engineering, knowledge engineering needs the integration of information of computer systems. In engineering, design information is an essential aspect. If the information is collected and held in the knowledge base, important cost and output gains may be accomplished. In a range of fields, information base content may be used as to reuse information in other ways for diverse goals, to employ knowledge to create smart systems capable of carrying out complicated design work. We shall disseminate knowledge to other individuals within an organization.

While the advantages of information capture and usage are obvious, it has long been known in the AI world that knowledge is challenging to access from specialists. Second, the specialists do not remember and describe "tacit knowledge" effectively, and this operates subconsciously and, where it is not impossible, is difficult to overcome the problems arising from several subject matters that they learn. To elaborate, they have to know what it is called. Third, there are various prospects and points of view which include aggregation to provide a coherent view. Last, professionals create abstract concepts and shortcuts for which they cannot communicate. The area of information technology was created some 25 years ago to address such problems, and the role of the knowledge engineer was born. Since then, computer engineers have developed a variety of principles, methods, and tools that have improved the acquisition, use, and implementation of knowledge considerably.

1.3 Knowledge Engineering as a Modelling Process

There is also a consensus that the KBS construction approach may be used as a modeling operation. It is not intended to construct a cognitively appropriate model, but to build a model that offers similar results for problem-solving problems in the area of concern as seen in Figure 1.2.

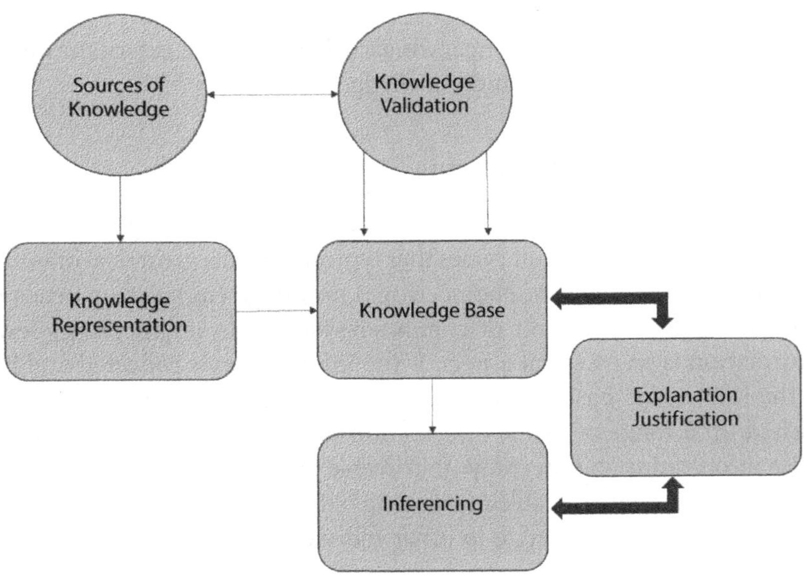

Figure 1.2 Knowledge as modelling process.

Building a KBS requires building a programming model to acquire problem-solving capabilities like those of a domain specialist. This material is not directly available; therefore, it needs to be created and arranged.

1.4 Tools

Intelligence engineers make the more efficient and less bogus use of dedicated computational tools for the acquisition, simulation, and handling of intelligence. PC PACK is a versatile compilation of this programmed, commercially available as a package of knowledge technology tools that are designed to be tested on a wide range of projects. The aim is to consider the key characteristics of the domain. The method simulates how anyone should label a text page with different colors such as green for suggestions and yellow for attributes. The labeled text would immediately be placed in the PCPACK database to be applied to all other resources when the user has highlighted a document. The MOKA and Popular KADS Methodologies are supported by the CFPACK. It is also fully compliant with information engineering approaches and techniques. The CFPACK is a software suite that includes the following

(i) Protocol tool: It enables the discovery, recognition, and definition of interview transcripts, conclusions, and documentation that may be included in the knowledge base.
(ii) Ladder Tool: This allows hierarchies of knowledge elements such as meanings, features, procedures, and specifications to be developed.
(iii) Chart Tool: This allows users to build mobile networks of connections between data elements, such as process maps, maps of ideas, and cutting-edge diagrams.
(iv) Matrix Tool: This allows grids that show the connection and attributes of the elements to be developed and edited.
(v) Annotation tools: This facilitates the creation of sophisticated HTML annotations, with links to other sites and other knowledge templates automatically generated in the CFPACK.
(vi) Tool publisher: This allows the creation from a knowledge base of a website or some other information resource using a model-driven approach to optimize re-usability. MOKA, CommonKADS, and the 47-step protocol provide approaches to run a project from beginning to completion, as well as maintaining best practice.

1.5 What are KBSs?

A knowledge-based framework is a system that utilizes AI tools in problem-solving systems to assist human decision-making, understanding, and intervention.

There are two core components of the KBSs:

- Information base (consists of a collection of details and a set of laws, structures, or procedures).
- Inference engine (Responsible for the extension of the information base to the issue at hand).

In contrast to human expertise, there are pros and cons to utilizing KBSs.

1.5.1 What is KBE?

It starts with a discussion about what KBE is in this book and begins with a simple definition: Knowledge-based Engineering (KBE) uses the knowledge of product and operation, which was collected and retained in specific software applications, to enable its direct usage and reuse in the development of new products and variants. KBE's implementation consists of applying a specific class of computing tools, called the KBE systems, which enable engineers to acquire and reuse engineering knowledge using methods and methodologies. The name of the KBE architecture is derived from the mixture of KBS and engineering, which are one of the major outcomes of AI. In the 1970s, KBE systems demonstrate the advancement of the KBS by applying the special engineering industry requirements. KBE systems combine KBS rule-based logic technologies with engineering data analysis and geometry like CAD.

For these reasons, a traditional KBE architecture provides the user with a programming language that is generally object-oriented and one (something more) embedded or closely connected CAD engine. The vocabulary of programming enables the user to capture and reuse rules and procedures in engineering, while the object-oriented approach of design corresponds well with how engineers see the world: systems abstract assets of objects, defined by parameters and behaviors, linked by relationships. Access and management by the programming language of the CAD engine meet the geometry handling requirements characteristic of the engineering architecture. The MIT AI laboratories and the Computer Vision CAD Group developed the first commercially available KBE system named ICAD1984 (now PTC). Fortunately, this asks the first inquiry you will hear in Figure 1.3 about KBE.

It will be useful at this stage to quickly clarify what we refer to as information and how we use this term to describe concepts other than data and evidence. Both terms are sometimes misused in the traditional spoken language; truth and knowledge are often interchangeably used. The hierarchy of data and intellect (and knowledge) is the subject of long-term disputes between epistemologists and IT experts. Since this subject goes well beyond the scope of this chapter, this is our definition of data. Data are objects that have no meaning before they are put in form like symbols, statistics, digits, and indications. The information consists of important processed data. The context in which the data are collected gives it meaning, importance, and intent. Human and electronic information can be collected, shared, and processed. The knowledge is encrypted by code, normally organized in a structure or size, and stored in hard or soft media to accomplish this. Awareness is the condition of information and awareness processing, which requires the chance to act.

New information may be produced as a result of the application of knowledge. The IGES file with a geometrical definition of the surface as a piece of information is an example of this. IGES files will be encoded with numbers and symbols (i.e., data) and will only provide knowledge useful if they understand the meaning (i.e., the fact they are the data of an

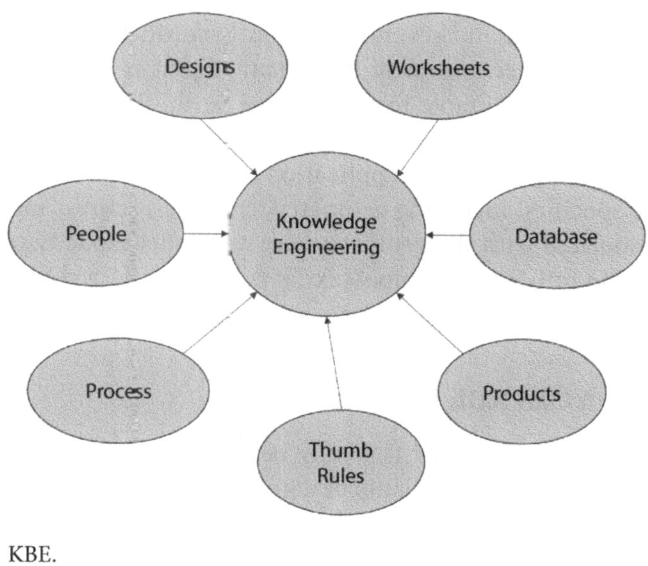

Figure 1.3 KBE.

IGES file). The information which can be collected with a KBE method is regarded as a simple example of the algorithm that reads such an IGES file, reconstructs a specified surface model, intersects it with a floor plane, and, if the crossroad is non-zero, calculates the length of the corresponding curve. It is also sensible to ask why geometry varies from the standard CAD paradigm and enables the creation and manipulation of geometry. Owing to the varying scopes of these systems, the differences are important. Digitized drawing systems, which allow programmers to catch their ideas have been designed to create CAD systems. They build and store the results using the CAD framework's geometry simulation functions. A set of points, lines, planes, and solids with reference and note are an almost all-inclusive link to the structure. These data provide enough information for the creation of a system that can be used to build a specification by production engineers. In doing so, creators store the specifics of "what," but they retain the "how" and "why." In a sense, the CAD approach can be considered a system "posterior," because before it can be moved to the system it is necessary to know what the principle is like. It can be argued that CAD is geometry or drawing/drawing engineering to distinguish this approach from KBE.

KBE-supported technology is different. Technology Instead of shifting "what," engineering experts are trying to move "how" and "why," encapsulating in the KBE process knowledge and thinking instead of geometry in the CAD framework. Not only does this work by manipulating geometric structures, but programming is needed rather than writing. The "how" and "why" in engineering are in some cases used in textbooks, databases, tip sheets, and several other outlets. Much of the knowledge is held up by engineers, mostly in a manner that is strongly compiled and not specifically suitable for translation to the KBE procedure. This experience should be sufficiently transparent to create a KBE program to be codified into a software application capable of producing all kinds of product specifics, including geometry templates, scores, and data that are not associated with the geometry. Because of its capacity to generate a specification rather than simply text, it is widely referred to as a generative model.

1.5.2 When Can KBE Be Used?

How easy is the use of KBE? It is only necessary to rapidly generate different configurations and variants of a given variable. In certain practical cases, this is not essential, so it may be a wrong expenditure to try to clarify details and program a KBE system. One-off prototypes and designs that need not be optimized or prototype versions for space travel are usually outside the scope of KBE implementation.

It explores the design field through the development of different design variants within the product family and evaluates its performance compared to previously tested versions with the multidisciplinary optimization (MDO) application. KBE will assist in many respects in this case.

It enables stable product parametric models to be generated which make topology changes and the freedom to make adaptation changes usually impossible for those built with a conventional CAD framework. This is important when considering broad variations like those which occur when a yacht manufacturer decides to accept one or more hull settings. It supports the integration into MDO through automation of the generation of necessary disciplinary abstractions of heterogeneous sets of analytical methods (low and high fidelity, in-house and off-shelf). It removes the optimizer from the challenge of managing the spatial integration constraints that generative models should guarantee. This is essential because the user does not need to specify constraints on configuration variables or restrictions to avoid intersection of two elements; or because a certain structural element does not need to remain beyond the same outer mold line; or because, during optimization, two products are expected to have a certain relative position apart.

KBE generative models can be a secret in producing MDO systems that are not multidisciplinary in return for adherence to science and that can handle complex problems reflecting actual industrial circumstances. We discuss the different models of current MDO systems and compare them to advanced KBE implementations in the next section to clarify this claim.

The third set of MDO structure implementation is available to overcome the weaknesses of the two approaches described earlier in this section by introducing generative models into the system. One advantage of this approach is that the exact geometry representations normally used for the use of high faithfulness analysis instruments may serve as a basis for the disciplinary study. It is therefore well adapted to the geometric nuances that are not included in a few general criteria of modern products. This geometry depiction is generated following individual tools of multidisciplinary system analyzers (BB SA) along with others, usually not schematic, product abstractions, and are systematically updated after each optimization loop. These MDO systems can fully resolve multidisciplinary cases without penalizing the degree of faithfulness and can contribute to the early stages of the design phase by addressing substantial changes in the shape and topology. They may also support a more sophisticated modeling method in which complex and accurate geometric models for high fidelity analysis are needed. These functions allow the early use of highly reliable testing approaches to be implemented with novel prototypes that do not have correct or unavailable semi-empirical and predictive technologies. The product modeling scheme which is the key feature of the MDO system is undermined by this approach.

1.5.3 CAD or KBE?

It is a mistake to know whether KBE is greater than CAD or vice versa. One is in the whole sense no bigger than the other, and we argue here that KBE should replace CAD. In certain circumstances, the KBE programming process is more suitable than the interactive application of the CAD platform, given that MDO supports one of the interests of this novel. This chapter is beyond the scope of a general debate about the suitability of one option for the next. The suggestions are as follows:

Where the focus is only on geometry development and manipulation; where considerations such as direct interaction with geometric models are important, graphical rendering and inspection are essential; if uniform, aesthetic, and heuristic design are the guides behind modeling, rather than engineering laws.

When it comes to design purposes, vocabulary is required instead of design results. The programming method of KBE systems offers the best solution in this case. Although CAD systems are committed to better documenting the results of the human design phase, KBE systems are designed to report the design procedure (i.e., the purpose of the design) and not just the results.

- A language is needed to promote automation while preserving continuity. Whenever the generative model is "played." The same protocol (i.e., the same rules and logic processes are applied) is constantly repeated with different appropriate inputs regardless of which operator and of how many replays. In some engineering design cases, one of which is the optimization of design, an obstacle to automation is placed in the loop (except process supervision).
- A vocabulary offers a competitive advantage when it comes to the ease of interaction with external modeling and simulation applications. Usually, both CAD and KBE systems link (to each other and) through standard data interchange formats including IGES and Move. In times of ad hoc interchange files that are dependent on ASCIIs, the most useful

approach to dedicated writers and parser production is full-function language programming. Also, the KBE system can detect and largely simplify these processes where the tool to be connected is required by complicated and knowledge-intensive pre-processing operations to schedule the input.

- Where there is an aesthetic facet of the architecture and details are produced, but at the same time an multidisciplinary research (MDA) and an optimization approach are used in the design and size of a given product, the best possible solution is provided by combined applications of CAD and KBE. In this case, the CAD process geometry would become the KBE application's feedback. This implementation will support the complex MDA structure and return the material (partially or fully) to the CAD system, where comprehensive work can take place more immersive.

At the end of the day, the heuristic and non-respecting, geometric or non-repeatable, one-off, and repetitive aspects of the design phase co-exist and are interlinked: Both CAD and KBE can contribute to this step, which must be the focus of both the creators of CADs and KBE's smooth integration.

1.6 Guided Random Search and Network Techniques

There are some methods designed to find suitable designs using techniques that avoid the use of the pursuit of gradients or almost gradients. A system that either uses random variations in design variables or avoids direct variations in design variables by the use of learning networks supplements the use of directional searches. We have selected the genetic algorithm (GA) as a representative in the first category, RAT (GRS), and we have selected the Artificial NERN (ANN) as a regular illustration in the second category, network-based learning methods.

1.6.1 Guide Random Search Techniques

Without stringent enumeration, guide random search technique (GRST) methods are attempting to seek a whole feasible design field and, in principle, have a global optimum. If this optimum is not suitable internationally, then traditional exploration procedures provide no underlying means to move away from the optimal local field to continue the search for the optimum global setting. However, it should be borne in mind that there can be no confidence that a GRST algorithm can solve a complex design problem globally and, as mentioned elsewhere in the book, no answer can be challenged to ensure that a global solution has been discovered. The methods, though, are rigorous and usually will include a solution that significantly improves on any initial concept put forward by the design team.

GRST methods can deal with problems with the architecture of undistinguished functions and with many local improvements. The ability to deal with non-differentiable functions makes it easy to address problems related to distinct design variables, which are common aspects of structural design. Many GRST methods are well adapted for parallel processing, in particular the evolutionary algorithms mentioned in the next section. The number of implementing variables would allow concurrent processing to be used to respond within a reasonable period if every MDO problem is resolved by the GRST method rather than trivial.

Evolutionary algorithms are a subset of GRST techniques that employ very special approaches that focus on evolutionary concepts seen in nature. This approach also exposes some designs to spontaneous variations and offers anyone with a practical advantage an increased opportunity to produce "spring" designs. There are a number and different methods to solving complicated optimization problems using the same straightforward probabilistic technique. We are concerned with GA, which may be the most popular evolutionary form of algorithms in-process libraries or in commercial MDO systems.

1.7 Genetic Algorithms

GA is a family of computational methods based upon the evolutionary theory of Darwinian/Russel Wallace, used to solve general problems through optimization. Caution is in order at this point! The use of biology-inspired terms in the GA system demonstrates genetics, which engineers considered several years ago as an optimization process. Since then, substantial progress in biology has shown that true genetic growth is many compounders but only a convenient metaphor remains the term "genetics" used in this book. Instead of looking from one design point to another in search of improved design, the GA shifts to a second population with a reduced meaning for the restricted purpose feature, from an existing set of design points called group. A replication and mutation process on the computer model of the plant design points achieves progress from generation to generation.

Although, the design team would like to see data from prior designs or preliminary studies in the application for engineering design from the original collections of design points. The question is not distinct from those used in the application of search methods. The objective design function and constraints at each design point of the population must be estimated. The experiments are independent such that the parallel treatment can be included. We now turn to the definition stage of data representation.

1.7.1 Design Point Data Structure

The architecture variables describing a specific design point are described by binary numbers and linked to a 0.1-bit string. Suppose, for example, that we create a solid cone with height and base diameter as design variables and then begin a design point with 4 m of height with a base diameter of 3 m (4, 3). This coordinate is a binary variant (100, 011) which is a concatenated string (100011). This string is named the chromosome of the structure that reflects its roots in genetics, and the individual sections are gene analogs. Therefore, there are multiple chromosomes in the population, equivalent to the number of design points that we intend to use in the field of design. Also, the chromosome number of digit slots (bits) should be sufficient to fulfill the software and the degree of precision of the various specification variable values.

1.7.2 Fitness Function

The problem with optimization was now reconfigured to a set of chromosomes which represent a century of designs with a special design for each chromosome. The AG encourages a "fittest survival" policy, which would eventually transfer chromosomes through generations before an optimum arrangement is found. This includes a chromosome recognizing or excluding process such that we monitor for the fitness to be included in the next generation of designs for the same chromosome. This is achieved by using a health function which is a metric of goodness common to all chromosome-based conception points with a separate meaning for each point. Why a penalty for a limitation violation is included in the exercise feature later is discussed.

Essentially, systems are designed to help design the next generation of chromosomes in the community and their well-being. To use the above example, the assessment approach is simple and better, but it can be pointed out that it reflects the mechanism of selection and that some might be used in business programs.

1.7.3 Constraints

The limitation will not be offset in the case of a GA by ensuring that the search algorithm does not traverse a non-feasible field by directly inserting the method limitations into the search direction. In the case of GA, limits are handled either using sanctions or by excluding ineffective chromosomes. This second approach should be implemented with care to prevent solutions from being rejected at the edge of a feasible field, where the solution is controlled by active limitations. However, side restrictions may also be added, for example, minimum gauges.

1.7.4 Hybrid Algorithms

GA has a reputation for being durable, meaning that it can usually deliver an overhaul of the initial design. But, for a particular design domain, they could not be the correct solution. A hybridization approach should be used to try to make the most of all the worlds to maximize their convergence rates in situations when more information is available and is not generated randomly (for example, where gradient details are available). Typically, a hill-climbing algorithm is used in the genetic code to allow everyone in the group to climb on the local hill. The system also encourages each offspring to climb a local hill, created at the breeding stage.

While the simple convergence of GA search algorithms associated with hybrid approaches is the common meaning, this term can also be used for a less straightforward hybridization, where GA and gradient search methods are employed in sequence. Use the GA to reverse the optimizing problem and then deliver the output to the conventional optimizer from this first stage to complete the operation. The first design approach would design the right initial layout for GA before moving on to the full design level, where the second stage optimization phase will begin with the use of classical search techniques. This can also be found in MDO implementations.

1.7.5 Considerations When Using a GA

- GA has the benefit of being able to handle a full variety of variability in a design. For example, in the preliminary design of an aircraft, the motor number and position must not be defined either in the wing configuration (i.e., monoplane, medium, mid-fuselage, and high), such that a selection algorithm can be used for the best combination.

- While GA tries to find the whole design space to find a global optimum, no guarantee exists that an algorithm finds this point and there are no certain parameters that show that the global optimum is achieved if fulfilled. Although the multi-minimum problems are similarly insufficient to deal with all alternate algorithms, the GA does not suffer. On the opposite, GA's potential not just to create an optimal design but an enhanced and feasible design population plays a major role for engineers since it allows them to make final design decisions by judgmental requirements that are outside of the formal framework of optimization. When a GA is regarded in an MDO procedure, the size of the design issue is significant. GA is suitable for parallel processing machines since the number of processors that allow for the calculation of multiple analyses can be simultaneously analyzed by all the people.
- A downside to these codes is that, after new structural modifications have been required, the engineer cannot provide any detail that can be used by the engineer. Taking account of these reasons, we propose that in the early phases of an MDO application in which the issue is relatively limited but the uncertainty is comparatively high, the GA has a valuable function to play.

1.7.6 Alternative to Genetic-Inspired Creation of Children

It emphasized its intrinsic malleability at various stages of definition with the introduction of GA. To explain this point further, we suggest an approach that differs considerably from the biological inspiration of GA and takes a method of child development that differs somewhat from gene exchange and crossover.

First, let us use the n-dimensional design area defined by a cartesian coordinate for two parents, created as described above and represented by points A and B. Now, draw a line between A and B and the location between A and B at that stage, according to the normal Gaussian middle point distribution, i.e., the midpoint is the most likely place. To begin with, construct a new n-dimensional coordinate system of Cartesian origin at O, whose coordinate axes parallel to the original system, extending from minus to further infinity. With a standard Gaussian distribution based on

O along each axis, n coordinates can be produced that define the design point of children A and B. This stage might be dropped off the AB side. This will yield more than one child for a couple of parents and allows multiple likelihood distributions rather than gaussian.

1.7.7 Alternatives to GA

In research papers and books on this topic, there is a wide range of GRST approaches. A lot of them are still under study and are still not sophisticated in making them attractive for developers or engineers of commercial devices that create an internal MDO structure. However, there are at least a few approaches in the "available" lists of methods utilized in publicity programs that are worth mentioning.

One is the principle of "simulated ringing" that allows atoms to form large crystals at an energy stage, at a minimum, given the differences present on the physical search pathway. The other approach is the ringing for steel and other metals. The search algorithm involves a way of randomly extracting inputs from neighboring designs and merge them according to a given set of laws as applied to the optimization issue. The key task of the virtual anneal is to have a small but not nil chance of flipping from an improved to a lower configuration. This makes it possible to break the local minimum trap at the cost of temporary design inferiority and, in the long run, pays off by going onto a new quest route that can optimize the probability of achieving an optimum overall.

The "particle swarm optimization" envisages designs in production rooms as a swarm of entities (the swarm of bees was inspired). In line with simple mathematical formulae, the swarm is moved into the design space to draw the position and velocity of all particles in the swarm to incorporate local and global information.

1.7.8 Closing Remarks for GA

The method class referred to in this section is still being created, based on its simplicity and compatibility with parallel technology. The manufacturing is available for the effort to produce innovative GA models. Changes may be made to allow the number of design points in the next generation to vary adaptively; control the distribution of these points to get them closer together to bring them closer to the points that have become the healthiest in the previous generation; and parent three-fold rather than parent peers, or, ultimately, a group of parents to produce in children.

1.8 Artificial Neural Networks

Let us switch now to artificial neural network (ANN) for memory. Again, it will have an overview of how these works are done; let the reader study alternative in-depth perspectives, and propose Raul Rojas' excellent text (1996). The section uses one aspect of a network and learning method that explains how business application vendors use a network, other network types, and learning processes. It is defined as having thousands of neurons, each of which is associated with more than a thousand other neurons in a rather simplified human brain model. Each neuron receives an electrical signal and transmits it to other brain network neurons. The neuron receives a signal from its associated neurons, and does not transmit the signal to other neurons immediately but waits until the concentration of the signal energy reaches level. In general, the brain learns by changing the amount of these connections and the signal thresholds.

ANN is constructed along identical lines except that node collections execute the location of neurons connected in the network, where a three-layer network is shown for ease. It has several layers defined as the input, hidden layers, and output of the neurons forming the interconnection network. The input neurons are the first information to deal with the problem, and the results and the solutions are in the output neurons. The hidden layer is an input and output layer network link. The diagram shows only one hidden layer, and we adhere for simplicity to one layer in this section, while there may be several such layers in some implementations.

The arrows in the image show the link between the neurons input n, the k hidden neurons, and the neurons in output m. Wisdom is seen as being fed on the left to right and is regarded as a feeding process. We undergo a back-breeding process in later portions. The way the network functions by its neurons has two major characteristics:

Neurons receive feedback from other neurons, however, the neuron also "flies" while the added neuron knowledge is of vital importance (a firing threshold). Information passing from one neuron to another is weighed by a variable that does not have a value affected by data within either neuron. The network is used to efficiently define alternatives to the issue by manipulating weighting variables.

1.9 Conclusion

One approach is to look for assistance using KBE approaches in solving the previous problem. KBE spans a wide range of engineering technologies and has tools that can capture and reuse the product and process knowledge to deliver individual users' or MDO environment details and data. A good connection between rule-based reasoning, object-oriented modeling, and geometric modeling inside the KBE frame makes certain measures in the MDO process easy to grab and automate. As seen in this section, the MDO approach includes direct cooperation between some testing, optimization, and other modules and codes that work along with several variables of nature for the enhanced product. Capturing and reusing knowledge facilities will help to deliver sustainable parametric models by adapting the data to the various discipline data models and allow adjustments to flow across them. A major benefit from this capability allows the integration of homogeneous sets across a range of simulation tools, such that data and concept information can be transmitted smoothly from low- to high-trust research models as the design progresses over time. It helps to integrate internally and acquired technical capabilities into the MDO environment and helps them operate in parallel with the evolution of data and architecture expertise and the acquisition of increasingly complicated data structures combined with dynamic data comfort.

References

1. Chan, P.K.M., A New Methodology for the Development of Simulation Workflows. Moving Beyond MOKA, Master of Science thesis, TU Delft, Delft, 2013.
2. Cooper, D.J. and Smith, D.F., A Timely Knowledge-Based Engineering Platform for Collaborative Engineering and Multidisciplinary Optimization of Robust Affordable Systems. *International Lisp Conference 2005*, Stanford University, Stanford, 2005.
3. Cottrell, J.A., Hughes, T.J.R., Basilevs, Y., *Isogeometric Analysis: Towards Integration of CAD and FEA*, John Wiley & Sons Inc, Chichester, 2009.
4. Graham, P., ANSI Common Lisp, Englewood Cliffs, NJ, Prentice Hall, 107, 384–389, 1995.
5. La Rocca, G., Knowledge Based Engineering: Between AI and CAD. Review of a Language Based Technology to Support Engineering Design. *Adv. Eng. Inform.*, *26*, 2, 159–179, 2012.
6. Lovett, J., Ingram, A., Bancroft, C.N., Knowledge Based Engineering for SMEs: A Methodology. *J. Mater. Process. Technol.*, *107*, 384–389, 2000.
7. Mcgoey, P. J., A Hitch-hikers Guide to: Knowledge-Based Engineering in Aerospace (& other Industries). INCOSE Enchantment Chapter, 2011. Available at: http://www.incose.org/. 1, 117–121.
8. Milton, N., *Knowledge Technologies*, Polimetrica, Monza, 2008.

2

A Framework for Big Data Knowledge Engineering

Devi T.[1]* and Ramachandran A.[2]

[1]*Department of Computer Science & Engineering, Saveetha School of Engineering, Saveetha Institute of Medical and Technical Sciences, Saveetha University, Chennai, India*

[2]*Department of Computer Science & Engineering, B.S. Abdur Rahman Crescent Institute of Science and Technology, Vandalur, Chennai, India*

Abstract

Analytics and analysis from a massive database using various approaches and techniques are experimented, and ongoing research brings its main focus toward the domain such as big data. Economic growth and technological growth combined with its data production are also highlighted in big data approaches. Data are analyzed from social media, online stock markets, healthcare data, etc., which can be collaborated with artificial intelligence by developing the automated learning algorithms and development in cloud computing as well. Data can be either discrete or continuous, which are independent of the various processes for understanding the decision-making that relies on knowledge engineering. The proposed work converges in transforming the observed sequential data analysis from weather forecasting dataset. These systems can perform the cognitive task in improving the performance along with integrity of data using the enhanced framework. The prediction of natural disasters is a challenge for customers accessing forecast data, since fluctuations in data occur frequently, which fail to update the localization, that are identified as sensor latitude and longitude that are updated as a sequence on regular intervals from various directions. These four hidden states are the features that differentiate the probability of distributions for calculating the best cognitive tasks. Improved Bayesian Hidden Markov Frameworks (IBHMFs) have been proposed to identify the exact flow of state and detect the high congestion, which leads to earthquakes, tremor, etc. As the data from the analysis are unsupervised and features are converted to discrete and sequential data (independent

**Corresponding author*: devi.janu@gmail.com

A. Jaya, K. Kalaiselvi, Dinesh Goyal and Dhiya Al-Jumeily (eds.) Handbook of Intelligent Healthcare Analytics: Knowledge Engineering with Big Data Analytics, (21–38) © 2022 Scrivener Publishing LLC

variables), IBHMF can utilize in increasing the performance and produce the accuracy results in state estimation.

Keywords: Artificial intelligence, big data, Improved Bayesian Hidden Markov Frameworks (IBHMF), hidden state, knowledge engineering, weather forecasting

2.1 Introduction

Catastrophic damage has been caused by natural hazards along with loss in a socioeconomic way, thereby depicting the increase in trend. Several disasters pose challenges to officials working in the disaster management field. These challenges may include resources unavailability and limited workforce, and these limitations force them from changing the policies toward managing the disasters [1].

The amount of data generated is huge in size including the real along with the simulation data. These data can be used in supporting disaster management. The technological advancement like data generated from social media as well as remote sensing is huge in size and also is real data. In certain times, these real data are scarce and lead us to usage of simulation data. Several computational models can be used in generation of simulation data that can be used in estimation of impact produced due to disaster. It is much necessary to acquire big data, manage it, and process within a short time span for effective management of disaster irrespective of the type of data being used. For this reason, artificial intelligence (AI) methods can be employed for analyzing the huge volume of data for extracting useful information. Such methods have gained popularity while they support the process of making decisions in case of disaster management [5, 6].

The usage of big data for managing disasters is still evolving. The main challenge for a scientist in today's technological world is handling the huge volume of information that is being generated during disaster time. When the volume of data is being increased, traditional systems (Figure 2.1) employed for storing and data processing are not able to perform better. The factors affecting their working include scalability along with data availability [8].

The storage systems at the present time are diverse in nature, and when it comes to collaboration, they provide much less scope. This leads to the necessity of methods that can be employed for integration, aggregation, and visualization of data. Also, the decisions taken need to be optimized as their quality is based on the available data quality. It is much necessary to

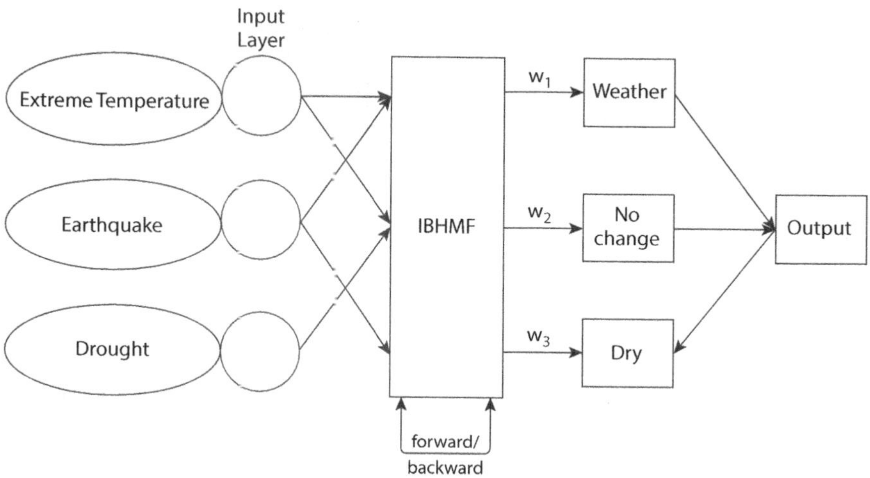

Figure 2.1 Traditional Bayesian Neural Network disaster prediction from the dataset.

organize data followed by storage and analysis of disaster data for further investigation [11].

2.1.1 Knowledge Engineering in AI and Its Techniques

AI can be useful for disaster management, wherein it can be further classified as the following categories: supervised model, unsupervised model, deep learning, and reinforcement learning along with optimization.

2.1.1.1 Supervised Model

Training is done using the human input on the data that is pre-existing in case of algorithms, representing the supervised model. Such models can represent a function by utilizing the methods such as classification for predicting the output value. This is due to the training data that is labeled and also the input as well as output pairs. The main advantages of these models include extraction of information and recognition of objects, patterns, and speech [17].

2.1.1.2 Unsupervised Model

Statistical methods are being used for extraction of hidden structure from the unlabeled data in unsupervised models. Human input is also absent here, and they are utilized in detection of abnormal data along with reduction of dimension of the data. Its applications include clustering as well problems such as data aggregation. The unlabeled data can be partitioned as several groups depending on the similarity feature. This recognition of patterns can be done using the clustering algorithms. On the other hand, the algorithms employed for reduction of dimension like PCA (Principal Component Analysis) play a significant role in data complexity reduction, which, in turn, helps in avoiding overfitting [18].

2.1.1.3 Deep Learning

Input data can be used for extracting the features by using the multiple layers and such algorithm classes constitute deep learning [20]. The learning performance is improved with a wide scope of application [3, 10]. The main disadvantage of using the algorithms in deep learning is that they take more time for training the data. These algorithms can be employed for solving problems such as assessment of damage, detecting motion, recognizing facial expressions, prediction of transportation, and also supports the management of disaster using natural language processing (NLP). Both Recursive and Recurrent Neural Networks can be used for applying along with NLP. At the same time, Convolutional Neural Networks can be used in image classification, text processing, and computer vision [19].

2.1.1.4 Deep Reinforcement Learning

Problems that are goal-oriented can be solved by taking decisions in the order of a sequence is performed by algorithms in reinforcement learning [11]. Here, learning can be carried out based on the reinforcement series, and modeling is based on decision processes such as Markov. These algorithms play a major role in finding solutions for the problems where the decisions to be made need to be in sequence and their applications extend to management of resources and controlling traffic light. Major problem in such algorithms is the preparation of the environment for training the data and performing tasks based on them. When a deep neural network (NN) is combined with reinforcement learning, it is referred to as deep reinforcement learning. The major goal of such algorithms is creation of software agents who learn themselves for establishing the policies in a successful manner. Additionally, they help in attaining the rewards pending in long term. In terms of performance, deep reinforcement learning solves the problems including sequential tasks that are complex such as robotics and computer vision. The need of more training data along with time for training for attaining better performance is a major drawback of deep reinforcement learning as they are computationally expensive. Data can be also be secured with usage of grids [6, 21, 24].

2.1.1.5 Optimization

Optimization helps in identifying the suitable model using objective function and is considered to be one of the important methods in case of disaster management. Several optimization methods are available, and based on their performance, they can be recommended for further applications [24].

2.1.2 Disaster Management

Disaster management starts with the phase of mitigation where activities are related to prevention of emergencies in the future including the consequences. It is the initial phase in managing the disasters. The activities related to mitigation includes enforcement of standards, providing hospital care along with shelters, and providing education for the public. This awareness can help people and also the stakeholders to deal with hazards and strategies for mitigation. The next phase is the preparedness, where the disaster is going to happen. These include the activities that can help in saving the lives of people along with helping the rescue operations like food stocking, providing emergency data, and evacuations. Following this phase comes the response stage. This stage takes place mainly at the time of disaster: evacuation of areas that are being threatened, fire fighting works, efforts such as search along with rescue, and management of shelter including assistance using humanitarians. When the disaster has occurred, the recovery stage deals with repair and the efforts related to reconstruction for returning the life to a normal level. Actions in case of recovery are cleaning the debris, assessment of damage, and reconstruction of infrastructure. It also includes assistance related to finance from the agencies of government or companies that provide insurance.

The objective of proposed work is as follows:

i) To develop the enhanced framework, which can perform the cognitive tasks to improve the performance of the weather forecasting;

ii) To reduce the fluctuations in data update from the entity column that can avoid the localization issues to update the directions;
iii) Using Improved Bayesian Hidden Markov Frameworks (IBHMFs), the performance can be predicted from the independent variable as sequence time series analysis data.

The chapter is structure as follows: Section 2.2 presents the big data in knowledge engineering. Section 2.3 presents the proposed system. Section 2.4 discusses the results obtained. Section 2.5 concludes the chapter.

2.2 Big Data in Knowledge Engineering

Information plays a major part in big data rules; the technologies for intervention can accumulate the information according to the needs. There are many computations mathematically performed for streaming data that are used as per requirements. The Internet provides many resources to learn and a platform for online learners. In such cases, big data that has multiple source providing ways can develop applications according to customers' expectations. There are many functions used for online learning techniques such as fragmented knowledge transfer that relies on big data models. Also, mining can extract the information based on demand that can create a framework that can be used according to the knowledge of engineering.

The software from IBM that was built with NN [13] optimized resources are also having certain rules to be achieved. They are applied to domains such as machine learning, computer vision, and NLP. Online learning not only used a fragmented knowledge method but also used the translation for various language learners. Those accessible translations are associated with libraries that support the online providers to reach the reader or learner in a fast or rapid manner. Mobile applications are also interconnected for knowledge services using big data learning procedures that can interpret the various users mode. Intelligent systems in AI such as robotics, automatic sensors, and latest technologies are all designed with the help of big data framework. There are four generations in knowledge engineering, which are the driving force of consuming data and massive volume of data access through the internet. To understand this generation, cognitive tasks are introduced, which have the sequential flow toward the customers' expectations.

2.2.1 Cognitive Tasks for Time Series Sequential Data

Time series [14] analysis such as frequent changes in the market, sensor updates in space, and instant observation maintenance are some of the examples for sequential data. The research work focuses on time series [22, 23] sequential data such as weather forecasting to predict the accuracy of natural disaster through the changes in sequential information. Since the longitudinal and latitude values from GPS are not always the same and the feed forward network is used to analyze the information which can have multiple hidden states. We have the various features from the Kaggle dataset for identifying the current state without a hidden model using a Bayesian Markov chain model, which can access the columns such as entity and economic changes due to various disasters [2].

Information processing are of two types: i) behavioral processes and ii) cognitive processes that are mostly used in knowledge engineering [15, 16]. Big data streaming data are through online processing via social users from Instagram, Twitter, etc., as video files, audio files, text data, and so on. They are organized according to the need of sequential changes. Transition based on current state and its path as directed graph are the most important analysis for maintaining the network flow. Applications such as weather forecast for detecting the severity through the temperature changes, climate changes, and complex tasks to predict the climate that brings a disaster such as earthquake and volcanic difference [12].

2.2.2 Neural Network for Analyzing the Weather Forecasting

NN is the most important concept in the AI system that can control the weather forecast system such that the natural disaster can be reduced. This work proposed a novel approach to predict the hidden state that updates the new changes in current state that can be directed graph and true to its transition matrix that can differentiate the modeling level that have performance analysis based on the range of economic difference that can step into the neural approach that can be in cognitive task [15] when there is a combination of multilevel longitudinal data that are from various directions from north, south, east, and west that are frequently changed. IBHMF's main task is to reduce the noise from the dataset, which takes particular columns to predict the performance rate and accuracy rate. Based on hidden NNs, the weather features can be variation, categorical data variables, etc., that are not stable instantly [13].

2.2.3 Improved Bayesian Hidden Markov Frameworks

Markov chain process is the type of technique to detect the chain represented in a random process. In this instance, the system can find the representations in disaster management dataset. In this Markov chain process, transition probabilities are described by the graph representations. The graph representations are the probability of moving the one state to another state. In state representation, the variable is declared as i and j, and the transitions are forward from i to j states.

In disaster management dataset, temperature, dry earthquake, and volcano are the attributes. The temperature attribute records the daily temperature in Celsius, which leads to disaster effects. The dry earthquake records the causes of earthquake and volcano records the hotness in lava and temperature in water vapor.

Hidden states are the layers, which are trained and tested in the model process. Hidden layers consist of weights and biases, which help in the training process. Weights help the input by multiplying and biases help to add the weighted input passed to the next hidden layer. These hidden layers are processed in forward and backward direction, which leads to reduce the error and loss rate and increase the accuracy of outcomes. In output layer, activation functions accessed the hidden layer output, which converts into binary format for the user understanding process.

For exponential time, executions are done by brute force solution. This brute force solution helps the model to efficiently approach exponential inputs. In this approach, sequences of observations of each state are executed using the Viterbi algorithm. The maximum probability of a sequence path of state and time are stored in observations of p. The observations of p are the property of Markov models. This process leads to the model in forward recursion.

$$P(Ot \; \varepsilon \; M)$$

Where,

Ot = probability, i = initial state, and j= hidden state of Markov.

In this model, Markov chain consists of four states; the transition probability between the temperature and volcano effects is 0.5; if it was high pressure today, then there is a 50% chance that it will cause disaster tomorrow. The invisible Markov chain processes each and every state that produces the random out for each transition. These transitions are stored in another variable of observations, which is named as m. These observations are visible to the user. There are four states of disaster management, and initial probabilities are transition probabilities of temperature and volcano probabilities matrix.

The parameter learning in the hidden Markov model is executed using the Baum-Welch algorithm. This Baum-Welch algorithm helps to find the probability of observations in local maxima. The following process shows the learning approach:

$$P(O/M)$$

where p = probability, O = converge of optimum, and M = parameters.

These chain properties are applied in discrete directions that are applied in each step. Let us consider the sequence of generated variables as follows:

$$S1, S2, S3, \dots, Sn \tag{2.1}$$

Equation (2.1) implies the Markov chain rule, which can illustrate the moments of update based on the sequence number. To get the distribution of probability that the next node relies on the current node, where the previous node stands idle that are not used for present distance calculation.

$$P_r(S_{n+1} = s|S_1{=}s_1{,}S_2{=}s_2{,}\dots{,}S_n{=}s_n) = P_r(S_{n+1} = s|S_n = s_n) \tag{2.2}$$

where the variables from Equation (2.2) can be defined as follows:

Pr = probability computation of input sequence, which describes the moving of state only to new state or the previous state that can derive the new next state [8].

Based on the visit of the node from weather forecasting according to the conditions that can be alphabets, numbers are accountable sets from the visited node space that are arbitrary in rules. Since the weather states are discrete time updates that are not stopped according to its application. IBHMF is the Markov chain applied for computational analysis using the variations and frequent update of directions that are noted as transitions [7]. Let us consider the transitions using two terms as node from the arrival as beginning S and s+1 for edge based relationship, which form the transition matrix using the Bayesian rules with its straight edges to measure the probability according to the matrices of visit from the current state that can be known from Equation (2.3)

$$s+1, P_r(S_{n+1} = s \mid S_n = s_n) \tag{2.3}$$

2.3 Proposed System

The proposed system deals with the improved IBHMF that was interconnected with probability of continuous data for classification in sequence. In IBHMF model, the weather forecasting features are in sequence, which gets frequent updates such that hidden nodes or states can monitor the variations. To identify the hidden node, there is inference included for decoding the features using the Viterbi techniques [4, 9]. Also, the transition matrix along with probability distribution is gathered from the likelihood formulation. Let us consider the features from the weather forecasting dataset downloaded from Kaggle. The features are entity, node, year, total economic damage from natural disasters (US$), and number of reported natural disasters (reported disasters). All natural disasters. Entities are observing the matrix that can be trained based on forward-backward rules. Table 2.1 defines the entity of the weather forecasting dataset [9].

Table 2.1 Entities from weather forecasting dataset.

❖	Natural disasters
❖	Drought
❖	Earthquake
❖	Wildfire
❖	Volcanic activity
❖	Dry
❖	Landslide
❖	Flood
❖	Extreme weather
❖	Extreme temperature

Daily weather can be defined, for instance, such that probability can be calculated {flood, landslide, volcanic activity} for processing the chain rule. These rules can find the instance according to Equation (2.1). The tokenized probability of each word that depends on the previous state is not updated with possibilities of visited information. To find the transition matrix and develop the graph above, illustrated states (three states) are considered between the probability rates in matrix format. Such a way, i and j are rows and columns that can build the matrix between transition form. In frequent interval, the time duration is also updated to know the last visited state along with the i-th entry of the possible probability toward the vector form of k-terms. So, in terms of year and hidden state, the climate changes and causes also differ.

Weather changes as non-probabilistic distance variation and then likelihood also become a problem such that maximization for change of directions remains same. The next state can be predicted using Markov chain model from the sequence of random generation of updates. Let us consider the state variables as state, followed with variables

Using the state estimation from Kaggle dataset data has been fed as an input layer; later, the hidden layer along with the weight (W) and bias (b) are initiated to classify the preprocessed data to predict the climatic change. The outputs expected from the reliable entity from a dataset such as extreme weather, dry, drought, and temperature change can be categorized using the value that creates the DAG form, which can avoid statelessness in nature. This model from the proposed work uses this stateless approach where the updates can memorize the information as analyzed from the buffer for unique classification on time series.

Weather forecasts always modify the day-wise improvement and random changes; Markov chain rule gives the priority to improve the challenge. Figure 2.2 shows the estimation of two different objects that are allowed from unique classes from the multilevel approach. The transition matrix from the level of updates can be prioritized, which can indicate the entity model id. The set of variables that are numeric and the updates that are based on the year-wise–dependent variable shows the economic difference. When the length and hidden states always vary based on the categories, the probability distribution are formulated. Later, covariation needs to be observed for probability distribution and rows are taken for x, y matrix, which are necessary to calculate the transition element wise. Rows and columns from the GPS are stochastic matrices, which are identified for performance as scalar value. Maximum likelihood has been generated to find the requirement of forecasting directions.

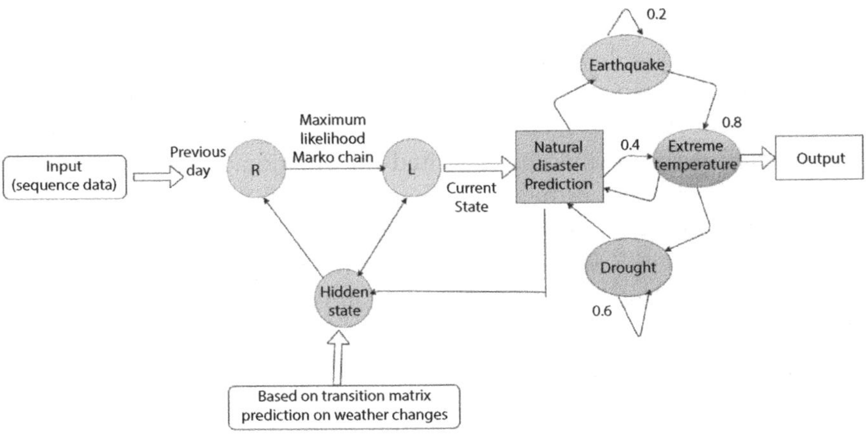

Figure 2.2 Proposed system for predicting disaster using improved Bayesian hidden Markov frameworks (IBHMF).

2.4 Results and Discussion

From the observations on each state moves, the probability with the power of the matrix is generated. Let us consider the steps as state-to-state transition.

Let A and B be the two states with K steps, which have many possibilities to initialize the time with constant value. When the iterations started from one to the i-th iteration, there will be the same number of vector probability ratio that are identified with three states such as natural disasters (Figure 2.4), dry, and drought. Figure 2.3 shows the variations of number of disasters.

Table 2.2 defines the dataset year-wise and its ratio when it is trained according to the NN. In addition, testing data with rating as well reported on a year basis are monitored from the parameters that match according to the testing and training accuracy.

Figure 2.4 shows the entities that are identified from the united states of America on various years that have the impact according to the climate changes. Prediction using this dataset based on the developed feature analysis can be performed. Each and every disaster along with the disasters reported year-wise with the count is also reported.

Natural disaster according to the disaster effects and probability of occurrence using stochastic matrix is described in the analysis. Figure 2.5 illustrates the understanding of probability ratio. It clearly indicates the year-wise analysis of the disaster along with the economic damage that

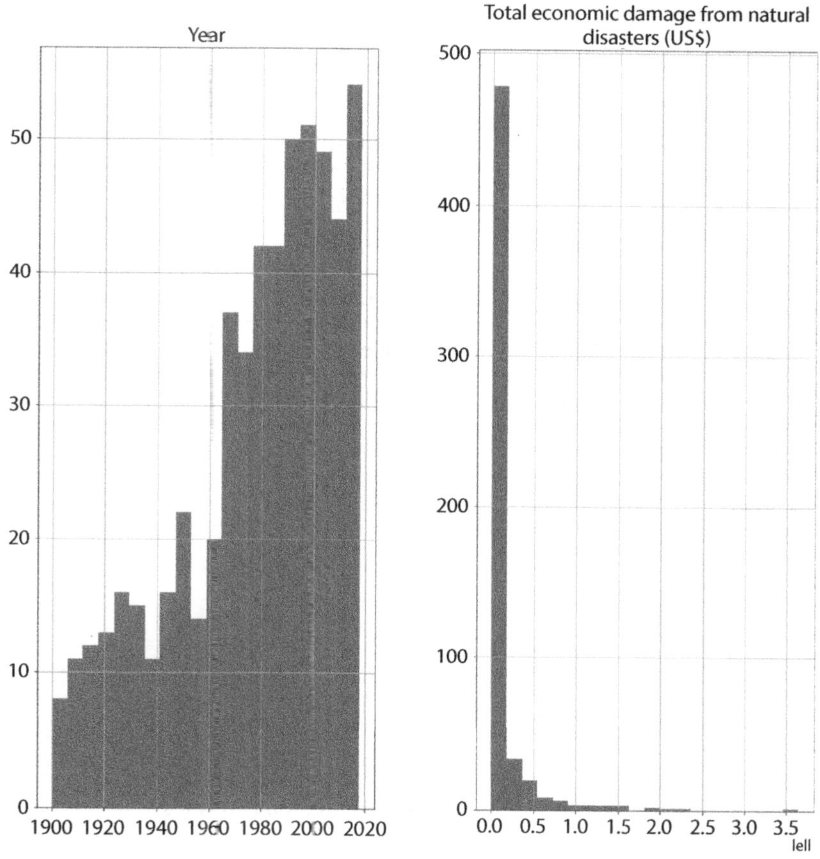

Figure 2.3 Total number of disaster analysis using improved Bayesian Markov chain model.

is caused by these damages. Figure 2.6 shows the boxplot view of natural disasters on various entities.

2.5 Conclusion

As observed from the Markov chain model that is developed, a framework for identifying random effects can create a disaster. Kaggle datasets are used to find the state transition matrix and occurrence that are predicting the natural changes in climate. Computation on the basis of frequency and the latent state changes are monitored based on the probability. By considering the likelihood from the part of observation that are identified from the matrix according to rows and column for assumption. Using the

Table 2.2 Sample dataset for predicting weather forecasting.

	Year	Total economic damage from natural disasters (US$)
count	561.000000	5.610000e+02
mean	1977.217469	1.146966e+10
std	30.399233	3.199525e+10
min	1900.000000	0.000000e+00
25%	1959.000000	6.50000e+07
50%	1984.000000	8.400000e+07
75%	2001.000000	5.444777e+09
max	2018.000000	3.640932e+11

Entity	Year	Number of reported natural disasters (reported disasters)
All natural disasters	2010	5
Drought	2000	2
Earthquake	1989	9
Wildfire	2002	8
Volcanic activity	2009	2
Dry	1999	4
Landslide	1976	17
Flood	2006	5
Extreme Weather	1998	4
Extreme temperature	1979	11
All natural disasters	2020	15

Figure 2.4 Changes from various impacts from natural disaster.

Entity	Year	Total economic damage from natural disasters (US$)
All natural disasters	1900	30000000
Extreme weather	1900	30000000
All natural disasters	1901	35000000
All natural disasters	1902	60000000
All natural disasters	1903	480000000
Flood	1903	480000000
All natural disasters	1904	0
All natural disasters	1905	100000000
All natural disasters	1906	650750000
Earthquake	1906	630750000
Extreme weather	1906	20000000
All natural disasters	1907	30000000
Earthquake	1907	30000000
All natural disasters	1908	116000000
Earthquake	1908	116000000
All natural disasters	1909	60000000
All natural disasters	1910	0
All natural disasters	1911	20000000
Extreme weather	1911	20000000

Figure 2.5 Economic damage changes a prediction analysis.

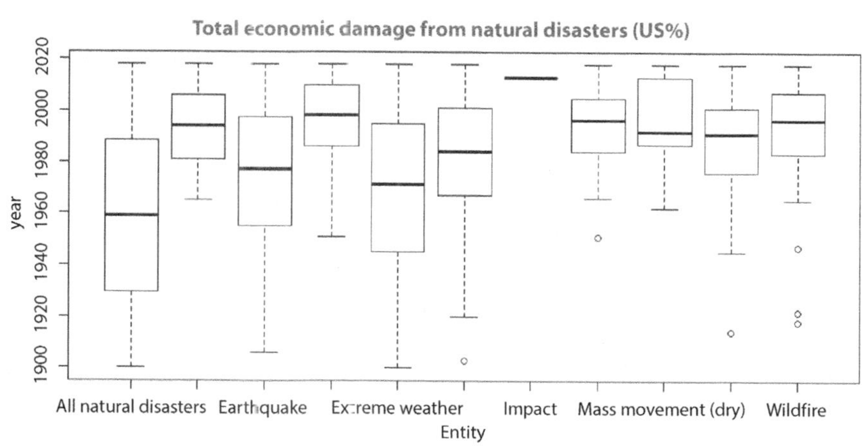

Figure 2.6 Boxplot view of natural disaster on various entity.

sequence numbering and entries on hidden state, the current sequence follows the Markov model and fixes the likelihood. Using the proposed framework, the tasks such as important performance based on weather forecast are achieved. The proposed algorithm IBHMF that produced a better performance and also independent variables that are factors of time series shows the exact transition matrix analysis to predict the forecast based on climatic change.

References

1. Baboo, S.S., Baboo, S.S., Shereef, I.K., An Efficient Weather Forecasting System Using Artificial Neural Network. *Int. J. Environ. Sci. Dev.*, 1, 321–326, 2010, https://doi.org/10.7763/ijesd.2010.v1.63.
2. *Cognitive Tasks and Learning*, n.d. SpringerReference, https://doi.org/10.1007/springerreference_226188.
3. Li, D. and Yu, D., Deep Learning: Methods and Applications. *Found. Trends Signal Process.*, 7, 3–4, 197–387, 2014.
4. Grinsted, A., Ditlevsen, P., Christensen, J.H., Normalized US Hurricane Damage Estimates Using Area of Total Destruction. *Proceedings of the National Academy of Sciences of the United States of America*, vol. 116, pp. 23942–46, 20191900-2018.
5. Anand, J. *et al.*, Efficient Security for Desktop Data Grid using Fault Resilient Content Distribution. *Int. J. Eng. Res. Ind. Appl.*, 2, 7, 301–313, 20092009.
6. Anand, J. *et al.*, Efficient Data Storage in Desktop Data-Grid Computing using Real-Time Parameters. *Int. J. Comput. Sci. Technol.*, 2, 3, 392–397, 2011.
7. Kuswanto, and Kuswanto, Bayesian Model Averaging with Markov Chain Monte Carlo for Calibrating Temperature Forecast from Combination of Time Series Models. *J. Math. Stat.*, 9, 349–356, 2013, https://doi.org/10.3844/jmssp.2013.349.356.
8. Minsley, B.J., A Trans-Dimensional Bayesian Markov Chain Monte Carlo Algorithm for Model Assessment Using Frequency-Domain Electromagnetic Data. *Geophys. J. Int.*, 187, 252–272 2011, https://doi.org/10.1111/j.1365-246x.2011.05165.x.
9. Padli, J., Habibullah, M.S., Baharom, A.H., The Impact of Human Development on Natural Disaster Fatalities and Damage: Panel Data Evidence. *Econ. Res.-Ekon. Istraž.*, 7, 1–17, 2018, https://doi.org/10.1080/1331677x.2018.1504689.
10. Pouyanfar, S., Sadiq, S., Yan, Y., Tian, H., Tao, Y., Reyes, M.P., Shyu, M.-L., Chen, S.-C., Iyengar, S.S., A Survey on Deep Learning. *ACM Comput. Surv.*, 22, 1–23, 2019, https://doi.org/10.1145/3234150.
11. R. Stuart, J., Stuart, J.R., Norvig, P., Davis, E., *Artificial Intelligence: A Modern Approach*, Prentice Hall, Springer, New Jersey, 2010.

12. Vista, A., Awwal, N., Care, E., Sequential Actions as Markers of Behavioural and Cognitive Processes: Extracting Empirical Pathways from Data Streams of Complex Tasks. *Comput. Educ.*, 92–93, 15–36, 2016, https://doi.org/10.1016/j.compedu.2015.10.009.
13. Wollsen, M.G. and Jørgensen, B.N., Improved Local Weather Forecasts Using Artificial Neural Networks. *Distributed Computing and Artificial Intelligence, 12th International Conference*, 2015, https://doi.org/10.1007/978-3-319-19638- 1_9.
14. On Sequential Spectral Analysis of Amplitude-Modulated Time Series, in: *Sequential Analysis*.
15. Habermann, A.N., Engineering Large Knowledge-Based Systems. *Data Knowl. Eng.*, 15, 105–117, 1990, https://doi.org/10.1016/0169-023x(90)90007-z.
16. Kellett, J.M., Winstanley, G., Boardman, J.T., A Methodology for Knowledge Engineering Using an Interactive Graphical Tool for Knowledge Modelling. *Artif. Intell. Eng.*, 4, 92–102, 1989, https://doi.org/10.1016/0954-1810(89)90004-6.
17. Knowledge Management and Knowledge Engineering. *Knowl. Manage.*, 152, 2001, https://doi.org/10.1201/9781420041125.ch1.
18. Mohamudally, N., Introductory Chapter: Time Series Analysis (TSA) for Anomaly Detection in IoT, in: *Time Series Analysis and Applications*, 2018.
19. Motta, E., THE KNOWLEDGE MODELING PARADIGM IN KNOWLEDGE ENGINEERING, in: *Handbook of Software Engineering and Knowledge Engineering*, 2001.
20. Niharika, M., Deepa, N., Devi., T., Image segmentation and detection for healthcare data in deep learning, in: *Test Engineering and Management*, vol. 81, pp. 5402–5407, 2019.
21. Vaishnavi, R. *et al.*, Efficient Security for Desktop Data Grid using Cryptographic Protocol. *Proceedings of IEEE International Conference on Control, Automation, Communication and Energy Conservation*, vol. 1, pp. 305–211, 2009.
22. Śmietanowski, M., Nonlinear Parameters Estimation from Sequential Short Time Data Series. *Auton. Neurosci.*, 90, 158–166, 2001, https://doi.org/10.1016/s1566- 0702(01)00283-1.
23. Zhang, Y., Zhong, M., Geng, N., Jiang., Y., Forecasting Electric Vehicles Sales with Univariate and Multivariate Time Series Models: The Case of China. *PloS One*, 12, 5, e0176729, 2017.
24. Gunn, S.W.A., The Scientific Basis of Disaster Management. *Disaster Prev. Manag.: An Int. J.*, 1, 446 1992, https://doi.org/10.1108/09653569210018672.

3

Big Data Knowledge System in Healthcare

P. Sujatha[1]*, K. Mahalakshmi[2] and P. Sripriya[3]

[1]Department of Information Technology, Vels Institute of Science, Technology & Advanced Studies, Pallavaram, Chennai, India
[2]School of Computing Sciences, Vels Institute of Science, Technology & Advanced Studies, Pallavaram, Chennai, India & Women's Christian College, Chennai, India
[3]Department of Computer Applications, Vels Institute of Science, Technology & Advanced Studies, Pallavaram, Chennai, India

Abstract

The present society is an Informational Society. The information is essential for industries for accurate results. Knowledge systems integrate the information from multiple sources to extract important insights for decision-making. Due to digitalization, the data available is abundant in all kinds of industry. In order to handle voluminous data and to predict the outcomes accurately, a new framework is evolved using big data and big data analytics. Big data has huge implications for knowledge management. Medical data in the healthcare industry is enormous. In recent years, the healthcare sector is changing from volume-based service industry into value-based service industry, because health is the most precious gift to human beings. Advanced technologies and its related analytical tools increase better healthcare practices and drive human beings into longer life spans. Big data knowledge system has a significant responsibility in the healthcare practices. Big data knowledge systems analyze the large volume of patients' electronic health records and transform the medical information into knowledge for assisting the process of clinical decisions, for recommendation of medicines, and for better diagnoses of diseases. In this chapter, we discuss big data, its applications, and challenges of the big data knowledge system in the healthcare sector

Keywords: Knowledge system, big data, electronic health records, healthcare, big data knowledge system, big data analytics, medical data, clinical decisions

**Corresponding author*: suja.research@gmail.com

A. Jaya, K. Kalaiselvi, Dinesh Goyal and Dhiya Al-Jumeily (eds.) Handbook of Intelligent Healthcare Analytics: Knowledge Engineering with Big Data Analytics, (39–66) © 2022 Scrivener Publishing LLC

3.1 Introduction

Since the last decade, there is a constant development in information technology. The most recent trends in technology like social networks, smart phones, computers, and smart wearable devices lead to the growth of data. This growth in the data has given embarks on this latest notion described as big data. Big data is a large data set, which is generated at rapid speed. This data set is very complex, which goes beyond the capacity of traditional database and processing tools for storing, processing, and analyzing. Big data refers to diverse large amount of information created by the industries, machines, and the individual person. The production and usage of big data is persistently increasing. In today's world, the volume of information far exceeds what the human mind can perceive. Big data requires sophisticated computer storage and analysis tools. In simple terms, big data is a huge amount of data in structured, semi-structured, and unstructured format. Day by day, the digital data volume is increasing steadily at a rapid pace. So, the big data is too large to process using the traditional technologies. Many companies realize the importance of the availability of big data. They use those large data to improve their decision-making and planning capabilities.

Healthcare big data is huge because of its abundant sources like electronic health record (EHR), the data from wearables and medical devices, genomic data, clinical research data, Internet of Things (IoT), the search engine data, and the social media data. It is very difficult to integrate big medical data with traditional information processing systems and databases because of its variety and volume. However, the tool and technology of big data is used for gathering, processing, and storing large volumes of medical data, which is diverse in nature. The role of big data in the healthcare industry is important due to availability and easy accessibility of vast amounts of medical data, increase in healthcare costs, and the need for personalized patient care. Big data analytics effectively accumulate, analyze, and take the knowledge from the medical data.

Big data knowledge systems in the healthcare industry are more important because all these data about patients are used for developing health recommender systems, clinical decision support system, disease prediction system, and knowledge discovery systems. The medical big data with knowledge system assists during the process of analyzing and extracting valuable knowledge. The extracted knowledge used by the healthcare professional for their clinical decisions. The big data knowledge system transforms the healthcare big data to useful information. The healthcare industries are providing better clinical decisions, for the quality patient care with lower cost using the healthcare data analytics.

Meaning of big data, various dimensions of big data, tools and technologies used in big data, process of creating value from big data, big data analytics, role of big data knowledge system in healthcare, healthcare big data analytics, the applications of big data, and the challenges faced by big data in the healthcare industry are presented and discussed in the following sections.

3.2 Overview of Big Data

3.2.1 Big Data: Definition

Every day, the organization is producing an enormous quantity of data. These huge volumes of data compose "big data". Big data, complex in nature, requires powerful technologies and advanced algorithms for its processing.

A formal definition of big data was given in [1]: "Big data is the information asset characterized by such a high volume, velocity, and variety to acquire specific technology and analytical methods for its transformation into value."

Data has been increasing constantly in an unpredicted way from the last decade due to the digitalization and the advancements in technology. In common, the big data is being generated from following sources [14]:

Social data: The social data refers to social media data. It is generated from social media such as YouTube and Twitter. This data is mainly used in market analysis. The analysis on Facebook likes and comments and tweets on Twitter provide the details about the consumer behavior.

Machine data: Machine data refers to the data generated by machines, such as wearable, sensor devices, web logs, and satellites.

Transactional data: The transactional data are generated as a result of the transactions. The transactions can be online or offline. Examples of the transactional data are the delivery receipts, order, invoices, etc.

Human generated data: The human generated data is extracted from the emails, electronic medical reports, messages, etc.

Search engine data: The search engine data are generated from the browsers.

All the abovementioned data are in diverse formats such as comments, videos, email, and sensor data, most of which are in unstructured format. Big data is the huge size of a data set that grows exponentially with time.

Examples of big data: Amazon product list, YouTube videos, Google search engine, and Jet engine data. Storing and processing of abovementioned big data is not possible with conventional databases because traditional databases can contain only gigabytes of data. But, the big data contains several petabytes of data. The big data solutions solve this entire problem with distributed storage and processing systems.

3.2.2 Big Data: Characteristics

The massive set of information generated with the utilization of the latest technologies is called big data. This large set of data is used for individual and organizational purposes. Previously, the information was generated, stored, and processed easily because of limited sources of data. The conventional database was the single supply of data. Most of the data in the conventional database was in structured format. Presently, data is in a wide variety of formats such as sensor data, email messages, images, audio, and video, which makes the big data. Most of this data is unstructured. The five common big data characteristics are volume, variety, veracity, velocity, and value. Apart from the abovementioned five dimensions of big data, many researchers also added new dimensions of big data. Other dimensions of big data are volatility, validity, visualization, and variability. Big data is characterized by following commonly adopted V's. Figure 3.1 and Table 3.1 represent the big data features in term of Vs [24].

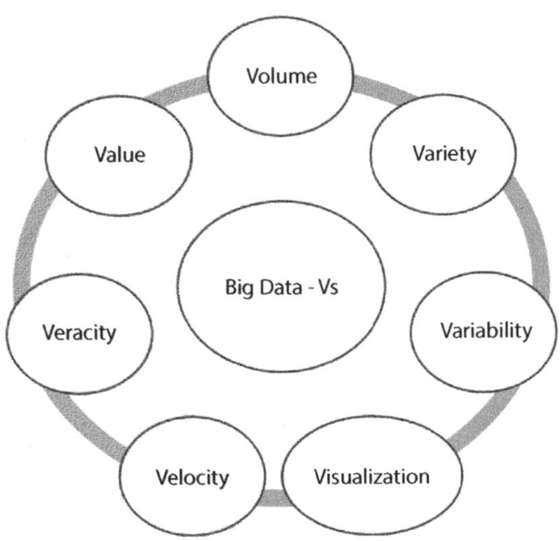

Figure 3.1 Dimensions of big data.

Table 3.1 Dimensions of big data.

Dimensions of big data	Description
Volume	In general, volume refers to quantity or amount. The data that contains gigantic larger data sets is called volume in big data. Big data is known for its voluminous size. The data is being produced on a daily basis from various places.
Variety	Meaning of variety is diversity. The data that are generated and collected from diversified sources is called a variety in big data. Usually, big data is in a variety of forms that comprises structured, semi-structured, and unstructured data.
Velocity	Velocity means speed. In big data context, the velocity denotes the speed of data creation. Big data is arriving at a faster rate, like a stream of water.
Veracity	The veracity is the credibility, reliability, and accuracy in big data and the quality of the data sources.
Variability	The variability is not the same as variety. The variability means data that are regularly changing. This is an important feature of big data.
Visualization	Big data is illustrated in the pictorial form using the data visualization tools.
Value	The potential derived from the big data is known as value.

3.3 Big Data Tools and Techniques

3.3.1 Big Data Value Chain

The value chain of big data is the process of creating potential value from the data. The following steps are the value chain process for creating knowledge insights from big data [12, 25, 29].

- Production of big data from various sources
- Collection of big data from various sources
- Transformation of raw big data into information for storing and processing

- Preprocessing of big data
- Storage of big data
- Data analytics

Production of big data from various sources

The production and generation of big data at various sources refers to big data generation. This is the initial step in the value chain process. Big data is known for its volume. The data is generated from various sources on a regular basis. Similarly, medical data is generated by the clinical notes, IoT, social media data, etc. Big data includes internal data of the industry and also includes internet and IoT data.

Big Data Collection

The second phase of the value chain process is the collection of data from a fusion of different places. Big data is generated at various sources. All these are in various formats like text, image, and email, which may be structured, semi-structured, and unstructured data. Big data collection process used to retrieve raw data from the various sources. The most common big data sources are computers, smartphones, and the Internet [14].

Big Data Transmission

Once the data collection is over, it should be transformed to the preprocessing infrastructure.

Big Data Pre-Processing

Big data is collected from a variety of sources. So, the data may consist of inconsistent, noisy, and redundant data. The data preprocessing phase improves the data quality by removing inconsistent, noisy, incomplete, and redundant data. This process improves the data integrity for data analysis.

Big Data Storage

The data should be stored for exploration. Big data refers to huge data that cannot be processed and stored using traditional information systems. The large amount of big data can be stored and handled easily with the help of big data storage technologies. All these technologies ensure data security and integrity.

Big Data Analysis

The final phase in the value chain process is analysis. Artificial intelligence, deep learning, machine learning algorithms, and other data analytical tools are used to analyze the data and to extract knowledge from a huge data set. Based on the data analysis process, the decision-makers can solve the problem and make good decisions.

3.3.2 Big Data Tools and Techniques

Technologies are used to analyze, process, and extract the useful insights available in complex raw data sets. The major technologies of big data are storage and analytics technologies. The most accepted big data tools are Apache Hadoop, Oracle NoSQL, etc. The technologies of big data are classified into operational and analytical.

Operational big data: This data is generated on a daily basis. Ticket booking, social media, online shopping, etc., are examples of operational big data.

Analytical big data: This data is nothing but analyzing the operational big data for real-time decisions. Examples of analytical big data are the medical field and weather forecast.

The technologies and tools used in big data are classified on basis of following steps [11]:

- Gathering of big data
- Processing the big data
- Storing the big data
- Analyzing the big data
- Visualizing the big data

Table 3.2 gives the detailed description about various big data technologies used in the process of creating a value from the raw data.

The Hadoop contains tools that are used to gather, preprocess, and investigate the huge volume of structured, semi-structured, and unstructured data. Hadoop is an open source, highly scalable, fault-tolerant, and cost-effective tool for faster big data processing.

3.4 Big Data Knowledge System in Healthcare

Knowledge means facts, information, and skills obtained in the course of experience. Knowledge is nothing but better understanding of a subject. Healthcare practice is a knowledge based process. Giving the right knowledge during the decision-making process is very important. In healthcare, the knowledge is used for the diagnosis of the disease, the prescription of medicine and the clinical management.

Table 3.2 Big data technologies [12, 14, 26].

Big data capability	Primary technologies	Developer	Description
Big data collection	Sqoop	Apache Software Foundation	The Apache Sqoop tool is used for transmitting the data among Apache Hadoop and structured databases like relational databases such as Oracle and MySQL. It exports and imports data among relational databases and the Hadoop file system.
	Flume	Apache Software Foundation	The Apache Flume is a service for collecting, aggregating, and moving huge volumes of log files to centralized data stores. The log files are collected from the internet and aggregated in the Hadoop Distributed File System for investigation.
	Kafka	Apache Software Foundation	The Apache Kafka is used to collect data and to perform real-time analysis. The Kafka is used by Uber, Netflix, and PayPal.
Big data storage	Hadoop Distributed File System (HDFS)	Apache Software Foundation	The distributed storage system for Hadoop application is Hadoop Distributed File System (HDFS). This HDFS is a storage system based on a distributed file system. It is a majorcomponent of the Apache Hadoop. The HDFS accommodates large data sets. HDFS framework used for storing files in distributed environments.

(Continued)

Table 3.2 Big data technologies [12, 14, 26]. (*Continued*)

Big data capability	Primary technologies	Developer	Description
	Oracle NoSQL	Oracle Corporation	NoSQL Database is also known as a non-SQL or non-relational database. It has a method for storing and retrieving data from other than tabular relational models. NoSQL is mainly used for storing big data. Unstructured data are managed by NoSQL using dynamic schemas.
	Apache HBase	Apache Software Foundation	Apache HBase is a distributed and non-relational database in the Hadoop framework. HBase provides a way to store sparse data sets. It dynamically distributes the table when the data set is too big to manage.
	Apache Cassandra	Apache Software Foundation	The Apache Cassandra database is a distributed database to manage huge amounts of data in several commodity software. Cassandra is a scalable, high-performance, NoSQL database.
Big data processing	MapReduce	Apache Software Foundation	The Apache MapReduce is used for the distributed processing of large amounts of data in a trustworthy, error-tolerant way. Hadoop MapReduce is a programming language. This framework is used to write applications for managing and processing huge data.

(*Continued*)

Table 3.2 Big data technologies [12, 14, 26]. (*Continued*)

Big data capability	**Primary technologies**	**Developer**	**Description**
	Apache Hadoop	Apache Software Foundation	The ApacheHadoop framework is used for storing and processing big data in the cluster of computers that is in a distributed environment instead of a single computer. Hadoop uses a cluster of computers for analyzing large data sets in parallel. The Hadoop consists of following main layers: • Processing/computational layer: MapReduce • Storage layer: HDFS Hadoop MapReduce, Yet Another Resource Negotiator, Hadoop Common, and HDFS are important modules of Apache Hadoop framework. Apache Yarn: Yet Another Resource Negotiator used for resource management and job scheduling. YARN allows processing of HDFS data with different data processing engines. Hadoop Common consists of common libraries and utilities. These libraries and utilities are necessary for supporting the other modules of Hadoop.

(*Continued*)

Table 3.2 Big data technologies [12, 14, 26]. (*Continued*)

Big data capability	Primary technologies	Developer	Description
Data integration capability	Oracle bigdata connector, Oracle data integrator	Oracle Corporation	The Oracle big data connector is software that combines the Apache Hadoop and oracle database. Apache Hadoop is used for data accumulation and pre-processing. The connector integrates data in the oracle database with Apache Hadoop for the analysis purpose.
Statistical analysis capability	R, Oracle R Enterprise	Oracle Corporation	R programming language is an open-source package for Statistical analysis. Oracle R enterprise refers to the oracle database for advanced analysis. It consists of R software and oracle database features. This feature is used by R users to access the database data without Structured Query Language.
Big data analytics	Apache Hive	Originally developed by Facebook and it's an Apache licensed software.	The Apache Hive is open-source software for data warehouses. This framework is used for storing, querying and analyzing the large data. Hive is built on Apache Hadoop. Hive framework is used to store and process large data sets quickly. Hive is used for summarizing the data, query and analyzing the data. Hive is written in java.

The complexity of massive data is the major problem with big data because it goes beyond the capacity of traditional data processing systems. Healthcare big data is the massive amount of patient's information because of the innovations in digital technologies. The roles of big data in the healthcare profession have the greatest impact on human lives. The healthcare organization is the major sector that produce enormous quantity of information such as administration data, laboratory test data, medicine data, sensor data, smart wearable device data, pharmacy data, EHR, doctor's appointment details, and health insurance. There is a need to analyze this healthcare data for the better treatment process and better healthcare services at lower costs and to increase the patients' satisfactions. The healthcare professionals, the hospitals, and many researchers are analyzing the medical data to understand the clinical perspective and to prevent health issues. The meaningful information is extracted from the massive data using the data analytics tool. Big data analytics helps to examine the patient's information quickly and to formulate better conclusions on the patient's treatment process. The disease prediction and detection, the clinical and drug research, the hospital administration, and the personalized patient healthcare services are a few of the major uses of big data analytics in the healthcare professions.

Figure 3.2 represents the process of creating knowledge from raw facts. A knowledge system objective in big healthcare data is to transform the information into a knowledge asset.

The big data sets are analyzed to get the valuable information. The information is transformed into actionable insights. The actionable insights are turned into knowledge using analytical tools. This knowledge can be used for processing the new information and for forecasting the future events [6].

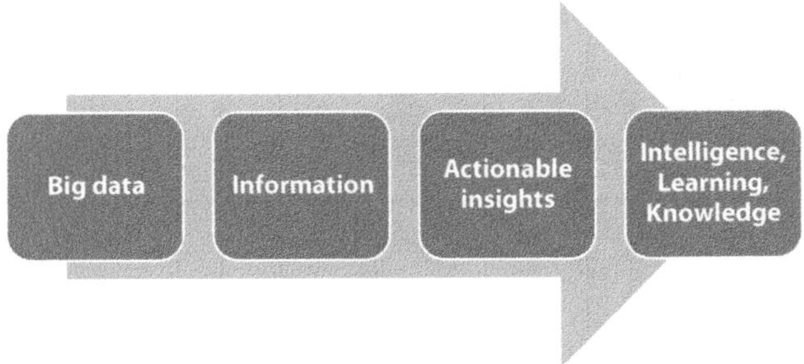

Figure 3.2 Big data value creation flow.

3.4.1 Sources of Medical Big Data

The medical care industries are producing a bulky amount of patient information. Figure 3.3 shows an important origin of the hospitals and patient data.

EHRs, electronic medical records (EMR), hospital reports, images of radiology, sensor devices, and IoT are the most important origins of the medical data. These data are used to diagnose and treat the diseases of human beings. The healthcare data are diverse in nature. It is a very difficult and challenging process to integrate the big data of healthcare industries. The technological innovation transforms the healthcare big data into useful information.

Internet of Things

The IoT consists of the physical devices and the sensors that are connected to the internet. All the interconnected devices gather and transfer the data over the internet. IoT device enable remote monitoring of the patients, which avoids patient's visit to hospital unnecessarily. The wearable device senses the human heart rate, blood pressure (BP), weight, stress level, physical activities, glucose level, etc. Because of the IoT-enabled devices, the people can self-track their health conditions. These sensor devices gather a large volume of health data continuously. Apart from wearable devices, there are applications in smartphones and other medical devices, which can gather information of patients and send it to the cloud on a continual basis.

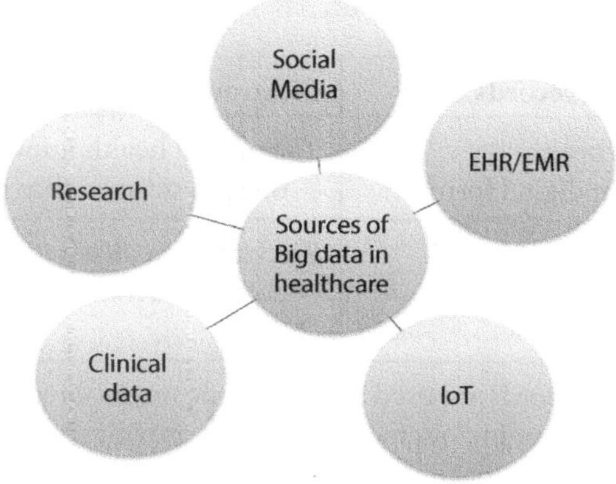

Figure 3.3 Different sources of healthcare data.

Electronic Medical Records and Electronic Health Records

EMRs and EHRs are the most widespread sources for big data in healthcare. The EHRs consist of health status information and data of people in digital format. This health information will be distributed among the diverse healthcare professionals, doctors, and hospitals. The EHRs of patients can be used for improving the healthcare process and enable the healthcare professional to treat the patients well.

The EHR consists of following information about patients:

- Patient's demographic information, medical history
- Billing data
- Insurance information
- Laboratory results
- Clinical information
- Medication records
- Medical imaging data

The EMRs contain data from a particular physician's office. The EMRs are private and confidential documents. The EMRs contain treatment history of the patients in any physician's office.

The difference between EHRs and EMRs are discussed in Table 3.3.

The main difference between EHR and EMR is that the EMR holds information of the patient's history by one provider. The EHR goes ahead of the information from one provider and contains wide-ranging patients' history, which can be shared between all providers.

Table 3.3 Difference between electronic health record and electronic medical record.

Electronic health records	Electronic medical records
All the patients' health information in digital form.	The patient health information form one physician's office in digital form.
Electronic health records are shared among medical professionals.	Electronic medical records are private and confidential documents.
Electronic health records are used by many numbers of providers for the diagnosis and for the better treatment.	Electronic medical records are used by one provider for the patient's treatment.

The EHR provides enormous data that enable doctors for better clinical decisions with advanced analytics.

Apart from IoT and EHR, the other sources of big data, in healthcare are as follows:

- Insurance data
- Other clinical data
- Healthcare-related research studies data
- Social media and search engine data

All these medical related data exist in many forms: laboratory results, the physician's clinical notes, the medical images, the sensor data, etc. There is no standard form for all these sources of medical data. The medical data is large in volume, of structured and unstructured data.

Structured data: Structured data means the information accumulated and presented in an organized way. Examples of the structured health data are height, weight and BP, blood group, and stages of a disease diagnosis [5, 8].

Unstructured data: Unstructured data is the data that is not organized with any standard format. Examples of the unstructured health data are the doctor's clinical notes, images, and the search engine data. The main problem of big data in healthcare fields is that around 80% of the healthcare data are in unstructured form [9]. Deep learning, artificial intelligence, cloud computing, data mining, machine learning algorithms, text analysis, and natural language processing tools are used for transforming the unstructured data into a meaningful knowledge insight. Table 3.4 summarizes the different sources of healthcare data.

3.4.2 Knowledge in Healthcare

In general, the knowledge is classified into the following two categories [5]:

- Explicit knowledge
- Tactic knowledge

Explicit knowledge: The explicit knowledge refers to clear and unambiguous information that is very simple to distribute, written down, and coherent. Studying the life cycle of a parasite, research on cancer, etc., are some of the examples of explicit knowledge. In healthcare, explicit knowledge is identifying the medication to treat a disease.

Table 3.4 Summary of different sources of healthcare data [13].

Sources of healthcare data	Details	Format
Medical records	The medical records consist of electronic health records (EHRs) and the electronic medical records (EMRs). These records offer information on laboratory tests, diagnoses, etc. The health records give all information of patients.	Structured and unstructured data
Human generated data	Clinical notes	Unstructured data
Machine generated data	Sensor devices data	Structured and unstructured data
Transaction data	Billing records and insurance claims	Structured and semi-structured data
Social Media data	Search engine data and social media post	Unstructured data

Tactic knowledge: The tactic knowledge is gained from the course of experience which is more difficult to express. It lies sub-consciously within the human and is improved by continuous practice.

In the healthcare industry, both the tactic knowledge and the explicit knowledge are used together for the diagnosis of many diseases.

In healthcare industry, knowledge is classified into three types:

Provider knowledge: The provider knowledge integrates both tactic and explicit knowledge. For example, in healthcare, a doctor should know the standard treatment for a particular disease. Doctors will gain internal knowledge through years of experience for better treatment.

Patient knowledge: Patient knowledge is tactic knowledge. "Patient's knowledge" is the knowledge of a patient about their medical condition. This information is important for the provider to treat the disease.

Organizational knowledge: The organizational knowledge is the data available for both provider and patient access.

The knowledge in healthcare is essential because healthcare industries are knowledge driven industries.

3.4.3 Big Data Knowledge Management Systems in Healthcare

The knowledge management system is used to generate, store, distribute, use, and reuse the valuable information, knowledge, and insights by means of information technologies.

The knowledge system in healthcare is to create, store, share, and reuse valuable information. Adoption of knowledge management systems in the healthcare industry is very challenging. Gathering of information from various sources and processing of that information needs the latest tools and technologies. Big data knowledge system refers to finding the knowledge from raw data. Big data analytics is the way to transform the raw big healthcare data into knowledge. Data, business intelligence, and machine learning are the most important in big data knowledge systems. The big raw healthcare data is transformed to insights using business intelligence, and then, insights are transformed to knowledge with machine learning and artificial intelligence. Figure 3.4 depicts the information flow and knowledge discovery process of healthcare big data.

Healthcare is based on knowledge such as medical knowledge, clinical knowledge, health service knowledge, and disease knowledge. So, healthcare services are knowledge-based services. All this knowledge enables the doctors to make better decisions in the treatment process with care. This knowledge is continuously evolving, based on new diseases and new drugs. The healthcare professional must learn constantly to provide up to date patient care.

The healthcare industry uses the "healthcare knowledge management systems", for better services to the patients with improved efficacy.

Advantages of knowledge management system in healthcare are as follows [16]:

- Improved patient services
- Medical error reduction
- Reduction of healthcare cost

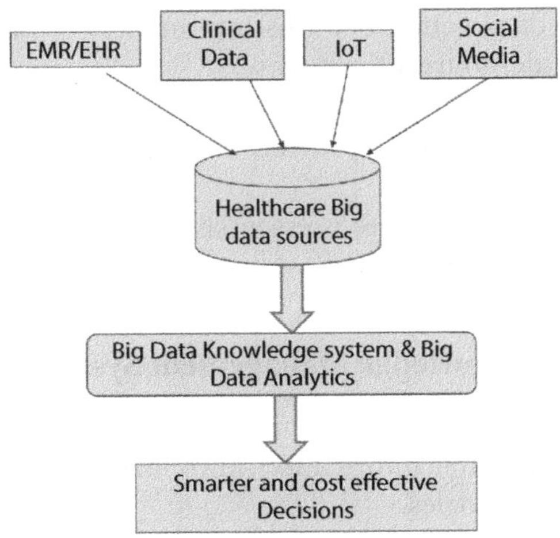

Figure 3.4 Knowledge discovery process of big data in healthcare.

The big data knowledge system transforms the raw data set into data insights, which lead to a better treatment at a lower cost.

Knowledge discovery is the method for transforming raw big data into useful information. The important elements of knowledge discovery are data, analytical tools, methods, and understanding of the domain [17]. Big data analytics is used to discover meaningful insights from the big data set. Machine learning, cloud computing, data science, natural language processing, text analytics, predictive analytics, statistical investigation, data mining, and artificial intelligence are advanced big data analytical techniques. This technique gains new knowledge from the data. All of the data is not readily usable in big data analytics. They have to undergo a "data cleansing process" to make it understandable. Understanding the details about, where the raw data come from and how they have to be treated before analyzing them, are also important. So, the data have to go through a process called "Extract, Transform, and Load" (ETL) before it can be analyzed. The data are harvested, that is, "Extracted". Then, the data is formatted to make it readable, by an application that is "Transformed". Then, the data is stored in the memory for reuse, that is, "Loaded". This is called the ETL process.

3.4.4 Big Data Analytics in Healthcare

Big data analytics is the process of discovering knowledge. The stages of healthcare big data analysis are descriptive, diagnostic, predictive, and prescriptive [7, 27]. These analytical stages give the answer for the following questions:

- What happened to the patient?—Descriptive Analytics
- Why did it happen and when?—Diagnostic Analytics
- What will happen in the future?—Predictive Analytics
- How can we make it happen or not?—Prescriptive Analytics

The following are the categories of the healthcare big data analytics [18–20].

Descriptive analytics: This analytics summarizes past data in an easily readable format. This is like reporting. This gives the information, like what happened and when. The descriptive analytics in healthcare data gives the answer for the following questions:

- How many patients were in the emergency ward in the last month?
- How many numbers of patients have been infected because of epidemic disease in the last three months?

The descriptive analytics give the details for the above question using data. The descriptive analytics gives answers in simple statistical measures such as count and percentages.

For example, how many people have been infected with COVID-19 in the last month? From descriptive analytics, the hospital management can discover the number of people infected with COVID-19 in the last month. For example, we can discover from the descriptive analytics that 10,000 people are infected with COVID-19 in the last month. Descriptive analytics looks at the last month's data and gives the answer in number. The government and the healthcare organization are able to identify the problem with current clinical aspects and get better healthcare practice, using the report of descriptive analysis. This is the first step of analyzing the big data that transforms the raw data into precious knowledge. Descriptive analytics is mainly for reporting, monitoring, and visualization.

Diagnostic analytics: This type of analytics gives information like what happened and why. The diagnostic analytics give explanation for the descriptive analytics data. The diagnostic analytics give answers for the following questions:

- Why were the patients infected with a particular disease last month?
- Why were the patients in the emergency ward last month?

The diagnostic analytics goes a little beyond reporting, that is, descriptive analytics. The above question can be answered using the data with the business intelligence for insight. For example, diagnostic analytics analyze the big healthcare data to discover the reason for why these patients are infected with COVID-19.

Predictive analytics: Predictive analytics uses machine learning and artificial Intelligence to analyze the present and the past data. This type of analytics is used to predict and forecast the future. The predictive analytics use descriptive data to forecast what will happen in the future possibly. This predictive analysis is the very significant analytics for predicting and diagnosing the disease. These types of analysis have capability for controlling and avoiding non communicable diseases like cancer, heart disease, stroke, and diabetes. Noncommunicable diseases are together accountable for most of the deaths worldwide. The predictive analytics use patient history and patient current health information to make medical decisions. The predictive analytics identify the following:

- Identify the patients who have the maximum possibility of getting diabetics, stroke, and heart attack?
- Identify the patients who have the maximum risk of hospitalization?

The above questions are answered using the data and machine learning for forecasting. In predictive analytics, the researchers and data analysts train a model using data mining algorithms, machine learning techniques, artificial intelligence, and the deep learning algorithms to predict the future events.

Prescriptive analytics: This analytics makes recommendations from the predicted output. The prescriptive analytics in healthcare is the last course of action in analyzing medical big data. This analysis has the capacity to suggest the action to overcome the problem in the healthcare organization. The imminent opportunity and challenges of medical big data is prescriptive analytics. Prescriptive analytics is the most advanced level of data analytics in medical big data. It is going to be realistic in the near future due to machine learning techniques, data science, cloud computing, deep learning, data engineering, and artificial intelligence.

These analytics give the recommendation or suggest action to change the prediction. Prescriptive analytics uses the data, the business intelligence for insight, and the machine learning for forecasting.

Big data analytics perform the above analysis using machine learning, artificial intelligence, and natural language processing to explore the unknown patterns and relationships among data.

3.5 Big Data Applications in the Healthcare Sector

Healthcare sector is the top most sector which generates large volumes of data. So, big data is having a huge impact in the healthcare industry. The EHR of patients is widely adopted and analyzed to get deeper insights on clinical knowledge and for enhanced knowledge about illness and disease. The healthcare organization uses big data that improves the efficiency of healthcare practice and care. Big data with the recent development of data mining techniques, machine learning algorithms, deep learning, artificial intelligence, and image processing used to find many important features. The major applications of big data in the healthcare industry are discussed in the following sections.

3.5.1 Real Time Healthcare Monitoring and Altering

Remote patient monitoring or telehealth is healthcare service for patients and doctors, which use IoT devices to track and analyze health status [25].

The work becomes easy for the doctors and nurses in the hospital because of the application of big data. According to researchers at IBM Watson [2], "Above one million gigabytes of medical associated data is likely to be generated by an average person in their lifetime." With the new innovations in technology, there is a prevalent use of wearable health monitors. The wearable devices collect data such as temperature, oxygen level, heartbeat, calories burned, BP, water intake, and distance walked. These data can be used by the healthcare professional for the clinical research, monitoring and tracking patients' health. The sensors are also used with in-patients to continuously monitor the patients. For example, the Fitbit smart watches are used by the patients for self-tracking of their health status. All these monitoring systems are connected to hospitals and physicians through the IoT concept for medical assistance. Table 3.5 lists various remote patient monitoring devices with its features.

3.5.2 Early Disease Prediction with Big Data

Predictive analytics is not new to the healthcare industry. But, it became more powerful with the emergence of big data and its tool to understand the gathered data. The predictive analytics in the healthcare sector helps the doctors to predict or for the early diagnosis of the disease like diabetes, heart disease, stroke, and cancer. Disease prediction is the big data automotive tool of the healthcare field. These analyses are used for developing

Table 3.5 Patient health checking devices.

Patient health checking devices	Features
Glucose checking meter	It used by diabetes patients for checking the glucose level.
Heart rate monitoring	It is used to monitor heart rate.
Hand hygiene monitoring	It is used in hospital to remind the people to sanitize their hands before entering hospital room.
Depression monitoring	It collects information like heart rate and BP and analyzes this information for a patient's mental health.
Blood pressure (BP) monitoring	This device monitors blood pressure. Some BP devices take multiple readings for the daily averages.
Pulse oximeter	This device is used to track the oxygen saturation level in blood and also the pulse rate of the patient.
Patient wearable	Patient wearable devices are used to sense heart rate, blood pressure, glucose levels, weight, and physical activities.
Maternity care monitoring	This remote monitoring reduces clinic visits for pregnant women.
Electrocardiography (ECG) devices	This device is used to diagnose cardiac abnormalities.

the disease diagnosing process which improves healthcare practices. The patients, medical professionals, pharmaceutical industry, and insurance companies can be benefitted from predictive analytics in healthcare using big data.

Predictive analysis mainly used by hospitals:

- To accurate diagnosis of disease
- To reduce healthcare costs
- For preventive medicine

Predictive analyses with medical data are used for determining the risk level of patients for disease based on lifestyle choices and health history.

One example for predictive analytics in healthcare is the predictive analytics algorithm by Kaiser Permanente [4]. This algorithm predicts the infants' illness using the infant mother's medical data and medical summary.

3.5.3 Patients Predictions for Improved Staffing

Main big data applications in the medical industry are the patient's predictions. The prediction of anticipated number of patient's visits to emergency will improve the staffing and provide a way to utilize the resources of the hospital properly. One of the major problems in hospitals is staffing, that is how many people (doctor, nurse, and other workers) are required at any given time period. Big data is used to solve this problem.

Intel and the Assistance Publique-Hôpitaux de Paris (AP-HP) worked toward developing a cloud based system for predicting the anticipated amount of patient's visits [3]. The overcrowding is an emerging problem in most of the hospitals. The cloud based big data analytics system will predict the patient visits and the admissions of patients in hospitals as inpatient or in emergency ward. These solutions make it simple for the hospital administration to productively use the resources (staff, beds, etc.) of the hospitals. The advantages of patient's prediction are improved patient services, reduced wait time, and reduced labor costs, and this will lead to better care with lower cost for patients.

3.5.4 Medical Imaging

Big data analytics for healthcare read the images using algorithms. These algorithms analyze the image by identifying patterns in the pixels and convert it into numbers. The result of the analysis is used by the doctor during the diagnosing process. Because of big data analytical tools, the doctor will analyze the outcomes of the algorithms.

Artificial intelligence is more efficient for the diagnosis of many diseases. One of the examples of AI is diagnosing breast cancer, which is developed by scientists of Google. This diagnosing system uses deep learning algorithm to predict the cancer cells which uses medical images as input [15]. Artificial Intelligence with medical imaging is used in early diagnosis of neurological disease, cardiovascular abnormalities, and cancer cells [30]. Medical imaging with the big data analytical tool and artificial Intelligence is mainly used in the treatment and disease diagnosis process in the early state.

The other big data applications in healthcare industries are as follows:

- Automation of hospital administration process
- Personalized healthcare service

- Medical research and the drug discovery
- Prevent health insurance fraud
- Prevent unnecessary emergency visits of patients
- Practice telemedicine
- Enhance the data security
- Improved strategic planning with the health data.

All the abovementioned applications of healthcare, in big data, can save human lives with more accurate and personalized patient care. The applications of healthcare in big data have a great impact on our lives. The researchers, hospitals, and the physicians are continuously analyzing the vast amount of healthcare data for developing new therapies and for finding new treatments. The innovation in healthcare application is evolving due to continuous development in technologies. The big data application in medical industries helps to prevent pandemic and disease with lower cost.

3.6 Challenges with Healthcare Big Data

3.6.1 Challenges of Big Data

The main challenge in big data is to handle and manage the huge amount of assorted, complex, and interconnected information. It is difficult to sort out the data and prioritize the data because of its volume and variety.

Major big data challenges are as follows:

- Lack of proper understanding of the big data
- Data growth
- Combining data from various sources
- Selection of big data tool to capture, store, process, and analysis
- Data security

3.6.2 Challenges of Healthcare Big Data

The healthcare data and its analytics have problems like data accuracy, data security, data sharing, data capturing, and data visualization. The following are the major challenges faced by healthcare organizations with big data [10, 21, 22, 28].

Data Capturing

The main problems in healthcare big data are capturing data. The qualities of the clinical decisions are based on how well the big data is captured. The healthcare organization improves their data capture method by prioritizing the data based on the values.

Data Cleaning

Data cleaning is essential as medical data is captured at a variety of data sources. The data cleaning process checks the data for accuracy and correctness and also ensures whether the data are relevant.

Data Storage

The volume of healthcare data grows exponentially with time. It is difficult for the healthcare organization to store the big healthcare data with the traditional database and in a single system.

Data Security

Data security is the major challenge in the healthcare industry because of the frequent hackings. The challenges of healthcare providers are data encryption, data masking, and other protection methods for the limited data access to the outside people.

Lack of Integration between Administrative Data and Clinical System

Sometimes, there is a gap between the administrative data and the clinical system. Example: variation in the treatment code and the care given to the patients.

Data Sharing

The standard of collecting and storing the big data is different from one healthcare organization with another healthcare organization. The sharing and integrating the different formats of medical records are the real problem with medical data.

Data Updating

The healthcare data is not fixed. Most of the patient data requires frequent updates. For better treatment the up to date and real-time patients' data is necessary. For example, the BP of the patient will vary every day. The old data will lead to wrong decisions in patient's healthcare practice.

The main challenge in healthcare data is maintaining balance between protections of patient's sensitive information, data integrity, and the usage of data for the knowledge discovery process [23].

All these challenges of healthcare data are not insoluble. The advanced technologies and new innovations will address all the abovementioned challenges of big data.

3.7 Conclusion

The objective of this chapter is to give a general idea about overview and characteristics of big data, different steps for deriving value from big data, big data technologies used for every step in the value chain process, use of knowledge systems in big data healthcare, various big data applications, and big data challenges in healthcare organizations. The healthcare organization needs to devote time and resources for implementing the big data value chain. Health is the precious gift of God to humans. A personalized and sustainable patient care service is the need of the hour. Big data analytics is capable of providing knowledge and valuable insights in medical data, which are useful for clinical decision, disease prediction, better treatment, etc. Healthcare big data is a permanent and continuously increasing phenomenon, which needs newer tools and technologies to analyze the data and to get knowledge for the present use and future research works.

References

1. De Mauro, A., Greco, M., Grimaldi, M., A formal definition of Big Data based on its essential features. *Library Rev.*, 65, 3, 122–135, 2016.
2. University of Illinois at Chicago. Big Data and Wearable Health Monitors: Harnessing the Benefits and Overcoming Challenges, 2019. https://healthinformatics.uic.edu/blog/big-data-and-wearable-health-monitors-harnessing-the-benefits-and-overcoming-challenges/.
3. white paper by Intel, in: *French Hospital Uses Trusted Analytics Platform to Predict Emergency Department Visits and Hospital Admissions.*
4. Kozlov, How to Improve EMR/EHR Using Predictive Analytics, 2020. https://www.romexsoft.com/blog/predictive-analytics-for-healthcare/.
5. Manogaran, G., Thota, C., Lopez, D., Vijayakumar, V., Abbas, K.M., Sundarsekar, R., Big Data Knowledge System in Healthcare, in: *Internet of Things and Big Data Technologies for Next Generation Healthcare*, pp. 133–157, 2017.
6. Cawthorne, J. Knowledge Management and Big Data: Strange Bedfellows? 2015. https://www.cmswire.com/social-business/knowledge-management-and-big-data-strange-bedfellows/.

7. Dash, S., Shakyawar, S.K., Sharma, M., Kaushik, S., Big data in healthcare: management, analysis and future prospects. *J. Big Data* 6, 54, 2019. https://doi.org/10.1186/s40537-019-0217-0
8. healthsnap. Structured and Unstructured Health Data: Challenges and Opportunities, 2019. https://healthsnap.io/structured-unstructured-health-data/.
9. Kong, H.J., *Managing Unstructured Big Data in Healthcare System. Healthcare Informatics Research*, vol. 25, 1, 1–2, 2019.
10. Krumholz, H.M., Big data and new knowledge in medicine: the thinking, training, and tools needed for a learning health system. Health affairs (Project Hope). 33, 7, 1163–1170, 2014. https://doi.org/10.1377/hlthaff.2014.0053.
11. Kiran, R., Top Big Data Technologies that you Need to know, 2021. https://www.edureka.co/blog/top-big-data-technologies/.
12. Bhadani, A., Jothimani, D., In Singh, M.K., Kumar, D.G., Big data: Challenges, opportunities and realities, in: *Effective Big Data Management and Opportunities for Implementation*, pp. 1–24, Pennsylvania, USA, IGI Global, 2016.
13. Fang, R., Pouyanfar, S., Yang, Y., Chen, S., Iyengar, S.S., Computational Health Informatics in the Big Data Age. *ACM Comput. Surv. (CSUR)*, 49, 1 – 36, 2016.
14. Bahri, S., Zoghlami, N., Abed, M., Tavares, J.M.R.S., Big Data for Healthcare: A Survey. *IEEE Access*, 7, 7397–7408, 2019.
15. https://missinglink.ai/guides/deep-learning-healthcare/ai-medical-imaging/
16. El Morr, C. and Subercaze, J., Knowledge Management in Healthcare, in: *Handbook of Research on Developments in E-Health and Telemedicine*, pp. 490–510, 2010.
17. Begoli, E. and Horey, J., Design Principles for Effective Knowledge Discovery from Big Data. *2012 Joint Working IEEE/IFIP Conference on Software Architecture and European Conference on Software Architecture*, 2012.
18. Bresnick, J., Healthcare Big Data Analytics: From Description to Prescription. Health IT Analytics, 2015. https://healthitanalytics.com/news/healthcare-big-data-analytics-from-description-to-prescription.
19. Usvyat, L., 4 Types of Healthcare Analytics to Use in Your Practice. Fresenius Medical Care, 2019. https://fmcna.com/insights/education/types-of-healthcare-analytics/.
20. Kaduk, T., 4 Stages of Data Analytics Maturity: Challenging Gartner's Model, 2016. Linkedin. https://www.linkedin.com/pulse/4-stages-data-analytics-maturity-challenging-gartners-taras-kaduk/.
21. SD Global. Big Data in Healthcare - Challenges and Opportunities, 2018. https://www.sdglobaltech.com/blog/big-data-in-healthcare-challenges-and-opportunities.
22. Bresnick, J., *Top 10 Challenges of Big Data Analytics in Healthcare. HealthITAnalytics*, 2019, https://healthitanalytics.com/news/top-10-challenges-of-big-data-analytics-in-healthcare.

23. White, S., A review of big data in healthcare: challenges and opportunities. *Open Access Bioinform.*, 6, 13–1, 2014, doi: 10.2147/oab.s50519.
24. Manogaran, G., Lopez, D., Thota, C., Abbas, K.M., Pyne, S., Sundarasekar, R., Big Data Analytics in Healthcare Internet of Things, in: *Innovative Healthcare Systems for the 21st Century*, pp. 263–284, 2017.
25. Ambigavathi, M. and Sridharan, D., Big Data Analytics in Healthcare. *Tenth International Conference on Advanced Computing (ICoAC)*, 2018, doi: 10.1109/icoac44903.2018.8939061.
26. Singh, S. and Singh, N., Big Data analytics. *International Conference on Communication, Information & Computing Technology (ICCICT)*, 2012, doi: 10.1109/iccict.2012.6398180.
27. Islam, M.S., Hasan, M.M., Wang, X., Germack, H.D., Noor-E-Alam, M., A Systematic Review on Healthcare Analytics: Application and Theoretical Perspective of Data Mining. *Healthcare (Basel, Switzerland)*, 6, 2, 54, 2018, doi: https://doi.org/10.3390/healthcare6020054.
28. Fatt, Q.K. and Ramadas, A., The Usefulness and Challenges of Big Data in Healthcare. *J. Healthc. Commun.*, 3, 2, 21, 2018. doi: 10.4172/2472-1654.100131.
29. Curry, E., The Big Data Value Chain: Definitions, Concepts, and Theoretical Approaches, in: *New Horizons for a Data-Driven Economy*, J. Cavanillas, E. Curry, W. Wahlster (Eds.), Springer, Cham, 2016, https://doi.org/10.1007/978-3-319-21569-3_3.
30. https://healthitanalytics.com/features/top-challenges-of-applying-artificial-intelligence-to-medical-imaging.

4

Big Data for Personalized Healthcare

Dhanalakshmi R.[1] **and Jose Anand**[2*]

[1]*Department of Computer Science & Engineering, KCG College of Technology, Chennai, India*
[2]*Department of Electronics and Communication Engineering, KCG College of Technology, Chennai, India*

Abstract

Big data has turned radically into a technological development that has been applied more often in biomedical healthcare research than earlier. The unparalleled developments in automated gathering of large-scale molecular and clinical data stand as main tasks to data analysis and interpretation, focusing for the growth of novel computational methods. As healthcare data cultivates forever additional heterogeneous and formless data becomes extra problematic to process, researchers have exposed new usages for big data in healthcare research and in the personalization of medication, which are created on numerous features. The formation of influential systems for the active use of biomedical big data in personal medication need noteworthy scientific and technical growths, including project management, infrastructure management, and engineering management. Big data has extensively been used in the business biosphere to achieve considerable amounts of data and forecast results. Since the arrival to maintain medical records electronically, healthcare earners and researchers have introduced big data as the resources to achieve the huge quantities of data made by patients in today's healthcare arena. The goal of the proposed system is to deploy healthcare system as a service over the cloud infrastructure. Most healthcare services are based on the premise that comprises handwritten assessment outcomes, non-digitized scan imageries, transcribed records, and distributed IT schemes. Maintenance and distribution of these medical data becomes difficult and sometimes impossible. Existing systems that are proposed to solve the above problem lack flexibility and dynamicity. This system provides a centralized repository for storing all the medical data and retrieve them anywhere at

**Corresponding author*: joseanandme@yahoo.co.in

A. Jaya, K. Kalaiselvi, Dinesh Goyal and Dhiya Al-Jumeily (eds.) Handbook of Intelligent Healthcare Analytics: Knowledge Engineering with Big Data Analytics, (67–92) © 2022 Scrivener Publishing LLC

any time. System provides diversified interface for patients, doctors, and hospital management. User's progressive medical records that are stored are used for personalized health analysis. On effective implementation of flexible, informative, dynamic, and scalable system, an easy relationship can be established among doctors, patients, and medical data.

Keywords: Big data, healthcare, cloud, personalized medicine, progressive web application, user authentication

4.1 Introduction

Beyond three decades, laptop structures are used considerably in scientific and healthcare structures. In documentation, garage, dispensation, evaluation, and demonstration of affected person's records garage gadgets and server structures are utilized in advanced nations currently. Yet, maximum healthcare structures are constructed on the premise that includes scientific statistics, handwritten outputs, non-digitized pictures, handwritten materials, and uneven IT structures. Distribution of records throughout carriers is incompetent, and unconfident and movability of facts could be very sporadic. Doctors and health practitioner rely upon the scientific staffs for affected person's facts, coordinating their care schedules and different administrative structures activities.

Cloud computing truly gives more than one advantages for agency computing environments; further healthcare cloud presents an infrastructure that lets in hospitals, scientific practices, and study centers to enhance their paintings ability in a first rate extent. All those approaches are bulky and time-consuming. So, there exist many extra demanding situations on this type of structures, wherein a big quantity of statistics is saved and facts necessities are ascendable, flexible, and clean to create, update, control, and get entry; however, safety has topmost priority.

Although richer fitness care and affected person's facts facilitates to make sure progressed and knowledgeable affected person care, catering to this developing facts is not a prosaic job. The facts generated have to now no longer best be saved however have to additionally be archived for numerous years primarily based on the scientific policies of the country.

4.1.1 Objectives

The system is deployed over the internet in cloud infrastructure as software as a service. Cloud provides access to manage huge data sets, and analyses are carried out for both single client and general purpose through parallel execution using the backend environment.

Most healthcare structures are constructed on the idea that includes paper scientific facts, handwritten outputs, non-digitized imageries, handwritten materials, and interrupted IT structures. The principal recognition is the affected person's facts series, garage, entry, evaluation, and presentation. The contemporary affected person's facts series strategies are time-intensive, incompetent, and exhausting for the staffs. Existing structures that are proposed to resolve the above hassle lack flexibility and dynamicity.

Cloud computing gives a couple of blessings for organization computing environment. Besides, facts contained in H-Cloud are highly analyzed and tracked (with the right facts governance) in order that fitness and remedy facts, costs, performance, and efficiency of research are analyzed and represented consequently. With the assistance of H-Clouds, the affected person's facts may be shared effortlessly and greater securely among legal physicians and hospitals. Thus, it presents greater well-timed entry to life-saving facts and decreases the want for replica testing. Evolving this sort of cloud primarily based on healthcare for rural fitness and on the occasion of screw ups is productive and worthy. Other hospitals and physicians are beginning to see cloud primarily based on the scientific facts and scientific photo-archiving services. Objective of this sort of answer is to lessen the burdensome challenge of the doctors, scientific staffs, etc., with progressed scientific structures.

The device is furnished as an innovative net utility that is deployed in cloud over the internet. Cloud-based and facts-pushed platform for the garage and evaluation of multisource healthcare facts is established.

Thus, on effective implementation of flexible, informative, dynamic, and scalable medical and healthcare environment, an easy relationship can be established and be provided to users over simple means to maintain a healthy society.

4.1.2 Motivation

Several works executed the use of the idea of universal and disbursed scientific processing field. A wearable and implantable sensor for disbursed cell tracking of the sufferers is represented [1]. Patients will bring sensors and that transmit the crucial records of sufferers to other devices for similar evaluation and process. This does not fulfill the needs such as scaling, security, and bendy records transmission. Telemedicine is an adaptable conversation for domestic tracking for heart sufferers. This facilitates faraway tracking and prognosis of the sufferers. There also are a few demanding situations on this form of structures such as heterogeneous, guidelines, database framework, formation of server, and records protection. Researchers on these days delivered a multitier telemedicine gadget, which can carry out real-time evaluation on scientific records, and it additionally presents remarks to customers and uploads records in to the server. It additionally suffers from the identical issues with records integration, privateness, and protection. Enabling Secure and Patient centric Access Control (ESPAC) gathers the records from sick people, sends it initially to the central server, and uploaded it into cloud environment. This is made with flexible and dynamic approach with the gadget. In case the local sever at hospital fails because of certain reasons, then the complete records will now no longer be sent. So, those varieties of structure lack protection and privateness of the records.

4.1.3 Domain Description

The amount of healthcare-related data generated day by day is increasing with huge volume. These data are very much necessary for obtaining decision-making and to provide best upkeep to the patients. For this, a simple cost-valued method such as cloud computed is used, in which real-time data collection is made with data storage facility and exchange the information between healthcare centers. Also, for proper utilization and analysis of medical records, it should be devoted to change healthcare data from old-style paper based on the digital setup.

Nonexistence of scalable, securable, and flexible data transfer infrastructure is an important issue. This particular problem has been solved by moving to a cloud platform. Since cloud is physically distributed and virtually presented as a collated resource, all underlying implementation and security assertions are provided and ensured by the cloud service provider. There are few difficulties in this method of schemes like scalable, storage of heterogeneous data, database administration, configuration of server, and data security. These problems are addressed by moving to a NoSQL database solution like MongoDB, where collection structure can be created as desired.

4.1.4 Organization of the Chapter

This chapter is structured as follows: Section 4.2 consists of related works to the proposed system for digitizing medical records and performs analysis using cloud infrastructure. Section 4.3 entails system analysis that describes the problem identification, component requirements, use cases, the proposed system architecture, and explanation on every module. Section 4.4 expresses the system implementation and the various algorithms. Section 4.5 discusses about the various output screenshots obtained and its discussion. Section 4.6 consists of conclusion and the future work for the project.

4.2 Related Literature

This section describes the related works in the field of big data in healthcare applications.

4.2.1 Healthcare Cyber Physical System Architecture

A cyber-physical gadget for patient-centric healthcare programs and services, known as Health-CPS, is constructed on cloud and huge statistics evaluation technology to offer a greater handy carrier and surroundings of healthcare. A clever fitness gadget assisted through cloud and huge statistics, which incorporates a unified statistics series layer for the mixing of public clinical sources and private fitness devices, a cloud-based and statistics-pushed base for multisource heterogeneous healthcare statistics garage and evaluation, and a unified API for builders and a unified interface for customers [2]. This structure has three layers, namely, statistics series layer, statistics control layer, and alertness carrier layer. Data series layer includes statistics nodes and adapters and presents a unified gadget that gets the right of entry to interface for multisource heterogeneous statistics from hospitals, internet, or user-generated information. Data control layer includes a Distributed File Storage (DFS) module and a Distributed Parallel Computing (DPC) module. Application carrier layer presents the primary visible statistics evaluation results to the customers. It additionally presents an open unified application programming interface for the builders at user-oriented software to offer rich, professional, and personalized healthcare services [3]. Cyber bodily structures are tested using a sequence of critical features to detail the anticipated capability of infusion and using STRIDE Threat Model to decompose the safety troubles and to deal with them [4].

4.2.2 Healthcare Cloud Architecture

Healthcare cloud structure is in which Wireless Sensor Network (WSN) is included with cloud computing to store patient's fitness facts in the cloud, and this structure is scalable and capable of save a big quantity of facts created through sensors [5]. As patient's healthcare facts are very touchy, the gadget is proposed to have the safety system in cloud to assure facts confidentially, integrity, protection, and high-quality entry to control. This model takes into consideration classes of users, healthcare professionals (doctors, nurses, and clinical staffs), and sufferers. The protection supervisor tests person authentication request for logging into the gadget through proper control approach, in which license key (person name and password) is used. The carrier supervisor is liable for taking person request and grants the essential offerings in line with the request. Cloud database is for garage of the facts, which might be uploaded from the WSN. It incorporates all updated fitness facts of the sufferers, which might be transmitted to it periodically [6].

4.2.3 User Authentication Management

Users can effortlessly get admission to their private records via cloud offerings and make it to be had to diverse offerings throughout the internet. An identification control system assists to authenticate customers in the cloud and log onto the cloud device [7]. Current authentication scheme is primarily based on passwords that have few inherent dilemmas and poses great risks. Thus, IDM device needs to constantly be capable of guard personal and personal records of the customers and processes. While customers engage with a provider furnished through the cloud, this provider would possibly want to assure that their identification is included from different offerings with which it interacts [8].

4.2.4 Healthcare as a Service (HaaS)

Health cloud is constructed on carrier-orientated structure and includes some of offerings for acting diverse healthcare operations. Irrespective of the whether or not the offerings are hosted on premise or on cloud, those offerings can be able to utilizing the cloud infrastructure. The offerings presented consist of garage and indexing of scientific records, photograph processing offerings, and record generation, charting, and fashion evaluation of healthcare records [9]. All of those offerings are secured with identification and get admission to manipulate functions that are able to utilize a hospital's present identification and authentication device like Lightweight Directory Access Protocol (LDAP) and Active Directory. The optimized records saved are indexed, configuring of the records backup, and archival is completed effortlessly and is executed at some stage in the configured intervals. In case of failures, the carrier additionally gives choice for geo-repetition of the scientific records. The carrier indexes the records; this is saved in cloud, and this enables in a quicker retrieval of records [10].

4.2.5 Reporting Services

Reporting offers a clean mechanism of sharing scientific reviews inside distinctive entities in a clinic/clinic organization and among distinctive fitness care providers. This facilitates to enhance affected person care immensely, lessen replication of scientific data, lessen scientific charges, and enhance collaboration among healthcare providers. For this optimized routes, genetic algorithm [11], evolutionary algorithms [12] and Q-learning-based approaches [13] are used.

4.2.6 Chart and Trend Analysis

The information accumulated in a health center in an afternoon is huge and is disparate. The charting and evaluation offerings assist to mix this information and carry out numerous forms of evaluation on it. These offerings enable to display the fitness of a affected person, enhance remedy efficiency, and take important guided choices on affected person's remedy beneath important situations [14]. The evaluation provider produces a wealth of data, which may be very beneficial for numerous functions like clinical research, preventive fitness care, and selling higher requirements in public fitness. This form of evaluation would require in-depth computing, and the cloud infrastructure affords on-demand and scalable compute resources [15].

4.2.7 Medical Data Analysis

Data mining or evaluation and sample prediction engine offer beneficial facts that aid selection, making through popular record mining techniques [16]. The version is aimed toward offering a platform with a view to properly use resources of the operations of HDOs in growing international locations through securing clinical information and offering facts with a view to help numerous stakeholders and coverage makers in selection making in the region. Healthcare groups continuously cope with massive quantity of digital information and virtual snapshots from which beneficial expertise approximately analysis of diseases, remedy administration, and managerial choices may be drawn [17]. The flaws in the end result are related to the orthodox processes to clinical offerings, and many lives had been misplaced, while numerous others have skilled severe deterioration of their fitness status. The saved information is wealthy treasured facts that may be used for enhancing the healthcare offerings supplied to the people. This version additionally offers a platform that decreases the boundaries for innovation, modernization of healthcare facts era systems, and different healthcare-associated applications [18].

4.2.8 Hospital Platform Based On Cloud Computing

In cloud-based e-health with the speedy financial and social development, the scientific statistics boom dramatically. In order to satisfy the desires of scientific statistics, clinical control has to constantly enhance the extent of modem statistics, which is control through the usage of superior technology and era innovation. At present, a few troubles in the conventional scientific control device are as follows: the garage capability and fine of digital report is restricted to hardware devices; with amount of statistics increased, the velocity of the device will gradual down due to channel get right of entry to mode and the bandwidth; due to the fact the contemporary backup mode is simple, it is not always healthy to the safety and long-time period garage of scientific files; and the conventional statistics garage mode is not always conducive to the sharing of resources [19].

4.2.9 Patient's Data Collection

A cloud computing answer for patient's facts series in fitness care establishments counseled the subsequent excessive degree necessities. It needs to enforce the strategies to collect, to procedure and to distribute patient's essential facts. It needs to be open, bendy, and extensible; that is, it needs to assist heterogeneous equipment in distinct numbers that may be delivered to device on ad hoc basis. It needs to be secure; that is, the device needs to assure the integrity and confidentiality of clinical facts. It needs to be manageable; it needs to offer manipulate over the myriad of computing gadgets related to the environment. It needs to be reliable; that is, it needs to assure device availability in spite of variations in operational situations and punctual issues. It needs to be scalable, to assist the deployment in huge fitness care environments and the combination of various establishments. It is optimal for computing resources; the software need to execute inexpensive on low profile computing gadgets [20].

4.2.10 H-Cloud Challenges

The minimized acceptance of cloud version in the fitness discipline is more often than not because of essential issues related to safety and interoperability. Those troubles want to be addressed so as to triumph over doubts while transferring to cloud and taking benefit of all of the answers and upgrades that it brings [21]. Security concerns in healthcare information not like different sort of information have strict confidential, privateness, and safety issues [22]. HIPAA compliance is the maximum essential requirement while transferring clinical statistics to the cloud. Migrating complete information storages to a third celebration agency is not always a clean undertaking to do, particularly while transferring touchy facts together with healthcare information. Even extra sturdy safety must be confident due to the fact that extra issues will stand up with get admission to controls, audit controls, authentication, authorization, transmission safety, and garage safety that allow you to keep away from exposing the facts to unauthorized entities [23].

4.2.11 Healthcare Information System and Cost

The courting among healthcare records device and price in health center. This examination has changed into purpose of examining the connection among Healthcare Information System (HIS) and price in health center. It additionally analyzed which organizational elements have an effect on the adoption of records device in hospitals. The examine covered 577 hospitals in the statistical analysis. The HIS changed into measured through three indicators, that is primarily based on the variety of software structures in three center health center functions (administration, management, and clinical). This examination determined that HIS measuring through the records device programs had courting with the decreased overall price [24].

4.3 System Analysis and Design

Most healthcare services are based on the premise that comprises handwritten test outcomes, non-digitized scan imageries, transcribed transcripts, and distributed IT services. Maintenance and sharing of these medical data becomes difficult and sometimes impossible. Existing systems that are proposed to solve the above problem lack flexibility and dynamicity.

4.3.1 Proposed Solution

The proposed system provides a centralized repository for storing all the medical data and to retrieve the data from anywhere and at any time. System provides diversified interface for patients, doctors, and hospital management. User's progressive medical records that are stored are used for personalized health analysis. On effective implementation of flexible, informative, dynamic, and scalable system, an easy relationship can be established among doctors, patients, and medical data.

Since cloud is physically distributed and virtually presented as a collated resources, all underlying implementation and security assertions are provided and ensured by the cloud service provider. There are few trials in this way of schemes like scalability, storage of heterogeneous data, database schema, data safeness, and server conformation. These problems are addressed by moving to a NoSQL database solution like MongoDB, where collection structure can be created as desired.

4.3.2 Software Components

The software components used in this project are Node.js, MongoDB, and Bootstrap. It also contains supporting APIs such as Mongoose, Service worker, Web push, MessageBird, Express, Chart.js, and Embedded JS.

4.3.3 System Design

The system offers a cloud-based service for digitizing the medical records and provides a personalized analysis for the patients, doctors, and the hospital management. For the system to be platform-independent and ease of access from any device, the entire system is delivered as a web application. The system has four different types of users; doctors, patients, hospital management, and pharmacists. The users are differentiated based on their login credentials. Each type of user has got their own interface that has a personalized dashboard and other operations related to them. The design of the system is modular and maintains consistency over the medical data stored.

Node.js is used for better code abstraction and provides modularity to system by easily integrating different functionalities without complexity. It handles all the GET and POST calls of each service and redirects it to appropriate APIs. MongoDB is used to reduce data storage and management overhead as it is unstructured database unlike a relational database, which is structured. Bootstrap is used to design a responsive and elegant interface such that it adapts to any viewing resolution. Chart.js is an open source library that is used to generate responsive charts, which scales up/down based on the data from the analysis service.

Security is ensured during every access to patient's medical records by enforcing a two-step one-time password (OTP) authentication in addition to the username and password. It is used for the login for authenticating into each type of user into the system. The system also provides a notification service that pushes notification to the user's device about the scheduled medications at appropriate time. This notification service is handled by a proxy server, called service worker, within the user's device. Entire design of the system is focused on modularity, scalability, robustness, and security.

4.3.4 Architecture Diagram

Various stakeholders and their interaction toward the system have been depicted below. The data providers for the system are hospitals, scan centers, and doctors themselves. With the given respective data of users, they are provided with analysis and consolidated storage and management. Effective chain of information sharing between various stakeholders is also achieved. Figure 4.1 shows the architecture of the system.

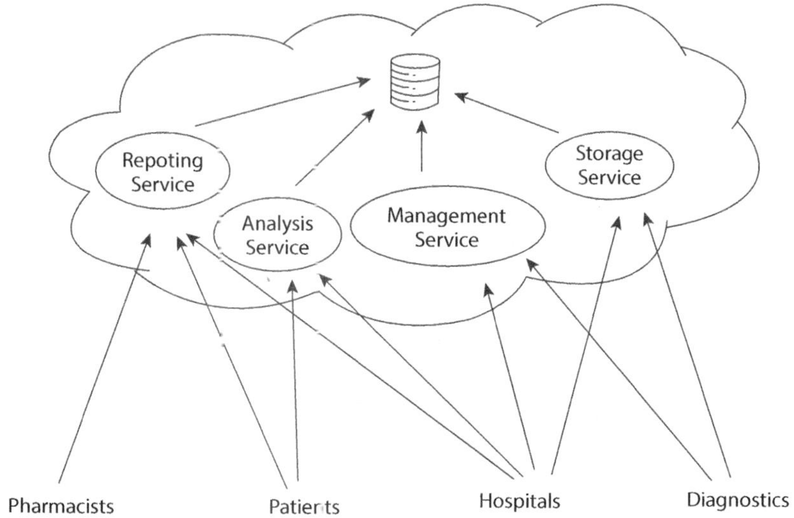

Figure 4.1 Architecture diagram.

4.3.5 List of Modules

The following are the list of modules in the system:

- Storage Service
- Reporting Service
- Analysis Service
- Management Service

Storage service is used for uploading patient's medical records provided by the doctor and diagnostic center to the database. Patients, doctors, and pharmacists can view the patient's medical records with the help of reporting service. Analysis service provides personalized analysis for the users based on the data stored in the database. Hospitals can manage the hospital, doctor and patient information with the help of management service. Figure 4.2 shows the functional parts of the system and the interaction between them. The different modules in the system are explained in the following.

Storage Service: The system provides centralized cloud storage for storing all types of medical records such as prescriptions, test results, and scan reports. The medical records are stored in digital format and raw documents. After authenticating the user, the scan center can able to upload the test results and scan reports to the database. It promotes scalability with less effort due to its well-structured storage of medical records. MongoDB cloud storage is used for storing medical records. Figure 4.3 shows the interaction flow between the doctor and the storage service.

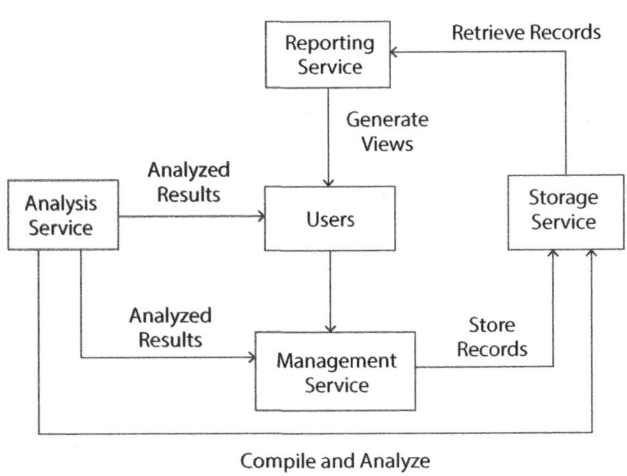

Figure 4.2 Functional block diagram.

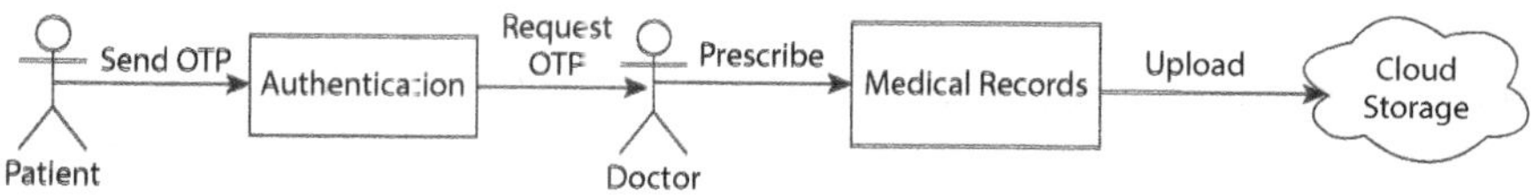

Figure 4.3 Storage block.

For storing files, GridFS specification is used, and the larger files such as test reports and scan reports are fragmented into chunks each having the size of 255 KB, while the last chunk will contain only the necessary, which may be less than or equal to 255 KB. GridFS involves two collections, namely, chunk and files, while chunk collection stores all the chunks and files; collection stores the files metadata, which is used to link all the chunks respective to the particular file. This provides additional security for patient reports since the file is fragmented into multiple chunks and stored in the cloud instead of single direct file. The medical records that are uploaded to the database are cannot be modified or deleted by any user.

Reporting Service: This service allows doctors, patients, pharmacists, and scan centers to view the patient's medical records. The doctors can use this service to review the patient's medical history to get clear knowledge about the patient. The patient can get an insight about the medical condition, history of visits, medications taken, ongoing medications, notifications, and laboratory reports. The pharmacists can view the prescriptions of the patient to provide required medicines. The scan centers can view the tests and scans prescribed by the doctor for the patient. Figure 4.4 illustrates the reporting service.

The simplified WebApp made up of Node.js is provided for the users, in which the users can view the stored medical records of the patients. The users can view the patient's medical record only after the authentication by the patient. User who wants to access the patient's medical record has to enter the patient's medical id or the registered mobile number of the patient, and the OTP is messaged to the patient's registered mobile number. The user should enter the OTP received on patient's mobile number within 150 seconds of it is received by the patient. After 150 seconds, the OTP will become invalid and the user should request for new OTP.

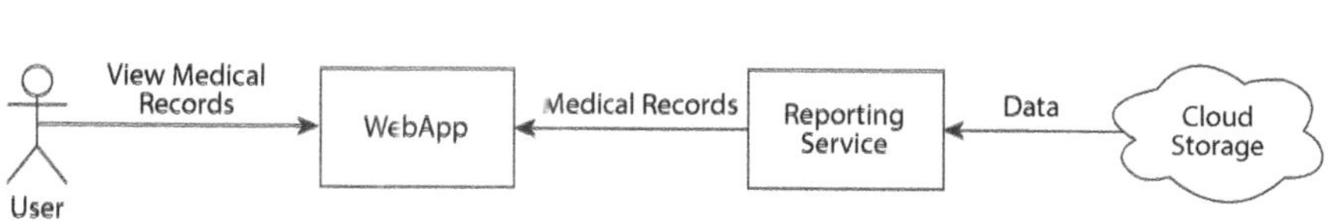

Figure 4.4 Reporting block.

The doctor can view the medical records prescribed by him/her without the authentication by the patient.

Analysis Service: Analysis service compiles and analyzes medical data of the patients. The compiled data is then used for providing analysis in form of various charts using Chart.js. Patients are given an analysis on laboratory reports and previous medical conditions to improve treatment efficiency. The data collected in a hospital is also used for trend analysis. Disease outbreak can be notified to promote better standards in public health. Doctors get the analysis based on the review and the ratings provided as feedback by the patients, which, in turn, improves their services. There are similar kind of various analyses that can be done for all patients, doctors, and hospitals. Figure 4.5 illustrates the analysis service.

Management Service: This service is an additional service that provides a dashboard view for the hospitals. In the database within the hospital collection, each hospital contains separate documents that consist of all the doctors working in that hospital along with their details. The dashboard contains status information of the doctor, numbers of inpatients and outpatients, admin operations, feedbacks, and in-hospital notices. When a patient is admitted, a new case is created for the patient, and case details such as case name, doctor's name, date of admission are provided. Figure 4.6 illustrates the management service.

This service also provides a function to add a newly joined doctor to the hospital and can keep track of the activities of that doctor. Notices created by the hospital are shown to all the doctors working in that hospital. During the admission of the inpatient, a new case is created by the hospital that contains the patient's name, reason for admission, doctor allocated to the patient, and date of admission. The status of the patient was kept as inpatient as long as patient stays within the hospital; once the patient gets discharge, the status changes to outpatient.

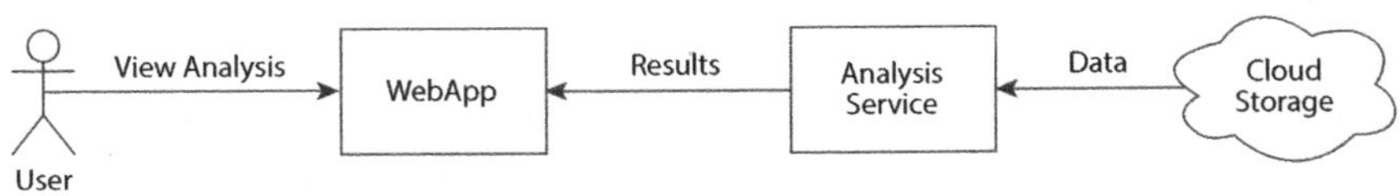

Figure 4.5 Analysis block.

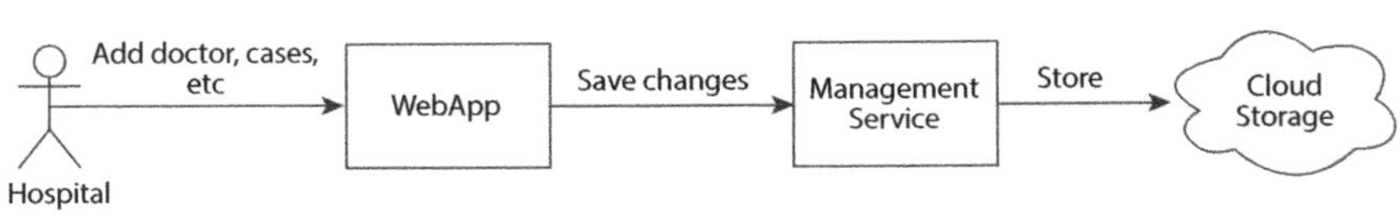

Figure 4.6 Management block.

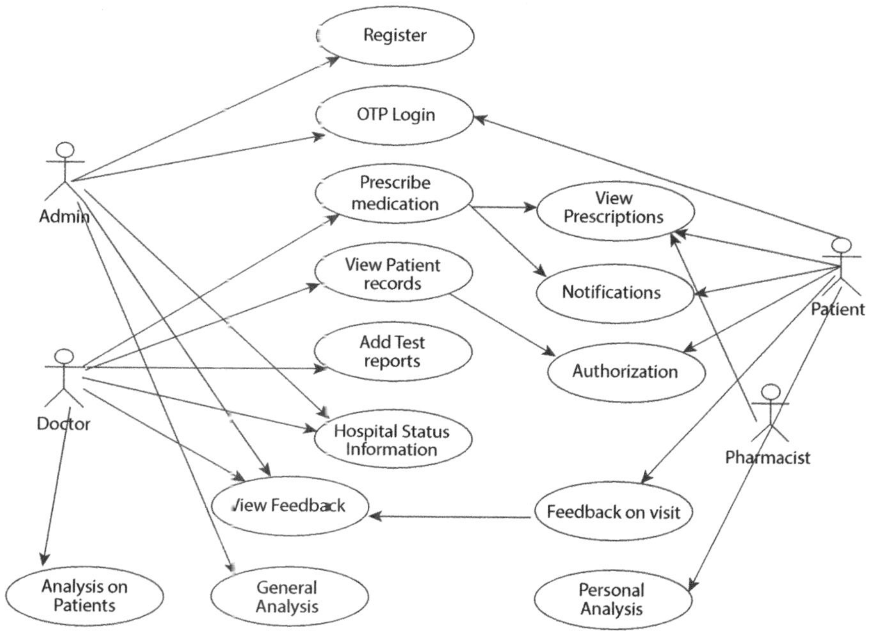

Figure 4.7 Use case diagram.

4.3.6 Use Case Diagram

Figure 4.7 shows the interaction among the patient, doctor, and other stakeholders with the system.

4.3.7 Sequence Diagram

Figure 4.8 illustrates that the admin is responsible of registering doctors, pharmacists, hospitals, and diagnostic centers. All stakeholders are provided with two-step login process. Doctors access patient's medical record only after users authenticate through an OTP initiated by the doctor. The same applies to both pharmacists and scan centers for viewing and reports uploading.

Upon authenticating, doctors will be able to prescribe medication for the respective patient. Patients are also provided with push notification for medicines based on the prescription. Patients will also be able to provide feedback on doctors and hospitals. Pharmacists will be able to only view prescriptions after authentication. With the data set generated, it is possible to provide various analyses and identify disease breakout.

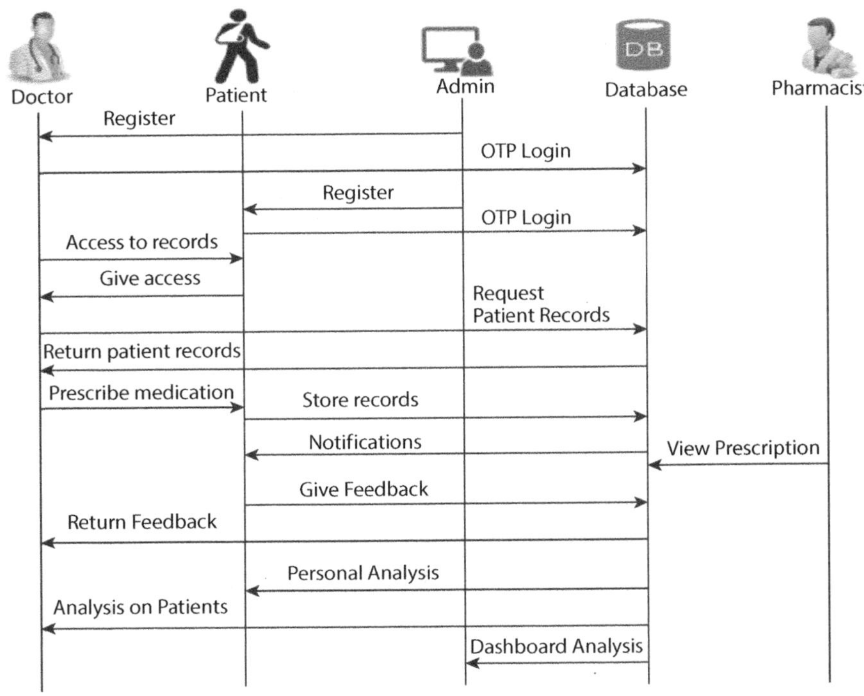

Figure 4.8 Sequence diagram.

4.3.8 Class Diagram

Figure 4.9 illustrates that there are four classes, namely, admin, doctor, express, and patient. The express class consists of all middleware; the system uses such as view renderer, push notifications, database connections, and authorization functions.

The patient class contains patient data and also related functions such as login, viewing records, and analysis. Admin class is responsible for registering doctors, registering the cases for the patient, providing authorization for doctors to hospitals, and also maintaining the overall system. Doctor class contains doctor information also functionalities like login, prescribe medication, and view user prescriptions after authenticated by the user. Express class acts as middleware for all the requests made by the client and response from the server for the respective request.

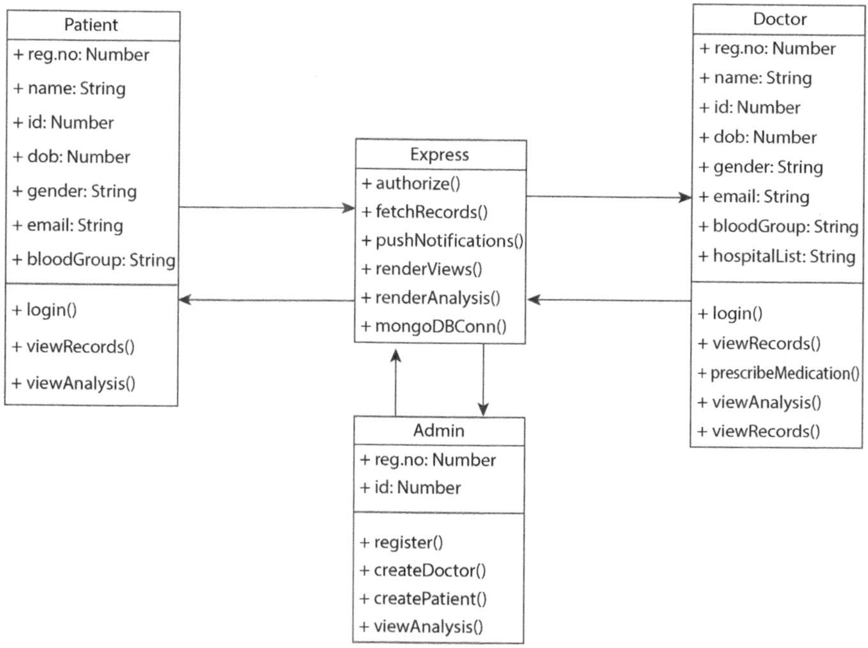

Figure 4.9 Class diagram.

4.4 System Implementation

4.4.1 User Interface

An interface is provided for users to interact with the system. The interface is designed using HTML, CSS, Bootstrap, and Node.js. While HTML is used to display contents in the user interface, CSS and Bootstrap are used for designing the user interface, and Node.js is used for handling HTTP request and response from the client and server and also used for sharing data within the web app.

NodeJS: Node.js is an open-supply, cross-platform JavaScript runtime surroundings that uses JavaScript code to browse, constructed on Chrome's V8 JavaScript engine as an asynchronous occasion pushed JavaScript runtime; Node is made to construct scalable community applications. Node takes the occasion version a piece further. It provides an occasion loop as a runtime assemble in preference to as a library. In Node, there may be no such start-the-occasion-loop call. Node surely enters the occasion loop after executing the enter script. Node exits in the occasion loop, while there are not any extra callbacks to perform. This conduct is like browser JavaScript—the occasion loop is hidden from the user.

Bootstrap: Bootstrap is an unfastened front-stop model for quicker and simpler internet applications. Bootstrap consists of HTML and CSS primarily based on the layout patterns for typography, forms, buttons, tables, navigation, modals, photo carousels, and plenty of other, in addition to non-compulsory JavaScript plugins. Bootstrap additionally offers the capacity to without difficulty create responsive designs. Responsive net layout is set growing net websites, which routinely regulate themselves to appearance correct on all devices, from small telephones to huge desktops.

Embedded JavaScript: EJS is an easy templating language, where we generate HTML markup with undeniable JavaScript. There is no religious a way to approximately arrange things. There is no reinvention of generation and control flow. It is simply undeniable JavaScript. Everyone recognize how rapid V8 and the alternative JavaScript runtimes have gotten. EJS caches the intermediate JS features for immediate execution.

Express: Express is a minimum and bendy Node.js net utility framework that offers a sturdy set of functions for net and cell applications. It is an open supply framework evolved and maintained with the aid of using the Node.js foundation. Express is an internet utility framework that offers with an easy API to construct websites, net apps, and lower back ends.

4.4.2 Storage Module

Doctors and diagnostic centers upload the patient's medical records to the database. MongoDB is used as a database to store patient's medical records and all other details necessary for system. GridFS is used to store all kinds of files provided by diagnostic centers.

MongoDB: MongoDB is a cross-platform and file-orientated database that provides excessive performance, excessive availability, and smooth scalability. It works on idea of series and file. Database is a bodily box for collections. Each database receives its very own set of documents at the record device. An unmarried MongoDB server usually has a couple of databases. Collection is a set of MongoDB files. It is the equal of an RDBMS table. A series exists inside an unmarried database. Collections do now no longer put in force a schema. Documents inside a set will have one-of-a-kind fields. Typically, all files in a set are of comparable or associated purpose. A file is a hard and fast of key-fee pairs. Documents have dynamic schema. Dynamic schema manner that files in an equal series do now no longer want to have the equal set of fields or structure, and not unusual place fields in a set's files can also additionally keep one-of-a-kind kinds of records.

GridFS: GridFS is the MongoDB specification for storing and retrieving massive documents including images, audio documents, video documents, etc. It is sort of a record device to save documents; however, its records are saved inside MongoDB collections. GridFS has the functionality to save documents even more than its file length restriction of 16 MB. GridFS divides a record into chunks and shops every bite of records in a separate file, every of most length 255 k. GridFS via way of means of default makes use of collections fs.documents and fs.chunks to save the record's metadata and the chunks. Each bite is recognized via way of means of its particular_id ObjectId area. The fs.documents serves as a figure file. The files_id area in the fs.chunks file hyperlinks the bite to its figure. For storing files, GridFS specification is used and the larger files such as test reports and scan reports are fragmented into chunks each having the size of 255 KB, while the last chunk will contain only the necessary which may be less than or equal to 255 KB. GridFS involves two collections, namely, chunk and files, where chunk collection stores all the chunks and files collection stores the files metadata that is used to link all the chunks respective to the particular file. This provides additional security for patient reports since the file is fragmented into multiple chunks and stored in the cloud instead of single direct file. The medical records that are uploaded to the database are cannot be modified or deleted by any user.

4.4.3 Notification Module

The personalized notifications for the patients are scheduled based on the medications provided by the doctor. The patients can get notification even without internet connection with the help of service worker.

Service Worker: A carrier employee is a script that the browser runs in the heritage, become independent from an internet web page, establishing the door to capabilities that do not want an internet web page or consumer interaction. Today, they already encompass capabilities like push notifications and heritage sync. It lets in guide offline experiences, giving builders entire manage over the experience. It is a JavaScript Worker, so it cannot get entry to the DOM directly. Instead, a carrier employee can speak with the pages it controls via way of means of responding to messages dispatched through the postMessage interface, and people pages can manage the DOM if needed. Service people make great use of promises. Service worker's functionalities are core for the proposed system. Only through service worker, users receive push notifications from server to registered devices. It also provides offline functionalities for the webapp.

Web-Push: It is helps in broadcasting short messages to the subscriber of the web application and help to support in user engagement with the application to notify every subscriber. It is really hard for any organization to send email notification for a short messaging event loop or broadcast messages like offers and updates of short messages, and it also increases the cost of server operations so the solution is Web Push Notification. Web push calls for that push messages caused from a backend be achieved through the Web Push Protocol; if one desires to ship facts with the frenzy message, then one ought to additionally encrypt that facts consistent with the Message Encryption for Web Push spec. This module makes it smooth to ship messages and also will manage legacy help for browsers counting on GCM for message sending delivery. The users are notified with prescribed medicines through web push. The data stored in the database is fetched by a scheduled task periodically, formatted, and then pushed to the user's registered devices. It has a timeout of 90 minutes in case of endpoint being offline. Message payloads are encrypted using VAPID keys generated at the server side.

4.4.4 Middleware

Middleware makes it less complicated for software program builders to put into effect conversation and input/output, so as to consciousness at the particular reason in their utility. The time period middleware is utilized in different contexts as well. Middleware is on occasion utilized in a comparable feel to a software program driver, an abstraction layer that hides element approximately hardware gadgets or different software program from a utility. Mongoose, bodyParser, and Express-Session are the middleware used in the system.

Mongoose: Mongoose gives a straightforward schema primarily based on the option to version the utility facts. It consists of integrated kind casting, validation, question building, commercial enterprise good judgment hooks, and more. Mongoose is an Object Data Modeling (ODM) library for MongoDB and Node.js. This achieves relations among facts, gives validation, and is translated among gadgets in code and the illustration of these gadgets in MongoDB. It is a schema-less, NoSQL report database. It can keep JSON files, and the shape of those files can range because it is not always enforced like SQL databases.

Body-Parser: bodyParser handles HTTP POST request in Express.js. bodyParser extracts the complete frame part of an incoming request move and exposes it on req.frame. It is a frame parsing middleware. The bodyParser item exposes numerous factories to create middleware. All middleware will populate the req.frame belongings with the parsed frame, while the Content-Type request header suits the kind option, or an empty item () if there has been no frame to parse, the Content-Type changed into now no longer matched, or mistakes occurred. This lets us to submit facts from one web page to the alternative or to the script that makes use of the facts in the question formation and execution of the equal.

Express-Session: HTTP is stateless; in an effort to companion a request to another request, one wishes a manner to keep person facts among HTTP requests. Cookies and URL parameters are each appropriate approaches to move facts among the customer and the server. But, they are each readable and at the customer side. Sessions resolve precisely this problem, assign the customer an ID, and make all in addition requests the usage of that ID. Information related to the customer is saved at the server connected to this ID. Whenever a request is crafted from the equal customer again, their consultation facts are saved (for the reason that the server changed into now no longer restarted), and the extra residences may be introduced to the consultation item.

4.4.5 OTP Module

A two-step authentication is required for the user to login to the system. The first step is to provide username and password, and the second step is to provide the OTP sent to the user registered mobile number during the time of login.

Message-Bird: MessageBird's APIs connect the website or application to mobile operators around the world and seamlessly integrate SMS, chat, and voice functionality. MessageBird's APIs use HTTP Verbs and a RESTful endpoint structure. Get admission to secret's used because the API Authorization framework. Request and reaction payloads are formatted as JSON (even though it affords a GET opportunity for requests): the use of UTF-8 encoding and URL encoded values. This API is used for two-step login verification and also for doctor authorization by patient to allow access to the medical records.

Analysis Module: The personalized analysis for the patients, doctors, and hospitals is provided with the help of data stored in the database using the Chart.js library.

ChartJS: Chart.js is a JavaScript library that permits drawing one of a kind variety of charts via the usage of the HTML5 canvas in detail. Chart.js is a clean manner to encompass animated and interactive graphs at the internet site for free. The charts are responsive, so they will adapt primarily based on the distance available. Chart.js charts are rendered on consumer supplied canvas elements. Thus far, the consumer creates the canvas detail in a manner that is on hand. The canvas detail has help in all browsers and could render on display; however, the canvas content material will now no longer be on hand to display readers. With canvas, the accessibility needs to be delivered with ARIA attributes at the canvas detail or delivered the usage of inner fallback content material located in the beginning and remaining canvas tags. Chart.js is being used to render various analyses in the form of graphs in the dashboard of respective stakeholders.

4.5 Results and Discussion

This section shows the screenshots taken from the system. In the login page for the user, the mobile number needs to be entered and an OTP is sent out to the user. When the correct OTP is submitted, it redirects the user to their corresponding dashboard. The next page shows a list of hospitals the doctor is associated. The doctor, on clicking the appropriate hospital, can navigate to the corresponding dashboard. For the doctors, an OTP authentication is also required to access the patient's medical records of the user. On successful verification, patient's medical records are brought into the view. The doctor can see all reports, prescriptions, and cases on further screens. Figure 4.10 demonstrates the patient's medical records grouped and organized for each case. Clicking on a case will show all the related prescriptions of that case.

Upon clicking, the prescription will display the medicines and other information provided in the prescription. Doctors can add medicines to prescriptions for patients. Doctors can add multiple medicines at a time and also provide relevant information, which aids the patient's awareness. Then, doctors grant notification permission for receiving medicine alerts

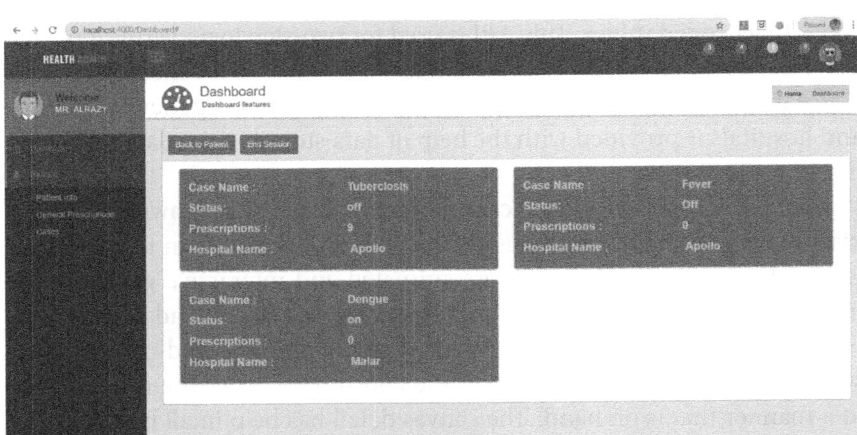

Figure 4.10 Cases of patients.

as notification on registered devices (endpoints). The endpoints or devices need to register for obtaining the alerts and viewing the information. Figure 4.11 shows the notification messages of medicines received on the registered endpoint respective to the user.

Figure 4.12 displays admin dashboard, which contains various analyses and also other functionalities that include adding doctors, adding patient, and creating notices.

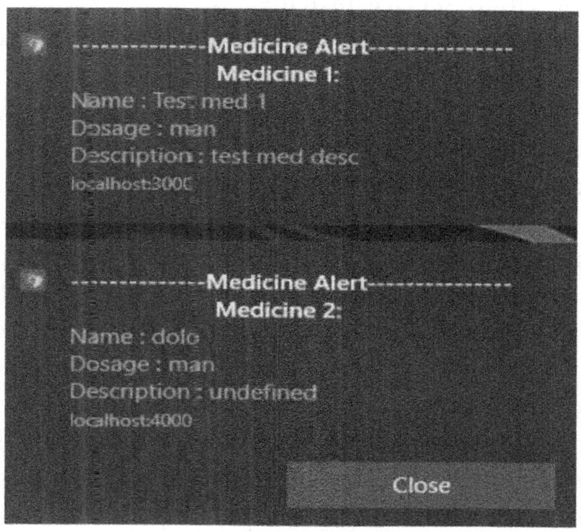

Figure 4.11 Notifications of medicines to endpoints.

Figure 4.12 Admin dashboard.

4.6 Conclusion

The system is deployed over the internet in cloud infrastructure as software as a service. Cloud provides access to manage huge data sets, and analyses are carried out for both single client and general purpose through parallel execution using the backend environment. Since cloud is physically distributed and virtually presented as a collated resource, all underlying implementation and security assertions are provided and ensured by the cloud service provider. Most healthcare structures are constructed on the premise that includes paper scientific records, hand-written check results, non-digitized images, handwritten notes, and fragmented IT structures. The essential consciousness is the affected person's records series, storage, access, analysis, and presentation. The modern affected person's record series strategies are time-consuming, inefficient, and onerous for the staffs. Users' progressive medical records like prescriptions test results and scan reports are stored and also provided with personalized health analysis from the stored data. On effective implementation of flexible, informative, dynamic, and scalable medical system, a digital healthcare environment is established and provided to users to maintain a healthy society.

In the future, the system can be integrated with functionalities for medical insurance verification and application while ensuring patient privacy. Pharmaceutical industries can get data of the patients anonymously with their consent and use it for enhancing production of the medicine. Nearby blood donors can be easily identified and notified at the time of emergency. The admin panel can be loaded with more options to manage resources of the hospital easily. When the application is used in large scale, it can be used for scheduling appointments with doctors. Since all the medical data are stored at on place, there is a good scope for machine learning to deliver and enhance better user experience.

References

1. Pijush K.D.P., Bijoy K.U., Saurabh P., and Tangos P., Chapter 1 – "Internet of Things, Smart Sensors, and Pervasive Systems: Enabling Connected and Pervasive Healthcare". ed(s): Nyanja D., Amira S. Ashour, Chintan Bhatt, and Simon James Fong,. In *Advances in ubiquitous sensing applications for healthcare, Healthcare Data Analytics and Management, Academic Press*, 1–58, 2019.

2. Zhang, Y., Qiu, M., Tsai, C.W., Hassan, M.M., Alamri, A., "Health-CPS: Healthcare Cyber-Physical System Assisted by Cloud and Big Data". *IEEE Syst. J.*, 11, 1, 88–95, 2017.
3. Gupta, S., Mukherjee, T., Varsamopoulos, G., Benergee, A., "Research Directions in Energy Sustainable Cyber Physical Systems". *Sustain. Comput. Inform. Sys.*, 1, 1 57–74, 2011.
4. Seifert, D. and Hassan, R., "A Security Analysis of Cyber Physical Systems Architecture for Healthcare". *Computers*, 5, 27, 1–24, 2016.
5. Onik, M.F.A., Anam, K., Rashid, N., "A Secured Cloud based Healthcare Data Management System". *Int. J. Comput. Appl.*, 49, 12, 24–30, 2012.
6. Monisha, V.A. and Jose, A , "Analysis of Dynamic Interference Constraints in Cognitive Radio Cloud Networks". *Int. Adv. Res. Sci. Commun. Technol.*, 6, 1, 815–823, 2021.
7. Subashini, S. and Kavitha, V., "A Survey on Security Issues in Service Delivery Models of Cloud Computing". *J. Netw. Comput. Appl.*, 34, 1–11, 2011.
8. Indu, I., Rubesh Anand, P.M., Bhaskar, V., "Identity and Access Management in Cloud Environment: Mechanisms and Challenges". *Eng. Sci. Technol., an Int. J.*, 21, 4, 574–588, 2018.
9. John, N. and Shenoy, S., "Health Cloud-Healthcare as a Service (HaaS)". *IEEE International Conference on Advances in Computing, Communications and Informatics*, pp. 1963–1966, 2014.
10. Aditya, R.R., Ajay, H., Balavanan, M., Lalit, R., Anand, J., "A Novel Cardiac Arrest Alerting System using IoT". *Int. J. Sci. Technol. Eng.*, 3, 10, 78–83, 2017.
11. Anand, J., Perinbam, J.R.P., Meganathan, D., "Design of GA-based Routing in Biomedical Wireless Sensor Networks". *Int. J. Appl. Eng. Res.*, 10, 4, 9281–9292, 2015.
12. Anand, J., Perinbam, J.R.P., Meganathan, D., "Performance of Optimized Routing in Biomedical Wireless Sensor Networks using Evolutionary Algorithms". *Comptes rendus l'Academie bulg. Des. Sci.*, 68, 8, 1049–1054, 2015.
13. Anand, J., Perinbam, J.R.P., Meganathan, D., "Q-Learning-based Optimized Routing in Biomedical Wireless Sensor Networks". *IETE J. Res.*, 63, 1, 89–97, 2017.
14. Islam, M.S., Hasan, M.M., Wang, X., Germack, H.D., Noor-E-Alam, M., "A Systematic Review on Healthcare Analytics: Application and Theoretical Perspective of Data Mining". *Healthcare*, 6, 54, 1–43, 2018.
15. Raghupathi, W. and Raghupathi, V., "Big Data Analytics in Healthcare: Promise and Potential". *Health Inf. Sci. Syst.*, 2, 3, 1–10, 2014.
16. Samuel, O.W., Omisore, M.O., Ojokoh, B.A., Atajeromavwo, E.J., "Enhanced Cloud based Model for Healthcare Delivery Organizations in Developing Countries". *Int. J. Comput. Appl.*, 74, 2, 34–39, 2012.
17. Johnson, M. and Dhanalakshmi, R., "Predictive analysis based Efficient Routing of Smart Garbage Bins for Effective Waste Management". *Int. J. Recent Technol. Eng.*, 8, 3, 733–5739, 2019.

18. Lavrac, N., Bohanec, M., Pur, A., Cestnik, B., Debeljak, M., Kobler, A., "Data Mining and Visualization for Decision Support and Modeling of Public Healthcare Resources". *J. Biomed. Inform.*, 40, 438–447, 2007.
19. Guo, L., Chen, F., Chen, L., Tang, X., "The Building of Cloud Computing Environment for e-health". *IEEE International Conference on E-Health Networking Digital Ecosystems and Technologies*, vol. 1, pp. 89–92, 2010.
20. Rolim, C.O., Koch, F.L., Westphall, C.B., Werner, J., Fracalossi, A., Salvador, G.S., "A Cloud Computing Solution for Patient's Data Collection in Healthcare Institutions". *IEEE Second Int. Conf. eHealth, Telemedicine, Soc. Med.*, 95–99, 2010.
21. Ahuja, S.P., Mani, S., Zambrano, J., "A Survey of the State of Cloud Computing in Healthcare". *Network Communication Technol.*, 1, 2, 12–19, 2012.
22. Mathew, S., "Implementation of Cloud Computing in Education a Revolution". *IJCTE*, 4, 473–475, 2012.
23. Rao, A.V.S.R. and Lakshmi, R.D., "A Survey on Challenges in Integrating Big Data". *Proceedings of 2nd International Conference on Intelligent Computing and Applications*, Springer, Singapore, pp. 571–581, 2017.
24. Lee, K., Wan, T.T., Kwon, H., "The Relationship Between Healthcare Information System and Cost in Hospital". *Pers. Ubiquit. Comput.*, 17, 7, 1395–1400, 2013.

5

Knowledge Engineering for AI in Healthcare

A. Thirumurthi Raja[1]* and B. Mahalakshmi[2]

[1]*Department of Computer Science, Vels Institute of Science, Technology & Advanced Studies (Formerly VELS University), Chennai, Tamil Nadu, India*

[2]*Department of BCA, Bhaktavatsalam Memorial College for Women, Chennai, Tamil Nadu, India*

Abstract

Artificial Intelligence plays a vital role in healthcare towards the way of providing easy assistance and also understanding the needs of the patients. KE tools and techniques that can be used in the construction of domain models reduce the work of AI towards the field of healthcare. The overall assumption denotes that the engineering process emulates the judgments and the behavior of human in terms of taking decisions. A simple schema method will represent some form of common sense or predefined knowledge which helps to solve the problems. Medical terms aim for clear images and after recognition, the perspective towards the image seems to be a major task. The most critical part of this is to identify the varieties of components that are present in that particular image. Task of analyzing can be achieved by Artificial Intelligence but the overall objective of the AI when ruling towards Knowledge Engineering is by through Computer Vision because this is the subpart of Artificial Intelligence. Deep observations of the images are easily and exactly identified by the AI mentored machines. This may have the chance to overlap with the concept of image processing but in contrary the result of images from the vision developed algorithms are far much better when compared with the image processing where the mechanisms used to enhance and produce the images are in a different level of understanding. The first interpretation indicates that it is the result of converting the images into pixels since being it as a picture element. Instead of considering as a big picture the information coming from those pixels are more important and it can be achieved by using the visionary concepts. The most interactive feature of this computer vision is that the images can be captured

**Corresponding author*: thirumurthiraja@gmail.com

A. Jaya, K. Kalaiselvi, Dinesh Goyal and Dhiya Al-Jumeily (eds.) Handbook of Intelligent Healthcare Analytics: Knowledge Engineering with Big Data Analytics, (93–114) © 2022 Scrivener Publishing LLC

in any number of ways such as medical scanners, video sequences and also from the multiple cameras the speculation follows from the artificial system which helps to extract information from the images.

Keywords: KE, schema, AI, deep learning, vision, extraction, AI mentored machines

5.1 Introduction

The new techniques are exciting and challenging in today's world in terms of technical and scientific. Out of all the perspectives, the notable one is the analysis of the human needs. Most important is in what way the information that are stored and produced to the end user plays a major role. Because the level of understanding between humans and the machines are having vast difference in the way of what a human cannot do, that can be done easily by the machines through some technology. Not everyone is familiar with the pseudo-codes or machine language instructions so a simplified vision and mission for the use of humans are in demand. The work of human has to be minimized starting from the understanding level and to the advanced level. In order to precede further levels, there should be a common area where the terms are fulfilled. The related fields should inculcate some of the basics such as the starting and the retrieval of the information. The advancements of these are the upcoming developments in all the respective areas. One of the areas of this is the challenging of the concept toward engineering, which will pave the way for the understanding of this toward the knowledge concepts. The general means says that the idea of the data will be taken on terms of how far it has been preaching to the normal people or to the machine learned objectives. Some of the data can be raw and some can be a well understood data but the ability of these needs a mechanism so that they can be tuned well in order to force toward the availability of these demands. The overall assumptions say that the vision toward the concept has to be simplified and analyzed, but the important factor has not been in terms of the machine or nor the humans. Here, the objective is how the interaction will takes place between the machine and the human, whether it is like a robotic or waved communication of the raw data. In order to achieve these simple terms, the only solution is to go to the heap of the problems arising in the area and then tries to arrive to the final stage. Since the stages are a different from the starting to the end, the overall view of this chapter will give a clear idea about what is the role of knowledge engineering (KE) in AI and how it is implemented in healthcare industry.

5.2 Overview

Knowledge aids to create a skill of what has to be done or familiarity gained by experiences of facts, data, and situations. KE is a related field of artificial intelligence (AI), where simple rules are created as a process of human thought to arrive at a decision. A variety of problem solving methods and knowledge are piled up to form a system where the methods could take the same data and sometimes arrives at the same conclusion. The step-by-step process seems to be an old method, and so, the knowledge process focuses on the system, which will provide the same result as how the older methods are served. The redefined model will be very closer to the human identified model, and finally, the sources will be the same with the overall concept. To develop or to build the expert systems, the engineering process will produce many customized methodologies. The problems associated with the real life are still under a process of finding out of the best solution along with the provided results.

5.2.1 Knowledge Representation

Humans at best use their knowledge to understanding, interpreting, and reasoning. Humans better analyze what is knowledge, and according to that, they perform various actions in the real world. In the same way, how machines do all these things comes under a verse part of engineering and monitoring. Knowledge representation comes under these categories:

- Knowledge representation and reasoning (KR, KRR) is the part of AI deals with AI agents thinking and how thoughts are contributed to intelligent behavior of agents.
- It is responsible for representing information about the real world so that a computer can understand and can utilize this knowledge to solve the complex real world problems such as diagnosis a medical condition or communicating with humans in natural language.
- It is also a way that describes how we can represent knowledge in AI. Knowledge representation is not just storing data into some database, but it also enables an intelligent machine to learn from that knowledge and experiences so that it can behave intelligently like a human.

The following are the kind of knowledge that needs to be represented in AI systems:

- Object: All the facts about objects in our world domain, e.g., guitars contain strings and trumpets are brass instruments.
- Events: Events and actions that occur in our world.
- Performance: This describes the behavior that involves knowledge about how to do things.
- Meta-knowledge: It is knowledge about what we know.
- Facts: These are the truths about the real world and what to represent.
- Knowledge-based: The central component of the knowledge-based agents is the knowledge base. It is represented as KB. The knowledge-based is a group of the sentences.

5.2.2 Types of Knowledge in Artificial Intelligence

A. Declarative knowledge:

- Declarative knowledge assists to know about something.
- This knowledge involves concepts, facts, and objects.
- It is also called descriptive knowledge and expressed in declarative sentences.

B. Procedural knowledge:

- It is also known as an imperative knowledge.
- This type of knowledge is responsible for knowing how to do something.
- It can be directly applied to any task.
- It includes rules, strategies, procedures, agendas, etc.
- Procedural knowledge depends on the task on which it can be applied.

C. Meta-knowledge:

- Knowledge about the other types of knowledge is called meta-knowledge.

D. Heuristic knowledge:

- Heuristic knowledge involves the knowledge about the subject.
- Heuristic knowledge is rules of thumb based on previous experiences and awareness of approaches and is good to work but not guaranteed.

E. Structural knowledge:

- Structural knowledge is basic knowledge to problem solving.

- It describes relationships between various concepts such as kind of, part of, and grouping of something.
- It describes the relationship that exists between concepts or objects.

5.2.3 Relation Between Knowledge and Intelligence

Knowledge and intelligence both plays a vital role in AI because this demonstrates intelligent behavior in AI agents. An agent is only able to accurately act on some input when he has some knowledge or experience about that input.

An AI system has the following components for displaying intelligent behavior:

- Perception
- Learning
- Knowledge representation and reasoning
- Planning
- Execution

5.2.4 Approaches to Knowledge Representation

There are mainly four approaches to knowledge representation, which are given as follows:

A. Simple relational knowledge:

- It is the simplest way of storing facts, which uses the relational method, and each fact stores data objects in columns.
- Relational knowledge works well in database systems where the relationship between different entities is represented and provides a little opportunity for inference.

B. Inheritable knowledge:

- In the inheritable knowledge approach, all data must be stored into a hierarchy of classes.
- All classes should be arranged in a generalized form of a hierarchical manner and their inherited values are inherited from other members of a class.
- Since instance and class are well represented, then it is termed as instance relation.
- Every individual frame can represent the collection of attributes and its value.

- In this approach, objects and values are represented in boxed nodes.
- We use arrows that point from objects to their values.

C. Inferential knowledge:

- Inferential knowledge approach represents knowledge in the form of formal logics.
- This approach can be used to derive more facts.
- It guaranteed correctness.

D. Procedural knowledge:

- Procedural knowledge approach is a step-by-step process.
- In this approach, one important rule is used which is If-Then rule.
- In this knowledge, we can use various coding languages such as LISP language and Prolog language.
- We can easily represent heuristic or domain-specific knowledge using this approach.
- But it is not necessary that we can represent all cases in this approach.

5.2.5 Requirements for Knowledge Representation System

A good knowledge representation system must possess the following properties:

A. Representational accuracy: KR systems are having the ability to represent all kind of required knowledge.
B. Inferential adequacy: KR system should have ability to manipulate the representational structures to produce new knowledge corresponding to existing structure.
C. Inferential efficiency: The ability to direct the inferential knowledge mechanism into the most productive directions by storing appropriate guides.
D. Acquisition efficiency: Acquisition of the new knowledge easily using automatic methods.

5.2.6 Techniques of Knowledge Representation

There are mainly four ways of knowledge representation, which are given as follows:

1. Logical representation
2. Semantic network representation
3. Frame representation
4. Production rules

5.2.6.1 Logical Representation

Logical representation is a language having strong rules, which has got propositions without ambiguities in their representation. Logical representation leads to a conclusion based on various conditions. This kind of representation lays down some important communication rules. It consists of precisely defined syntax and semantics, which supports the sound inference; by this, all the sentences can be translated into logics using syntax and semantics.

5.2.6.2 Semantic Network Representation

Semantic networks provide alternate of predicate logic to represent the knowledge. In semantic networks, the knowledge is shown as graphical networks. This network consists of nodes describes objects and arcs which describe the relationship between those objects. Semantic networks can categorize the object in different forms and can also link those objects. These networks are easy to understand and extensible.

This representation consists of mainly two types of relations:

a. IS-A relation (inheritance)
b. Kind-of-relation

5.2.6.3 Frame Representation

A frame is a record which has got collection of attributes and its values to describe an entity in the world. These are the part of AI data structure which divides knowledge into substructures by the use of stereotypes situations. Frame representation is a collection of slots and slot values and can be of any types which have names and values which are called facets.

Facets are features of frames where the constraints are forced on the frames. Example: IF-NEEDED facts are called when data of any particular slot is needed. In AI frame can be represented as slot–filter knowledge representation. A frame may consist of any number of slots, and a slot may include any number of facets and facets may have any number of values.

The basic for semantic networks are frames and later switched over into our modern-day classes and objects. Collections of frames are normally connected to each other. Knowledge about an object or event can be stored together in the knowledge base inside a frame. The frame is a type of technology which is widely used in various applications including Natural language processing (NLP) and machine visions.

5.2.6.4 Production Rules

Production rules system consist of (condition, action) pairs which mean, "If condition, then action". It has mainly three parts:

- The set of production rules
- Working memory
- The recognize-act cycle

In production rules, agent checks for the condition, and, if the condition exists, then production rule fires and corresponding action is carried out. The condition part of the rule determines which rule may be applied to a problem. In addition, the action part carries out the associated problem-solving steps. This complete process is called a recognize-act cycle.

The working memory contains the description of the current state of problem solving and rule can write knowledge to the working memory. This knowledge match and may fire other rules. If there is a new situation (state) generates, then multiple production rules will be fired together; this is called conflict set. In this situation, the agent needs to select a rule from these sets, and it is called a conflict resolution.

The focus on the healthcare is needed to facilitate the transmission of knowledge across the research and the healthcare is always interleaved. Important aspect toward this will be how far this will reach the healthcare related terms and achieve the goal of full understanding. Besides, all these decision support toward the health systems has to be highly concentrated and it is also very difficult because of the inherent complexity and the structures involved. Most of the research methods in medicine are mainly based on the restrictive experimental approach [12].

Healthcare systems involve very complex structures and high-dimensional interactions among multiple factors, where the meaning of the measurements becomes crucial for proper interpretation, and where phenomena are very difficult to model in their globalist under classical approaches. Paradoxically, the complexity and non-linearity of health systems has not been acknowledged until very recently by international organizations such as WHO.

Although very accurate models can be obtained under classical approaches while restricting the scope of the model to a particular aspect of the problem, other methods are needed to model phenomena grasping their whole complexity [14].

The healthcare specifics and the settings are precluded to the feasibility of a mono health record, and how far the interoperability between systems is achieved, this is the primary challenge faced by health informatics researchers. The transfer process has been left behind in favor of a modeling process. Instead of this step-by-step process of a decision, KE is focused on a system, which will invoke the same results as the expert without being inherited the same path [15].

Some of the issues are eliminated by tracking down the knowledge being used for nonlinear thinking, as the people doing it are often not aware of the information they are pulling on. As long as the conclusions are comparable, the model works. Once a model is consistently coming close to the human expert, it can then be refined. Bad conclusions can be traced back and debugged, and processes that are creating equivalent or improved conclusions can be encouraged. KE is already integrated into decision support software.

Specialized knowledge engineers are employed in diverse fields that are advancing human-like functions, including the ability of machines to recognize a face or parse what a person says for meaning. As the complexity of the model grows, the knowledge engineers may not fully understand how conclusions are being reached. Eventually, the field of KE will go from creating systems that solve problems as well as a human to one that does it quantitatively better than humans. KE sought to transfer the expertise of problem-solving human experts into a program that could take in the same data and come to the same conclusion.

This approach is referred to as the transfer process, and it dominated early KE attempts. It fell out of favor; however, as scientists and programmers realized that the knowledge being used by humans in decision-making is not always explicit Programmers realized that the knowledge being used by humans in decision-making is not always reaching up to the extent [13].

The automation of the various information nowadays has been reaching high stages due to the fact that an expertise will give a better and higher consumption toward the varieties of fields. Apart from these steps, the most focusing step toward that will be the highest range of problem solving.

5.2.7 Process of Knowledge Engineering

Basic relationship between the raw data which is producing in the standard format and the real data lies in the core. The next process is of extracting in a deeper format and finally how that has been transformed into relationships using various available, and newest algorithms are seeking more informative in all the ways, and this is how the engineering process is going on. The best explanation is described in Figure 5.1.

In other words, we can explain the KE domain as how the information are understood and represented using the knowledge of humans as data structures; some models like semantic models and, in general, collection of rules that come to a solution for each and every problem the domain is facing called as Heuristics. Most common examples that form the basis for the applications are expert systems and algorithms. Here, the acquisition of the knowledge can be collateral one, where the demands are the major aspect.

Now, the concentration is changing toward that, when technology and its standards have assisted in integrating data, it makes it accessible for the knowledge engineered process. Expert systems are there but needs to be modeled more where the problem and the solution should meet in a proper way. More the way, depth of knowledge is understandable, higher the way it is going into all problems and its related process.

Sometimes, the artifacts are made easy too: the way of representing the model because the conceptual data of reuse was there. The most common aspect of the KE in this larger multidisciplinary area was to build the models from the ground up using knowledge-based technology, the layer-by-layer processed data are seems to be the prioritized one. Best examples for the KE are as follows:

- Integration with the higher technologies is easier, e.g., CAD-based software
- Collaboration with the supporting team like automation, design artifacts, and the products simulation.

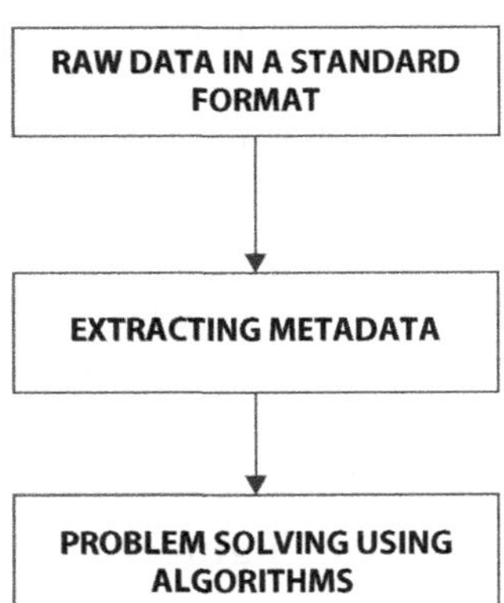

Figure 5.1 Process of knowledge engineering.

- Design of a model is easier.
- The technology can be made innovative when the knowledge, and its intensive activity are inherited together.
- Re-usability saves time; by this method, the sub-classing can be very easy to create new types of artifacts and the process to start with an existing class by adding a new subclass, forming a kind of inheritance.
- The learning health system can be best provided by the healthcare quality organizations, review organizations, researchers, and individual healthcare professionals.

It is particularly important to employ KE techniques when an organization employs a range of knowledge management approaches. This is becoming common in larger organizations, which already use a multiplicity of information systems tied into an intranet and see a multifaceted knowledge management system as normal. For example, such a knowledge management system might include a capability management system, discussion forums, a document management system, and several lessons-learned knowledge bases. In such cases, the key challenge becomes knowledge integration—linking the various sources at the knowledge-content level.

In knowledge process, the research is focused on three things mainly. These are as follows:

- Analysis.
- Design.
- Management or supervision.

The abovementioned aspects are applicable for both healthcare and for the engineering. If the research is more specific, then the efforts are involved more in the analysis, the design, and the development of the agents more specifically called as intelligent agents. The coalition setting dynamics and the structures which are socialized, new model construction, information process design, argumentation techniques based on the experience, and several statistical method have been solved and are still solving because of the processed information.

Not only in these fields but also in the application, expertise which leads to the development of Environmental Decision Support System (EDSS) and everything has been programmed to represent the knowledge explicitly, which unites all the knowledge-based system in one common theme.

More focus and more descriptive will be the self-assessment, which are needed for the patients not about the in-depth knowledge, but only in few terms of the basic information that can be shared. The patient-provided data are essential to know the symptomatically and for the care evaluation. The aim of this chapter is to introduce some basic details about how the patients can analyze themselves when they are injured through the applications. The combination of the process and the technology can be acquired. The initial stages are the best one to follow where the representation and the acquisition are happened to be in the same place.

Processed information proceeded with the basic knowledge where both the doctor and the patient has to be focused on the information representation but in the second stage, the data which are filled up are handled and extracted in the form of the metadata, whereas the last informative phase can be redesigned as the unprocessed algorithms and the results expected to produce a simple valuable statement can be termed as unprocessed. The combination of the processed and unprocessed together plays a vital role in the development of KE.

The human practice supporting systems can be the next generation development. The continuous innovation and improvement toward the science, informatics, incentives, with best practices seamlessly embedded in the delivery process and new knowledge captured as an integral by-product of the delivery experience.

The scenario will arise where the data and the knowledge meets (Figure 5.2).

When the data are simplified, there is no need for the detailed explanation to be given. Raw data will be collected on the basic term and the way how it is processed can be explained in terms of an application. The smaller the number of information and the basic idea between the unprocessed one is a big dilemma still, and so, the process will continue to reach on time

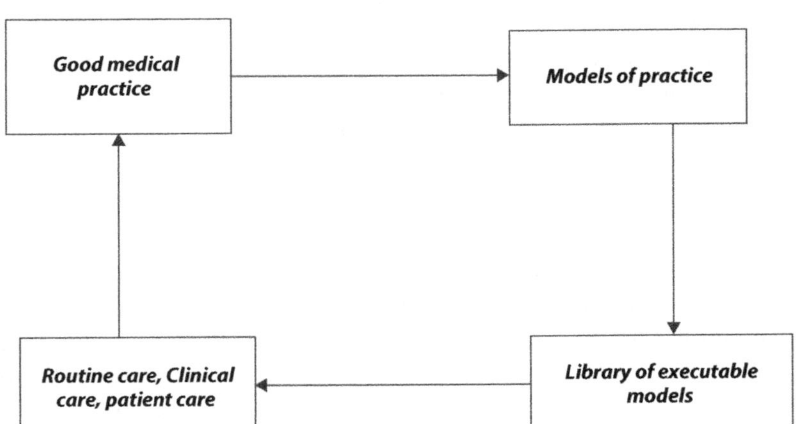

Figure 5.2 Data science and knowledge engineering.

from the smallest injury to the high rate of injuries. When an application helps to solve this problem, the solution can be made simpler, and so, the cost of the treatments can be reduced [9].

The use of traditional knowledge acquisition processes, such as schema-based, or frame-based, methods; with the protocol, analysis becomes problematic, because the knowledge engineer may not have the desired technical expertise in communication with the designer. As like in chain reaction, they may not be able to clause all the necessary collection. The original multilevel modeling approaches are being studied to deal with the complexity to develop applications, which will provide a specific environment for the healthcare situations.

The two common V's of KE are verification and validation because it offers the way of how to produce judge the success for knowledge managed process. If the verification from the patient side, then the validation is from the doctor side and so vice versa.

Already, techniques like CIT and CDM have been employed successfully to elicit information about human activities, export human knowledge, and model making along with the domain because both the methods are semi-structured format worked along with the resilience engineering, which in terms describes an ability to adapt or absorb disturbance, and even the change. All these questions are based to check how people are actually co-operating for the application. The focus again goes with the critical and non-routine events. This involves the adaptation of the safety and the recovery [1].

The semi-structured data allows the narrative response which helps in getting of the contextual detail according to the situation. From the functional point of view a suitable healthcare knowledge management services that aims to assist even the stake holders. Another perspective to be taken care is the exponential growth of the number of parameters has to be controlled by using the conditional probability tables which are in existence. Continual improvements are needed where everyone in the organization understands the system of work and it aims to work together to build new challenge and the knowledge in even daily works.

Many of the applications are implemented using the a KM framework, which starts from the scratch like knowledge creation, assets, inertia of the knowledge, the history about the KE, systems thinking, artificial engineering systems, and even the strategy management systems. The questions will be ask, whether the systems are knowledgeable? Or it is accessible for the specific domain problem? [2]

The reliability of software needed by the healthcare professionals rely on the simulation process, either it can be for the purpose of training or for decision-making and again the concept of the validation comes into existence, which paves the way for the error free software in demands. This is the most challenging one where no question can be raised whether the CAD has been worked properly for the construction setting and the parameters used are appropriate. Testing simulation software is the most challenging in a way that the program has to be written in a proper way. The aim of developing validated software and the simulation from that are to be suited in a proper way [3].

Once the data are collected properly, the way how it is getting accessible has to be recorded. With the current situation, explosion of data, retrieving of data, and integration of data are a major problem because the work of multi-database systems has begun to address the problem, but it has primarily focused on methods for the communication needed between database and a new database system, which are in demand. The collaboration of the medical records can be deeply utilized by the local mining and the learning approaches; basically, they are tightly linked reinforced also mutually. In some cases, we can even follow the global mining [5].

5.2.8 Knowledge Discovery Process

The emerging applied science field helps in the knowledge discovery process where the analytic systems are evaluated only for the domain experts. The discovery process can come under the following categories:

- Domain understanding.
- Data understanding.
- Physician and patient bridge gap.
- Data preprocessing.
- Data analytic techniques.
- Evaluation of the model.
- Deployment of the model.

5.3 Applications of Knowledge Engineering in AI for Healthcare

The applications of AI in healthcare hold the promise of affordable healthcare, improved success rates, efficient clinical trials, and a better quality of life. Mostly, we are familiar with AI related to robots, self-driving vehicles, etc. Now, we are gently accepting the potential of the clinical applications of AI. The stable rise in the impact of AI in healthcare can be demonstrated by looking at the following sectors of the healthcare industry [6].

5.3.1 AI Supports in Clinical Decisions

The ability of AI in healthcare is helping people to stay healthy. In the advent with the Internet of Medical Things (IoMT) in consumer health applications, the applications are growing in leaps and bounds from the last decade onward. Healthcare apps encourage healthy behavior among individuals. Most of the mainstream companies like Apple now making tracking cardiac health, fall detection, and emergency SOS major products of their portfolio, and it looks like IoMT is here to stay.

It is more unpredictable for the health professionals to account all crucial information into consideration while diagnosing patients. By this result, the shift toward various complicated unstructured records. A single mistake in keeping track of even a single relevant record, the life of a patient could be. For the innovation in NLP, it provides more convenient way for the doctors to narrow down all relevant information from patient reports.

AI holds the ability to store and process large sets of data, which can provide knowledge databases and facilitate examination and recommendation individually for each patient, thus helping to enhance clinical decision support. (Various ML algorithms are deployed in order to decipher medical data; you may find similar application as "Bioinformatics".) [12].

This technology can be relied upon by doctors for aid in detecting risk factors through unstructured notes. An interesting example of this is IBM's Watson, which examined the key indicators of congestive heart failure in almost 21 million patients over a span of 6 weeks, detecting the ones at risk of gaining the condition within a year in about 85% of the cases.

5.3.2 AI-Assisted Robotic Surgery

In terms of practice, AI helps in surgical performance. Typically, the outcome of a surgery, particularly, a new or complex procedure, can vary with the surgeons' skill. Utilization of AI can reduce the case-to-case variations and even help improve the efficiency of even the best surgeons.

For example, AI controlled robots can provide a three-dimensional (3D) magnification for articulation and perform with more precision and miniaturization. AI-enabled robots can perform basic acts of precision cutting and stitching. In 2017, we witnessed surgeons using AI-assisted robotics to suture extremely narrow blood vessels—0.03 to 0.08 mm across—at the Maastricht University Medical Center, Netherlands [11].

AI and collaborative robots have revolutionized surgeries in terms of their speed and depth while making delicate incisions.

AI machines are capable of employing data from past operations to develop new surgical methods. The preciseness of these machines reduces the possibilities of tremors or any unintended or accidental movements during the surgeries.

5.3.3 Enhance Primary Care and Triage

People have a tendency of booking appointments with their GP at the slightest threat or medical issue, which could often turn out to be a false alarm or something which could be cured of self-treatment.

AI assists in enabling smooth flow and automation of primary care, allowing doctors to stress over more crucial and dire cases.

Saving money on avoidable trips to the doctor, patients can benefit from medical chat bots, which is an AI-powered service, incorporated with smart algorithms that provide patients with instant answers to all their health-related queries and concerns while also guiding them on how to deal with any potential problems. These chat bots are 24/7 available and have the capacity to deal with multiple patients at the same time.

5.3.4 Clinical Judgments or Diagnosis

AI has the capacity to surpass human doctors and help them detect, predict, and diagnose diseases more accurately and at a faster rate. Likewise, AI algorithms have proved to be not only accurate and precise at specialist-level diagnostics but also cost-effective in terms of detecting diabetic retinopathy.

For instance, PathAI is developing machine learning technology to aid pathologists in making more accurate diagnoses. The company's current goals include reducing error in cancer diagnosis and developing methods for individualized medical treatment. (Other than medical and healthcare applications, machine learning also has these seven applications in daily life.)

AI can and is already improving the early detection of diseases including cancer and retinopathies. The use of AI in analyses and review of mammogram and radiology images can speed up the process up to 30 times and with 99% accuracy. In 2017, Stanford University published a study describing successful use of AI algorithms to detect skin cancer against the diagnosis of 21 dermatologists [10].

This year, Google's DeepMind technology successfully trained a neural network to detect more than 50 types of eye disease, by analyzing 3D retinal scans, in a collaborative study with joint research with Moorefield's Eye

Hospital, London, UK. The biggest advancement of this study over its predecessors is the explain ability of how the computer arrived at some of the interpretations. Overcoming a previous black box of data interpretation and inference reinforces the reliability and trust in this application.

Most importantly, it is important to acknowledge and emphasize the benefit of combining the powers of the AI algorithms with the powers of the physicians. At the International Symposium on Biomedical Imaging, a competition was held for computational systems that were programmed to detect metastatic breast cancer from biopsy images. While the winning program made the diagnosis with a 92.5% success rate, combining it with human pathologists' opinion and expertise increased that number to 99.5% success [7].

5.3.5 Precision Medicine

One of the most valuable examples of AI in healthcare is precision medicine, which is currently touted as the paradigm-shifting healthcare practice. The foundation of precision medicine relies on the copious amounts of data collected from many disruptive technological innovations, including health sensors patients use at home, cheap genome sequencing, and advanced biotechnology. Precision medicine refers to "tailoring of medical treatment to the individual characteristics of each patient". Medical practices are now rapidly shifting from making decisions based on few seemingly overlapping features among patients, to adopting a more personalized format.

Precision medicine depends on advanced super-computing algorithms with deep learning and thus uses the cognitive capabilities of physicians at a new scale. In this day and age of easy access to genomic data, one of the challenges is to plough through to identify genetic variants that increase disease risk. Intel, in collaboration with the Scrip's Research Institute, CA, USA, has developed a deep learning algorithm that could detect 23 patients with increased cardiovascular disease risk, unidentified by conventional statistic methods, with 85% accuracy. IBM Watson and Google DeepMind are the leaders in mining medical records, with the ultimate objective of creating a "cognitive assistant" equipped with a range of clinical knowledge, and analytical and reasoning capabilities, alongside [8].

5.3.6 Drug Discovery

The current format of clinical trials takes decades of research and costs billions of dollars. According to the California Biomedical Research Association, "Only five in 5,000 of the drugs that begin preclinical testing ever make it to human testing and just one of these five is ever approved for human usage". The use of AI in drug discovery can help pharmaceutical companies to streamline drug discovery and drug re-purposing. Many pharmacy giants, including Pfizer, Sanofi, and Genentech, are now partnering with AI service providers—IBM Watson, Exscientia's AI, and GNS Healthcare, respectively, to drive their oncology drug discovery programs. AI can pinpoint to previously unknown causes of various diseases and enable testing of more compounds with higher accuracy and reproducibility.

Using AI for drug discovery would enable us to get rid of the traditional trial and error approach and embrace a more patient-driven biology via using more data-derived predictive hypotheses. Atom wise, a drug development company used AI to analyze if existing medicines could be redesigned to target the Ebola virus in 2016. An analysis that would have normally taken months or years by the conventional means was instead accomplished in a single day, resulting in two potential hits. While the in silica modeling techniques grow more important in modern drug research and development (R&D), they are far from replacing the standard pharmaceutical-industry R&D productivity [11].

While the above said practices are some of the "cooler" AI applications, what goes unnoticed behind the scenes is the administrative workflow assistance by AI with an estimated value of $18 billion. AI has mobilized the automation of administrative jobs that can help doctors and nurses in their routine jobs of data collection, recording, and long-term storage. By getting rid of the repetitive parts of a physician's job, AI might assist the caretaker's availability for the patient on a regular basis [5].

AI describes the ability of a machine to learn the way a human learns, for example, through image recognition and discovering patterns in complex situations. AI in healthcare is changing the way information gets collected, analyzed, and developed for patient care.

The people who work in patient diagnoses, health logistics, radiology, and administration are increasingly vital to the operation of any sizable healthcare provider. Professionals who earn a Master of Science in Health Informatics are uniquely well positioned to step into important roles in these areas, and their services are shaping the ways AI in healthcare is changing the patient experience [14].

5.3.7 Deep Learning to Diagnose Diseases

One of the areas where AI in healthcare has shown the most promise is in diagnostics. Early diagnosis is one of the most important factors in the ultimate outcome of a patient's care. AI deep learning algorithms are being used to shave down the time it takes to diagnose serious illnesses. The way AI rapidly processes large amounts of information and arrives at likely causes for symptoms can drastically reduce the diagnosis-treatment-recovery cycle for many patients. The effects of this are already being felt in several areas.

In the article "Assessment of Machine Learning Detection of Environmental Enteropathy and Celiac Disease in Children" published in the Journal of the American Medical Association, the authors describe a machine learning tool developed at the University of Virginia School of Medicine that can rapidly analyze thousands of images from children's biopsies and distinguish between environmental enteropathy and celiac disease more reliably than human doctors, at least at the earliest stages when the two disorders are easily confused. It is hoped this can speed up children's treatment and reduce rates of stunted growth that can be the result of delayed diagnosis [4].

5.3.8 Automating Administrative Tasks

While patient care is always the top priority of healthcare providers, they still have to operate as businesses, which mean expensive and time-consuming overhead and administrative costs.

AI in healthcare administration is drastically reducing these expenses, and it is helping healthcare providers devote more of their limited resources to the patient care that really matters.

The American Hospital Association estimates that hospitals spend $39 billion annually to comply with the vast requirements of healthcare laws. Regulatory compliance is a full-time job for an average of 59 employees per medium-sized hospital, and meeting regulatory burdens occupies roughly 25% of doctors' and nurses' working hours. AI in healthcare is radically shifting that time and money back toward patient care. Some examples are intelligent checklists that automatically enforce patient privacy rules and automate the production of disclosures and check-in systems that remind hospital workers to wash their hands and replace equipment on legally mandated schedules.

More than 25% of healthcare costs in the United States go toward billing, usually for insurance. Even in countries with public health insurance, such as Canada, the cost of medical billers can be as high as 12% of the total healthcare budget. AI in healthcare billing applications uses smart algorithms to analyze and assign costs, as well as to correctly structure invoice requests and even negotiate with some insurers. Applications such as 1Desk can coordinate the workflow between patients, insurance companies, hospitals, specialists, and banks to keep human intervention to a bare minimum in the billing process [15].

Streamlining the billing process may be one of the most important, if underrated, benefits promised by the increasing use of AI in healthcare. Billing AI, such as 1Desk, is potentially very attractive for recent graduates in health informatics. The generally nonclinical nature of the program lets practitioners apply their skills virtually anywhere that uses automated billing programs, such as insurance and government offices [16].

5.3.9 Reducing Operational Costs

The operational costs of running a healthcare organization can be staggering. This is as true for private practice providers as it is for sprawling HMOs with tens of thousands of employees. AI in healthcare is cutting those costs and letting providers at every level extend their budget further. This results in more resources for the practice, more time for the patient, and significantly improved outcomes for many.

One example of how AI is improving healthcare delivery by cutting operating costs is in the field of joint replacement therapy. An intelligent program known as PeerWell helps patients prep for total joint replacement by guiding the course of pre-op physical therapy. As reported in Healthcare Finance, a study in the Annals of Translational Medicine reported that patients using the AI saw a reduction in surgery costs of $1,215 [7].

Even better, the intelligent prep work the program recommended got so many patients into such good shape prior to their surgeries that the test group saw a reduction of 25% in the time patients spent in the hospital, 80% in the need for post-op home care, and a staggering 91% in the number of patients discharged to nursing home care after their operations.

The providers saw similar reductions in operating costs because patients no longer required pre-op care from clinicians. This not only saves time for doctors and expense for patients and their insurers but also dramatically reduces demand for post-op and rehab facilities, which frees up resources for the remaining patients and improves outcomes for nearly everyone.

Advances of AI in healthcare are very exciting, especially for health informatics graduates starting out in the field. New graduates with the right credentials can be on the ground floor of a revolution in healthcare that has hardly begun.

Interested students can learn more about this rapidly developing career field and the many opportunities new graduates can have to make a difference for patients from the University of Illinois at Chicago's Master of Science in Health Informatics program page.

5.3.10 Virtual Nursing Assistants

AI systems facilitate virtual nursing assistants that can perform a range of tasks from conversing with patients to directing them to the best and effective care unit [17]. These virtual nurses are available 24/7 and can respond to queries as well as examine patients and provide instant solutions. Presently, many AI-powered applications of virtual nursing assistants presently enable more regular interactions between patients and care providers between office visits to avoid any unnecessary hospital visits. The world's first virtual nurse assistant, Care Angel, can even facilitate wellness checks through voice and AI.

5.4 Conclusion

AI no doubt will bring new efficiency for the outcome of healthcare. The impact of these systems depends on the problem identification along with their accurate solutions. There is a mild gap between how the technology is reaching in tremendous areas rather than the remote areas. The challenging information is how the digital world is raging the information between the user and the developers who are not blind ones to expect. Nowadays, information can be passed through various ways even though the hindrances are on the side. It is recommended that, before using the application, it will be good; if the norms are understood clearly, by then the flaws can be eradicated to a few extents. The issues related to privacy, misuse, and other factors can be decided in various ways but require far-reaching consideration before the safest healthcare systems are delivered. According to the content mentioned, AI developed systems are safer and easily adaptable to fair healthcare systems. The essential in getting the expertise from the domain is by the use of KE where the entire knowledge-based systems serve as the building blocks; by this, the automated learning techniques are more useful for acquiring the domain knowledge. Giving the experience to face the new challenges toward technology is gaining good aspects for the overall growth of the healthcare systems. In today's systematic world, novel practices are needed, and the future world is going to be the best of important practices that are going to be handheld. The systems are designed to imitate the human capabilities of thinking and sensing. The development of these functions normally associated with human intelligence, in the forms of reasoning, learning, and problem solving.

References

1. Sudeep H., Hettinger A.Z., Fairbanks R.J., Wreathall, J., Wears R.L., Knowledge Elicitation for Resilience Engineering in Healthcare, *Proceedings of the Human Factors and Ergonomics Society Annual Meeting*, December 20, 2016.
2. Knowledge management technologies and applications—Literature reviews from 1995 to 2002 on Effective Testing of Healthcare Simulation Software, Expert systems with Applications, 25, 2, 155–164.
3. https://www.kolabtree.com/blog/5-real-world-examples-of-ai-in-healthcare
4. https://www.javatpoint.com/knowledge-representation-in-ai
5. https://www.javatpoint.com/ai-techniques-of-knowledge-representation
6. https://healthinformatics.uic.edu/blog/ai-in-healthcare-4-examples-in-health-informatics.
7. https://www.analyticssteps.com/blogs/artificial-intelligence-healthcare-applications-and-threats.
8. Glymour, C. and Cooper, G.F., *Computation, Causation, and Discovery*, AAAI Press, USA, 1999.
9. Huang, The Sensitivity of Belief Networks to Imprecise Probabilities: An Experimental Investigation. *Artif. Intell.*, 85, 1–2, 363–397, 1996.
10. Druzdzel, M.J. and Simon, H.A., Causality in Bayesian Belief Networks, in: *Proceedings of the Ninth Annual Conference on Uncertainty in Artificial Intelligence (UAI-93)*, Morgan Kaufmann Publishers, Inc, 1993.
11. Cumberpatch, J. and Heaps, H.S., A disease-conscious method for sequential diagnosis by use of disease probabilities without assumption of symptom independence. *Int. J. Biomed. Comput.*, 7, 61–78, 1976.
12. Davis, R., Interactive transfer of expertise: Acquisition of new inference rules. *Proc. of 5th Int. Joint Conf. Artific. Intell.*, 1977.
13. Fagan, L.M., *Knowledge engineering for dynamic clinical settings: Giving advice in the intensive care unit*, 16, Feigenbaum, E.A., School of Humanities and Sciences Stanford University, IJCAI I, Stanford, California, USA, 1979.
14. Feigenbaum, E.A., The art of artificial intelligence: Themes and case studies of knowledge engineering. *AFIPS Conf. Proc*, .
15. Feinstein, A.R., Quality of data in the medical record. *Comput. Biomed. Res.*, 3, 426–435, 1970.
16. Gleser, M.A. and Collen, M.F., Towards automated medical decisions. *Comput. Biomed. Res.*, 5, 180–189, 1972.
17. Kanal, L.N., Patterns in pattern recognition. *IEEE Trans. Inform. Theory*, IT-20, 6, 1968–1974, 1974.

6

Business Intelligence and Analytics from Big Data to Healthcare

Maheswari P.[1]* A. Jaya[1] and João Manuel R. S. Tavares[2]

[1]*B.S. Abdur Rahman Crescent Institute of Science and Technology, G.S.T Road, Vandalur, Chennai, India*

[2]*Instituto de Ciência e Inovação em Engenharia Mecânica e Engenharia Industrial, Departamento de Engenharia Mecânica, Faculdade de Engenharia, Universidade do Porto, Porto, Portugal*

Abstract

Healthcare is one of the largest and complex sectors in the stock market. It comprises a broad range of companies including hospitals, healthcare providers, selling of medical devices, drugs, and insurance. When the coronavirus unexpectedly comes into sight, the entire world economy has stagger. This has decreased the surgeries, outpatient department footfall, international patients, medical device pharmaceutical, and healthcare commodities. Medical device industry has worst affected the export of medical devices and the critical raw materials are disturbed due to the restriction on movement, social distancing, travel, and transport. Healthcare industry challenged a burden such as 100% of alertness for the protection in hospitals and further investment of manpower, equipment, consumables, etc.

Health and wealth are the two main components of well-being in life. Almost everything requires money from food to education to health services. During this pandemic situation, we have to take care of both health and wealth. Healthcare industry is one of the world's major and fastest emergent enterprises. Nowadays, with the advancement of expertise like analytics, business intelligence (BI), and artificial intelligence (AI), the prediction of stock value has improved and benefits the investors to make the right decisions. This study aimed to predict the hospital and healthcare services stocks in the Indian Stock Market Index (Nifty). Companies like Apollo Hospitals Enterprise Limited, Cadila Healthcare Ltd., Dr. Reddy's Laboratories, Fortis Healthcare Limited, Max Healthcare Institute

**Corresponding author*: pmaheswari2011@gmail.com

A. Jaya, K. Kalaiselvi, Dinesh Goyal and Dhiya Al-Jumeily (eds.) Handbook of Intelligent Healthcare Analytics: Knowledge Engineering with Big Data Analytics, (115–146) © 2022 Scrivener Publishing LLC

Limited, Opto Circuits Limited, Panacea Biotec, Poly Medicure Ltd., Thyrocare Technologies Limited, and Zydus Wellness Ltd. were used in this study to predict healthcare stocks. Hospital and healthcare service stocks were predicted using linear regression (LR), autoregressive integrated moving average (ARIMA), and long short-term memory (LSTM).

Keywords: Healthcare, hospital, business intelligence, linear regression, long short-term memory (LSTM), analytics, big data, autoregressive integrated moving average (ARIMA)

6.1 Introduction

"Health is wealth" is the famous traditional proverb prevailing in almost each and every language in one or another way, which is concerned about the health. No one will disagree with the proverb because our health is one of the biggest treasures that we have. Once our health is lost its value, the wealth in life will also mean nothing to us. Health and wealth are the two main components of well-being in life. Nowadays, financial security is crucial. Almost everything requires money from food to education to health services. During this pandemic situation, we have to take care of both health and wealth. Healthcare industry is one of the world's major and fastest emergent enterprises.

The healthcare industry, medical industry, or health economy offers goods and amenities to treat patients with therapeutic, precautionary, rehabilitative, and relaxing care. The healthcare sector in India is emergent at abrupt speed due to its reinforcement treatment, facilities, and cumulative outflow by the public as well private enterprises[1]. India has ranked 145th place in terms of quality and accessibility of healthcare among 195 countries.

6.1.1 Impact of Healthcare Industry on Economy

The healthcare sector is the largest and complex sector in the stock market. Healthcare is a vast part of a country's economy. In most developed nations consuming over 10% of gross domestic product (GDP), a correlation exists between expenditure and incomes on healthcare in various countries. For example, industrial economic countries like the United States, Switzerland, Germany, and France spend more on the healthcare industry compared to the large developing economical countries like India, China, and Brazil that spend less on healthcare. Healthcare sectors include healthcare services, hospital activities, manufacturing medical equipment, devices, medical supplies, insurance, as well as healthcare providers, pharmaceuticals, biotechnology, and related segments.

6.1.2 Coronavirus Impact on the Healthcare Industry

Once the coronavirus appeared unprecedented, the entire world economy has fallen all of the sudden. Pandemic has decreased the surgeries, out-patient department footfall, international patients, etc., Medical device industry has worst affected the export of medical devices, and critical raw materials are disturbed due to the restriction on movement, social distancing, travel, and transport. Healthcare industry faced a burden such as 100% alertness for the protection in hospitals and additional investment of man-power, equipment, consumables, etc.

Nowadays, with the advancement of expertise like analytics, business intelligence (BI), and artificial intelligence (AI), the prediction of stock value has improved and benefits the investors to make the right decisions, which shows a vital role in decision-making and improves productivity faster and better decisions. BI can support the efficiency of operational rules like monetary and budgeting in an energetic commercial background. BI is used to collect and examine data using predictive analytics [23].

6.1.3 Objective of the Study

Investors are stressed to predict the stock market. Many are already betting that healthcare shares would not lead the way. Predicting the stock market is very complex during this pandemic situation. Earlier this year, during the market turbulence, the sector offered shelter investors followed by an outsize rebound. Healthcare stocks suddenly seemed to offer big opportunities for gains of several companies scrambling to make primary effective corona vaccine or treatment. This study aims to predict the healthcare sector stock prices using linear regression (LR), autoregressive integrated moving average (ARIMA), and long and short-term memory (LSTM). Hence, algorithms were applied to predict the preeminent result, maximize the profits, and provide the optimal results to the investors.

6.1.4 Limitations of the Study

The stock market is highly volatile and predicting stock value is a very critical job. Financial investors are facing challenges when to buy the stocks, when to sell the stocks, or else hold the stocks. Stocks are fluctuating rapidly based on the events that occurred in the world. These research works are based on a generic stock index. As a future direction, this research would like to include various factors like news, sentiments of the markets, the psychology of the investors, supply and demand, and results of the companies to predict the preeminent result.

The rest of this chapter consists of the following sections: Section 6.2 reviewed the related works of the LR, ARIMA, and LSTM. Section 6.3 explains the conceptual healthcare stock prediction system. Section 6.4 describes the implementation and results from the discussion of various healthcare stocks. Comparisons of the healthcare stock prediction framework are discussed in Section 6.5. Finally, in Section 6.6, conclusion and future enhancement of the chapter are given.

6.2 Related Works

In [1], the author had used Nifty 100 of selected pharmaceutical companies that predicted the share prices using the ARIMA model. Research work in [2] discussed the big data analytics methods, tools, etc., in the healthcare domain. The authors focused on their research major disciplines like medical image processing, bioinformatics, and clinical informatics.

A system was proposed in [3] to predict the foreign exchange like USDJPY. The authors have used RNN and ARIMA for prediction. They have achieved good accuracy compared to the previously published limitations. In the future, they have planned for another neural network to extract the underlying features. In [4], the proposed system uses deep AR models that provide more accurate results. After applying the purposed arrangement on continuous association information, the framework can acquire 92.38% correctnesses for the stock regarding intellectual demand predicting.

Studies concentrate on petroleum; basic metals for the forthcoming expectation of Tehran stock market were discussed in [5]. The authors used four metrics in each technique that were predicted. In this work, all algorithms were used and compared with the four metrics. LSTM model has the highest model fitting ability and shows more accurate results.

Research work proposed in [6] is an associated network model. LSTM model and deep RNN model are compared with the associated network. The proposed associated network provides greater accuracy than the other two models. The proposed system provides a well traditional statistical approach for predicting. They have used ARIMA for predicting. The prediction of the proposed model and previously available consistency models are compared in [7].

The ARIMA model has forecasted values that are accurate for medium-term or long-term possibilities. The authors were used past values of two indices of "S&P BSE Sensex" and "S&P BSE IT". Their results expose that both indices are positive, and results have vital recommendations in [8]. The study provides the novel hybrid recurrent network of ARFIMA-LSTM. Their model reduces the unpredictability problem and also overcomes the overfitting problem of neural networks. They have obtained 80% accuracy on RMSE in [9].

The authors have designed a novel LSTM-ARIMA hybrid model in [10]. Preprocessing requires less in the Prophet forecasting library by Facebook. The authors observed to adopt a scientific and crucial evaluation of approximately 122 relevant studies works mentioned in instructional journals for 11 years from 2007 to 2018 with inside the region of inventory change prediction the usage of the system getting to know. The diverse strategies recognized from those reviews have been collected into three categories: technical, fundamental, and mixed analyses. Support vector systems and synthetic neural community have been determined to be the most used system getting to know algorithms for inventory marketplace prediction in [11].

The author [12] incorporates management practices; their study recommends that both BI and BA used, based on the information needs, a different level of uncertainty versus uncertainty characteristic of the management practice. In [13], their proposed work uses ANN and RF techniques for forecasting the closing value for the subsequent day for five concerns in diverse sectors. They estimated by using RMSE and MAPE. The two indicators have showed the prototypes are effective in forecasting closing stock value.

In [14], the research work used LR, 3MMA, and Exponential Smoothing algorithms to predict the stock prices in Amazon (AMZN) and Apple (AAPL) datasets. From the prediction, they have concluded that Exponential Smoothing is best rather than the other two approaches. In [15], authors have assumed that news would affect the stock market price. In their model, they have selected the top news of each day, which is evaluated as positive or negative using machine learning algorithms. They have applied different algorithms, and the results were predicted.

The study was confirmed that investors utilized the logistic regression version as compared to LR in [16]. The authors had reviewed 122 appropriate research works from the prediction of the stock market. From various techniques that are identified, support vector machine (SVM) and artificial neural network (ANN) were established to be the foremost machine learning (ML) algorithms for predicting stock market in [17].

The authors proposed a work that used deep learning architectures for predicting the "National Stock Exchange (NSE)" and "New York Stock Exchange (NYSE)". They have used four networks. These are MLP, RNN, LSTM, and CNN. From the result, they have concluded that CNN has performed better in [18]. The author has investigated the tweeted messages that are used to predict the Dow Jones particular companies listed. They find the trends of the stocks for the future. They have employed more than 800,000 tweets and showed up to 82% of success in [19].

The authors examined the modern stock market data and make the administrative power resultant from the raw data and supports with inside the improvement of enterprise specialists who can exploit effectiveness through powerful pattern of BI in [20]. In [21], authors offered a widespread opinion of the interrelationships among BI, BD, and BDA. The analytics makes the system measured as one of the high organization capabilities, which are the system highlights. The [22] authors compared the ARIMA model; to calculate the performance of the system, mean average percentage error (MAPE) is used. BI can support the efficiency of operational rules like monetary and budgeting in an energetic commercial background. BI is used to collect and examine data using predictive analytics [23].

The ideas behind BI and analytics from big data to healthcare are discussed, and the conceptual healthcare stock prediction has been implemented.

6.3 Conceptual Healthcare Stock Prediction System

The proposed healthcare stock prediction system is comprised of three phases (Figure 6.1). The first phase of the healthcare stock prediction is the data source. Data are retrieved from the Yahoo finance website using panda reader in python libraries. The data that are retrieved from Yahoo finance consists of 6 months of data with a daily frequency of chronological

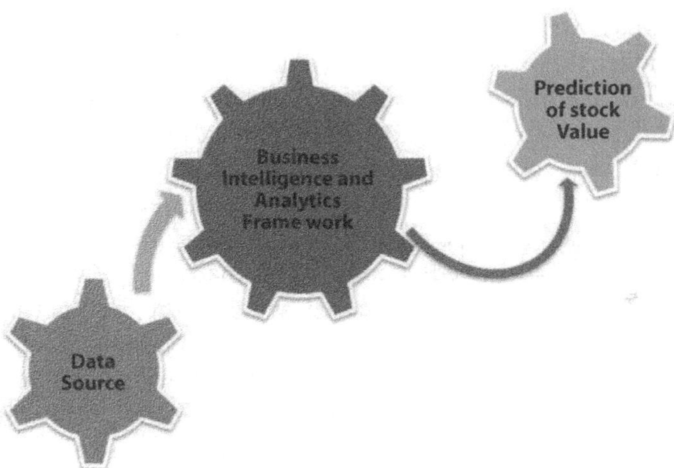

Figure 6.1 Conceptual healthcare stock prediction system.

data. The second phase comprises of BI and analytics framework, which is an overview of BI and analytics framework in Figure 6.2. The third phase of the healthcare stock prediction is the prediction of stock value. Investors can able to predict whether to buy, sell, or hold the stock.

Figure 6.2 depicts that the BI and analytics framework comprise simple machine learning (SML) models, time series forecasting (TSF), and complex deep neural network (CDNN).

The proposed healthcare stock prediction system predicts the closing price of the healthcare stocks using BI and analytical framework. Figure 6.3 depicts the illustration of the healthcare stock prediction system.

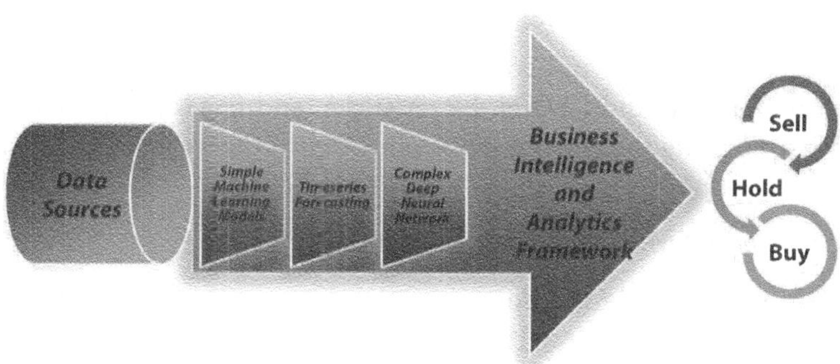

Figure 6.2 Overview of business intelligence and analytics framework.

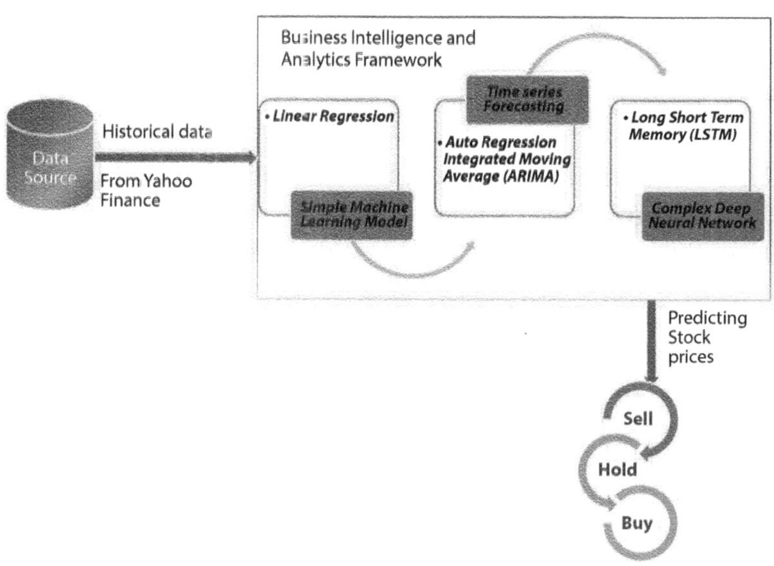

Figure 6.3 Illustration of healthcare stock prediction system.

6.3.1 Data Source

The chronological prices are collected from Yahoo finance using the python library pandas_reader. The daily frequency of the past 6 month's data is retrieved from Yahoo finance. Data source consists of various elements like date, open, low, close, and high. This research works predict the healthcare company stocks like Apollo Hospitals Enterprise Limited, Cadila Healthcare Ltd., Dr. Reddy's Laboratories, Fortis Healthcare Limited, Max Healthcare Institute Limited, Opto Circuits Limited, Panacea Biotec, Poly Medicure Ltd., Thyrocare Technologies Limited, and Zydus Wellness Ltd.

6.3.2 Business Intelligence and Analytics Framework

BI and analytics framework comprise of SML model, TSF, and CDNN. In a SML model, TSF, and CDNN, LR, ARIMA, and LSTM, respectively, are used for predicting the closing stock value of healthcare stocks.

6.3.2.1 Simple Machine Learning Model

A SML model is used for recognizing the data configuration and appropriating the data into model. Machine learning algorithm allows computers to train the data that are input, and statistical analysis is used for the values that are output within an exact range[2]. LR is used in SML. LR is supervised learning, and it is a measure of regression analysis used to find out the association between attributes of input and the objective variables. In this model, 80% of training data and 20% of testing data are used.

6.3.2.1.1 Algorithm to Forecast the Stock Price using Linear Regression (LR)

The algorithm describes predicting the closing stock price using LR.

Input: Date, Open, Close, Low, High, and Volume of the particular stocks retrieved from Yahoo finance.
Step 1: Start
Step 2: Data is loaded.
Step 3: Declaring the x and y variables.
Step 4: Split 80% of training data and 20% of testing data.

Step 5: Using x train and y train regressor is fitted.
Step 6: Regression coefficient is calculated.
Step 7: Predict the test data.
Step 8: Calculating the MSE and RMSE.
Step 9: Actual and predicted stock prices are visualized.
Step 10: Stop the execution.
Output: The closing price of the given stock is predicted.

6.3.2.2 *Time Series Forecasting*

The TSF models predict the future value based on the past value. ARIMA model is used to predict the closing price of the healthcare stocks. ARIMA is the predicting algorithm based on the past values of the time series data future value that is predicted. ARIMA is the blend of autoregressive (p), integrated (d), and moving average (q).

6.3.2.2.1 Algorithm to Forecast the Closing Stock Price using ARIMA

TThe algorithm describes predicting the closing stock price using ARIMA.

Input: Date, Open, Close, Low, High, and Volume of the particular stocks retrieved from Yahoo finance.
Step 1: Start
Step 2: Time series data is visualized.
Step 3: Data is transformed into a logarithmic format.
Step 4: Check whether the data is stationary or non-stationary.
Step 5: If the data is non-stationary convert by taking the difference.
Step 6: Plot the correlation and autocorrelation charts.
Step 7: Construct the ARIMA model based on the data.
Step 8: The model is used for prediction.
Step 9: Stop the Execution.
Output: The closing price of the given stock is predicted.

6.3.2.3 *Complex Deep Neural Network*

The CDNN is a process of the deep learning module that uses LSTM to predict the healthcare stocks LSTM resides of cell states, input gate, hidden state, forget gate, and output gate.

6.3.2.3.1 Algorithm to Forecast the Stock Price using Long Short-Term Memory

The algorithm describes predicting the closing stock price using LSTM.

Input: Date, Open, Close, Low, High, and Volume of the particular stocks retrieved from Yahoo finance.
Step 1: Start
Step 2: Dataset is loaded.
Step 3: Features are loaded from feature extraction into the neural network.
Step 4: Training neural network comprises Optimizer, Regularization, and Dropouts.
Step 5: Optimizer increase the accurateness of the model, Adaptive Movement Estimation (Adam) is used.
Step 6: Regularization develops the performance of the model and shrinks overfitting. Tikhonov regularization is used.
Step 7: In dropouts calculate the trends based on the input and recurrent connections to LSTM units.
Step 8: The output layer of RNN with the objective value is generated by Output generation.
Step 9: Actual and predicted values are visualized.
Step 10: Stop the execution.
Output: The closing price of the given stock is predicted.

6.3.3 Predicting the Stock Price

Predicted healthcare stock prices are compared using MSE and RMSE. The model that provides the minimum error is considered to be the preeminent model to forecast the stock prices. With the help of this prediction, investors can predict and make decisions about the stock price in healthcare stocks whether to sell, hold, or buy the stocks.

6.4 Implementation and Result Discussion

The healthcare stock prediction framework uses LR, ARIMA, and LSTM. The proposed framework uses the data from Yahoo finance. The data source consists of the date, open, close, low, and high of the healthcare domain. The proposed framework predicts the closing price of the healthcare stocks of Apollo Hospitals Enterprise Limited, Cadila Healthcare Ltd., Dr. Reddy's Laboratories, Fortis Healthcare Limited, Max Healthcare Institute Limited, Opto Circuits Limited, Panacea Biotec, Poly Medicure

Ltd., Thyrocare Technologies Limited, and Zydus Wellness Ltd. using LR, ARIMA, and LSTM.

6.4.1 Apollo Hospitals Enterprise Limited

Apollo Hospital is the first corporate healthcare in India. Prathap C Reddy is the founder of the hospital in 1983. Its first branch was opened in Chennai. The healthcare network including hospitals, pharmacies, primary care, and diagnostic clinics is powerful[3].

Figures 6.4, 6.5, and 6.6 depict the prediction of the closing price using LR, ARIMA, and LSTM, respectively.

6.4.2 Cadila Healthcare Ltd

Cadila Pharmaceuticals Ltd., also known as Zydus Cadila, is one of the leading pharmaceutical enterprises in India. They are developing and manufacturing medicines for an affordable cost to patients around the world. The company also got permission from the DCGI to conduct the human experimental from July 2020 for the COVID-19 vaccine named ZyCoV-D[4].

Figures 6.7, 6.8, and 6.9 depict the prediction of the closing price using LR, ARIMA, and LSTM, respectively.

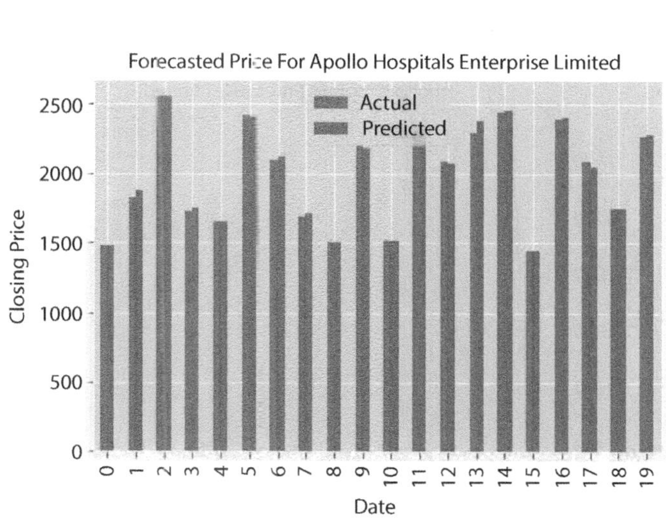

Figure 6.4 Prediction of the closing price using LR.

```
==============================================================================
Dep. Variable:                    D.y   No. Observations:                  126
Model:                 ARIMA(0, 1, 0)   Log Likelihood                 296.242
Method:                           css   S.D. of innovations              0.023
Date:                Sun, 10 Jan 2021   AIC                           -588.483
Time:                        15:36:20   BIC                           -582.811
Sample:                             1   HQIC                          -586.179

==============================================================================
                 coef    std err          z      P>|z|      [0.025      0.975]
------------------------------------------------------------------------------
const          0.0048      0.002      2.316      0.021       0.001       0.009
==============================================================================
```

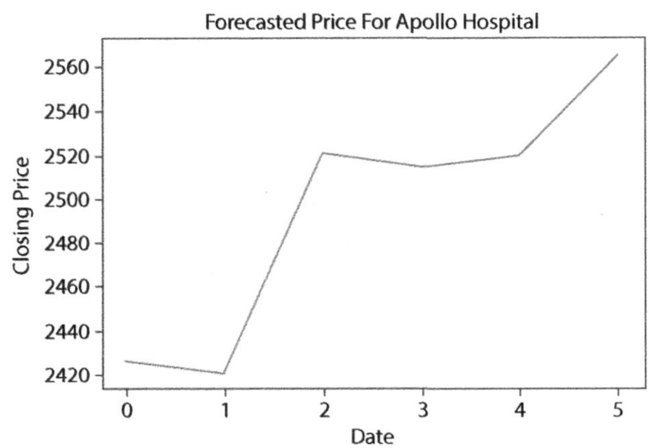

Figure 6.5 Prediction of the closing price using ARIMA.

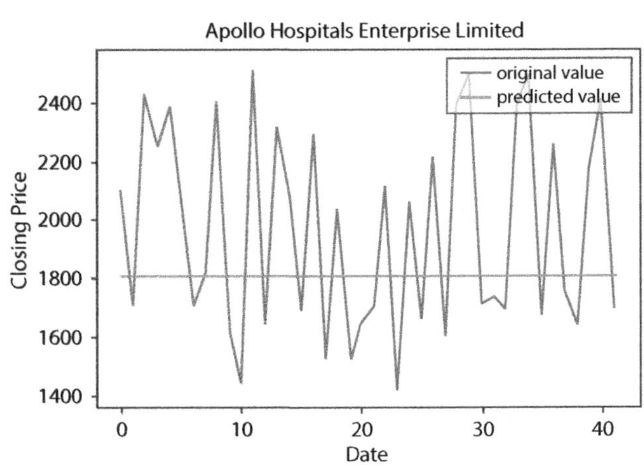

Figure 6.6 Prediction of the closing price using LSTM.

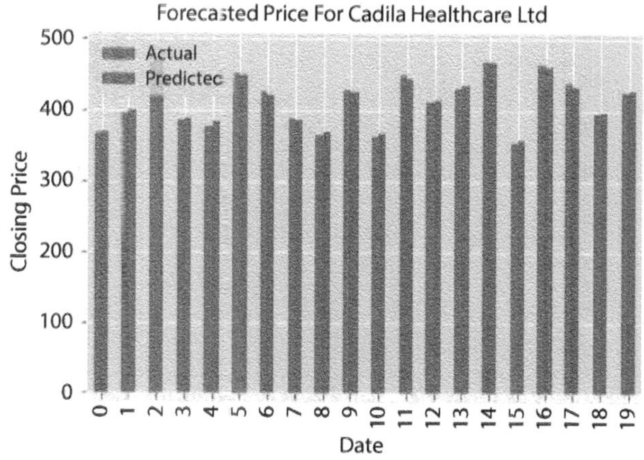

Figure 6.7 Prediction of the closing price using LR.

```
                              ARIMA Model Results
==============================================================================
Dep. Variable:                    D.y   No. Observations:                  126
Model:                 ARIMA(0, 1, 0)   Log Likelihood                 307.507
Method:                           css   S.D. of innovations              0.021
Date:                Sun, 10 Jan 2021   AIC                           -611.013
Time:                        15:56:34   BIC                           -605.341
Sample:                             1   HQIC                          -608.709

==============================================================================
                 coef    std err          z      P>|z|      [0.025      0.975]
------------------------------------------------------------------------------
const          0.0024      0.002      1.303      0.192      -0.001       0.006
==============================================================================
```

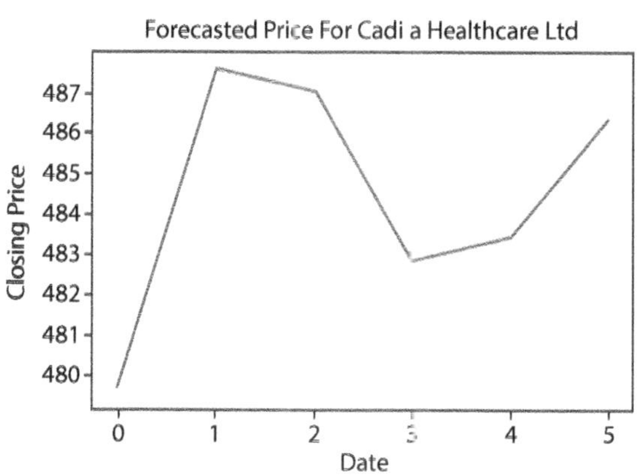

Figure 6.8 Prediction of the closing price using ARIMA.

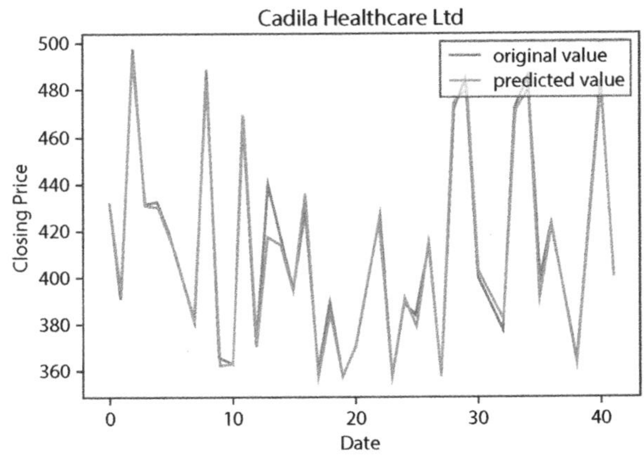

Figure 6.9 Prediction of the closing price using LSTM.

6.4.3 Dr. Reddy's Laboratories

Dr. Reddy's Laboratories is a worldwide pharmaceutical completed nearly 190 medications and 60 APIs for manufacturing drug. It is founded by Anji Reddy. The company has a contract for the final-stage trials of Russia's SputnikV-COVID-19 vaccine in India[5].

Figures 6.10, 6.11, and 6.12 depict the prediction of closing price using LR, ARIMA, and LSTM, respectively.

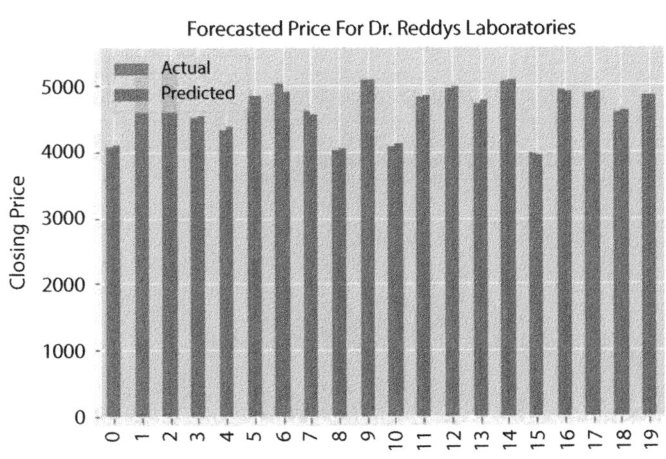

Figure 6.10 Prediction of the closing price using LR.

```
==============================================================================
Dep. Variable:                    D.y   No. Observations:                  126
Model:                 ARIMA(0, 1, 0)   Log Likelihood                 324.117
Method:                           css   S.D. of innovations              0.018
Date:                Sun, 10 Jan 2021   AIC                           -644.235
Time:                        20:06:27   BIC                           -638.562
Sample:                             1   HQIC                          -641.930

==============================================================================
                 coef    std err          z      P>|z|      [0.025      0.975]
------------------------------------------------------------------------------
const          0.0025      0.002      1.507      0.132      -0.001       0.006
==============================================================================
```

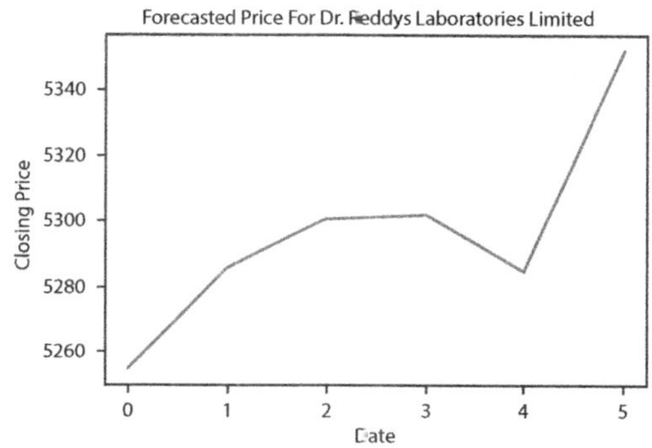

Figure 6.11 Prediction of the closing price using ARIMA.

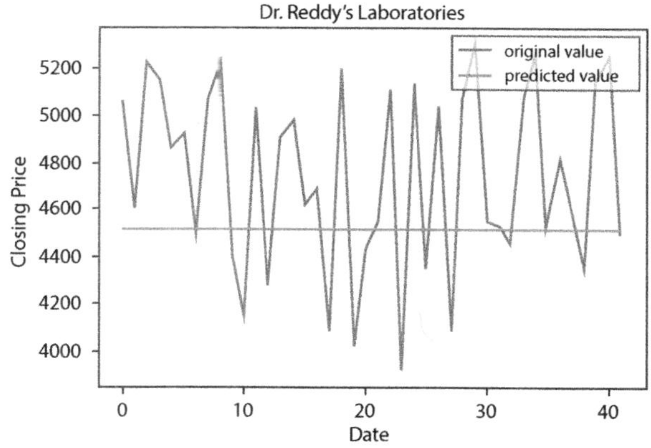

Figure 6.12 Prediction of the closing price using LSTM.

6.4.4 Fortis Healthcare Limited

India's leading incorporated healthcare distribution service provider is Fortis Healthcare Limited. The medical amenities verticals of the institute involve fundamentally emergency health center, diagnostics, and day care specialty amenities. There are 36 healthcare facilities and 415 diagnostics centers in India, Nepal, Dubai, and Sri Lanka[6].

Figures 6.13, 6.14, and 6.15 depict the prediction of the closing price using LR, ARIMA, and LSTM, respectively.

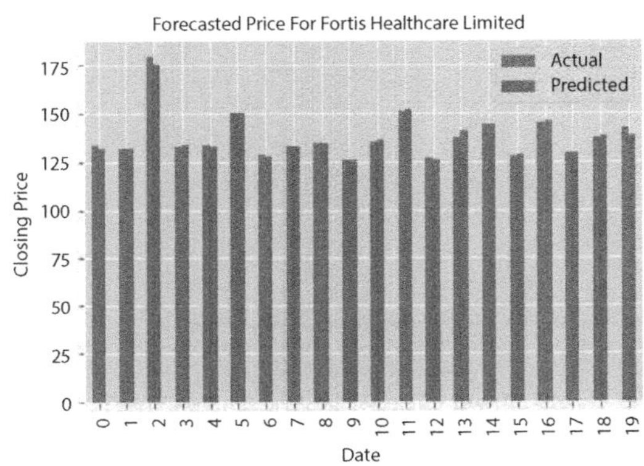

Figure 6.13 Prediction of the closing price using LR.

```
==============================================================================
Dep. Variable:                    D.y   No. Observations:                  126
Model:                 ARIMA(0, 1, 0)   Log Likelihood                 322.284
Method:                           css   S.D. of innovations              0.019
Date:                Sun, 10 Jan 2021   AIC                           -640.567
Time:                        20:10:49   BIC                           -634.895
Sample:                             1   HQIC                          -638.263

==============================================================================
                 coef    std err          z      P>|z|      [0.025      0.975]
------------------------------------------------------------------------------
const          0.0031      0.002      1.856      0.063      -0.000       0.006
==============================================================================
```

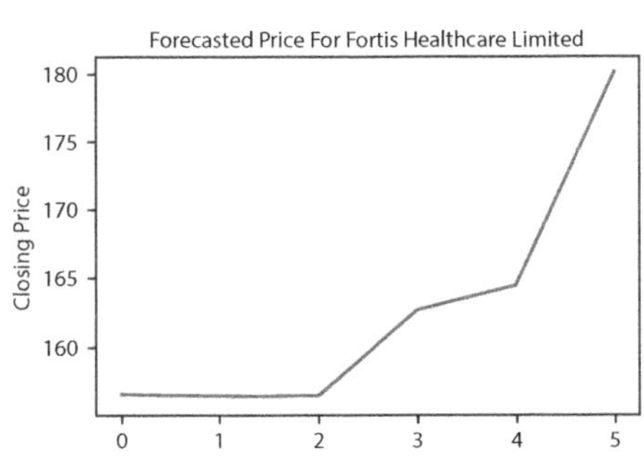

Figure 6.14 Prediction of the closing price using ARIMA.

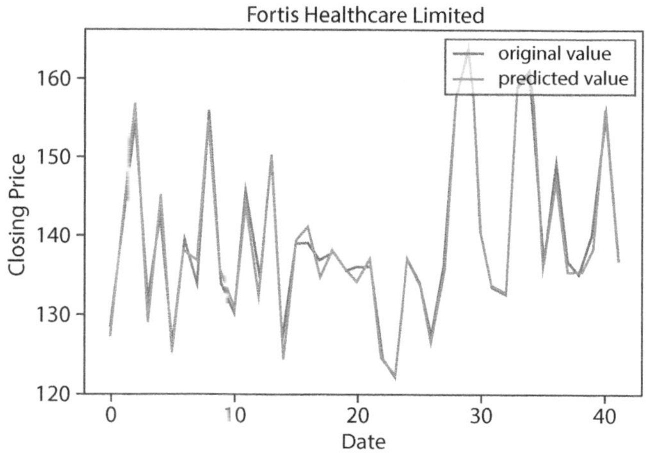

Figure 6.15 Prediction of the closing price using LSTM.

6.4.5 Max Healthcare Institute Limited

Max Healthcare is the second biggest medical care chain in India by income. It has 16 healthcare facilities. Max Healthcare retains and works on medical care offices across North India, the National Capital Region (NCR) of Delhi, and the western port city of Mumbai. Max Healthcare united with Radiant Lifecare, which worked at the notable Nanavati Hospital in Mumbai and BLK Hospital in Central Delhi to arise as the second-biggest healthcare amenities organization (by income) in 2020[7].

Figures 6.16, 6.17, and 6.18 depict the prediction of the closing price using LR, ARIMA, and LSTM, respectively.

6.4.6 Opto Circuits Limited

Opto Circuits Limited (OCI) is a vertically incorporated worldwide clinical innovation Group that has some expertise in essential, intense, and basic consideration items for the worldwide business sectors. OCI creates and makes mechanically progressed clinical gear and gadgets for present-day medical service rehearses. OCI represents considerable authority in heart and indispensable signs checking, crisis cardiovascular consideration, vascular medicines, and detecting advancements[8].

Figures 6.19, 6.20, and 6.21 depict the prediction of the closing price using LR, ARIMA, and LSTM, respectively.

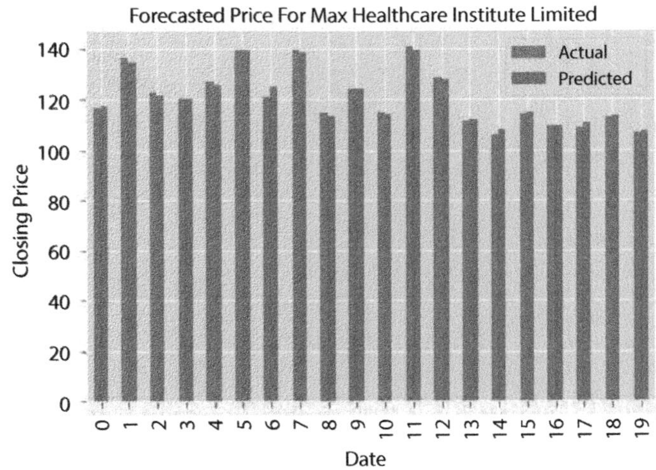

Figure 6.16 Prediction of the closing price using LR.

```
                             ARIMA Model Results
==============================================================================
Dep. Variable:                    D.y   No. Observations:                   96
Model:                 ARIMA(0, 1, 0)   Log Likelihood                 189.047
Method:                           css   S.D. of innovations              0.034
Date:                Sun, 10 Jan 2021   AIC                           -374.093
Time:                        20:13:10   BIC                           -368.964
Sample:                             1   HQIC                          -372.020

==============================================================================
                 coef    std err          z      P>|z|      [0.025      0.975]
------------------------------------------------------------------------------
const          0.0034      0.003      0.975      0.330      -0.003       0.010
==============================================================================
```

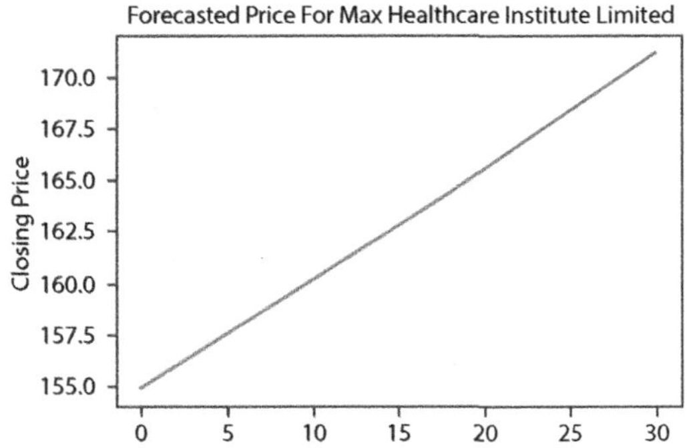

Figure 6.17 Prediction of the closing price using ARIMA.

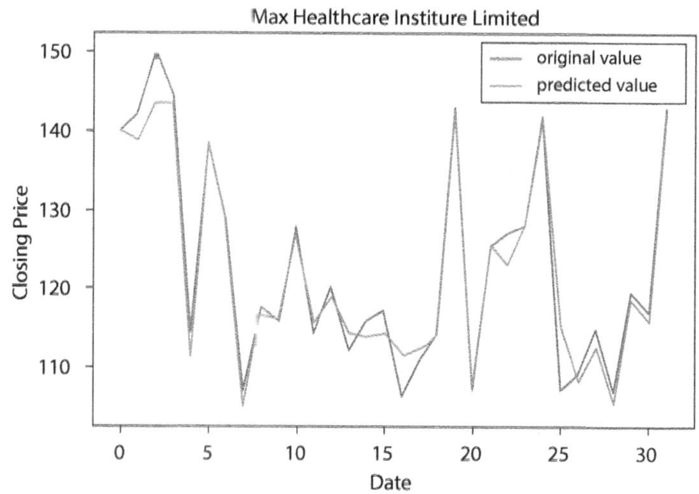

Figure 6.18 Prediction of the closing price using LSTM.

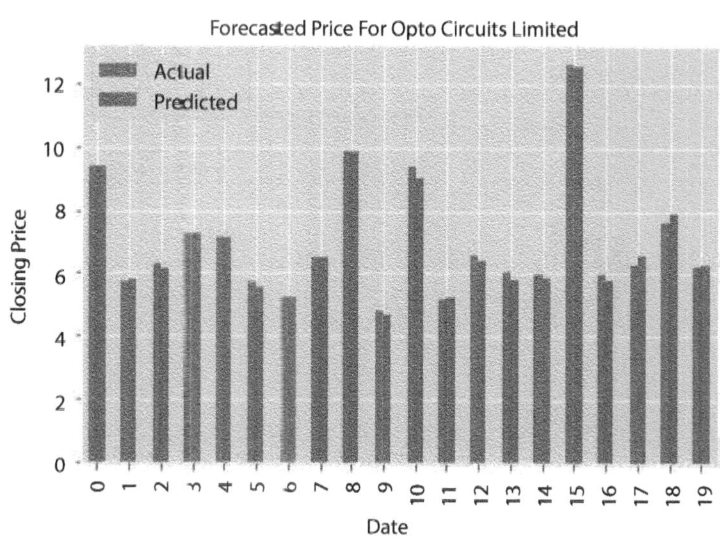

Figure 6.19 Prediction of the closing price using LR.

```
                              ARIMA Model Results
==============================================================================
Dep. Variable:                    D.y   No. Observations:                  126
Model:                 ARIMA(0, 1, 0)   Log Likelihood                 233.503
Method:                           css   S.D. of innovations              0.038
Date:                Sun, 10 Jan 2021   AIC                           -463.006
Time:                        20:17:50   BIC                           -457.333
Sample:                             1   HQIC                          -460.701

==============================================================================
                 coef    std err          z      P>|z|      [0.025      0.975]
------------------------------------------------------------------------------
const         -0.0063      0.003     -1.878      0.060      -0.013       0.000
==============================================================================
```

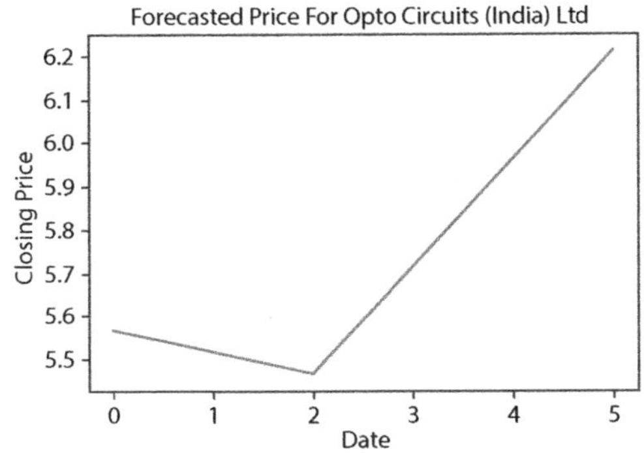

Figure 6.20 Prediction of the closing price using ARIMA.

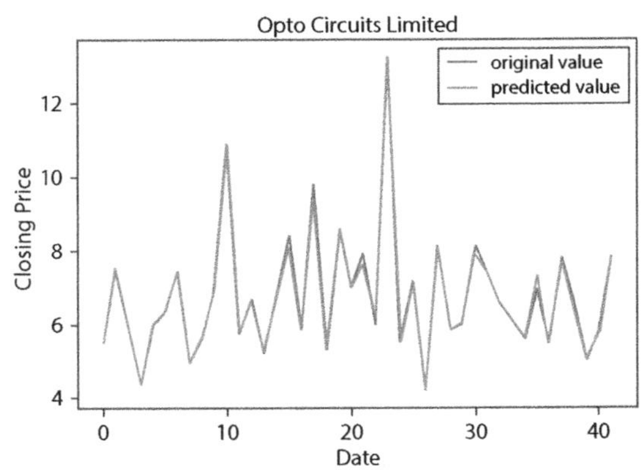

Figure 6.21 Prediction of the closing price using LSTM.

6.4.7 Panacea Biotec

Panacea Biotec is an innovation-driven biotechnology association. The company has a strong background in exploration and growth in the field of biotechnology, manufacturing, and vaccines. The company has established a joint endeavor firm in Ireland with US-based Refana Inc. to build up an antibody for COVID-19[9].

Figures 6.22, 6.23, and 6.24 depict the prediction of the closing price using LR, ARIMA, and LSTM, respectively.

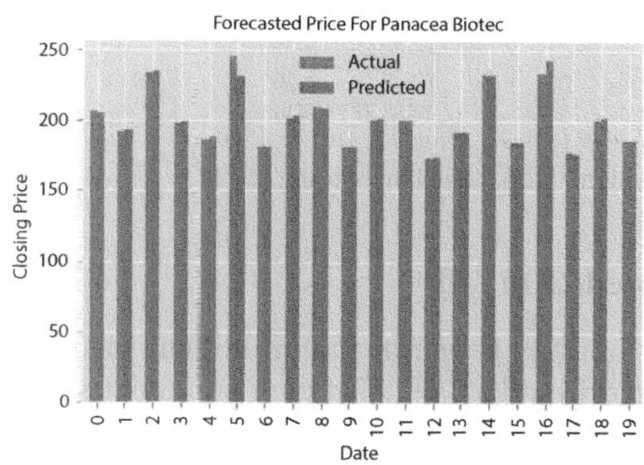

Figure 6.22 Prediction of the closing price using LR.

```
==============================================================================
Dep. Variable:                    D.y   No. Observations:                  126
Model:                 ARIMA(0, 1, 0)   Log Likelihood                 251.708
Method:                           css   S.D. of innovations              0.033
Date:                Sun, 10 Jan 2021   AIC                           -499.416
Time:                        17:21:13   BIC                           -493.743
Sample:                             1   HQIC                          -497.111

==============================================================================
                 coef    std err          z      P>|z|      [0.025      0.975]
------------------------------------------------------------------------------
const          0.0016      0.003      0.547      0.584      -0.004      0.007
==============================================================================
```

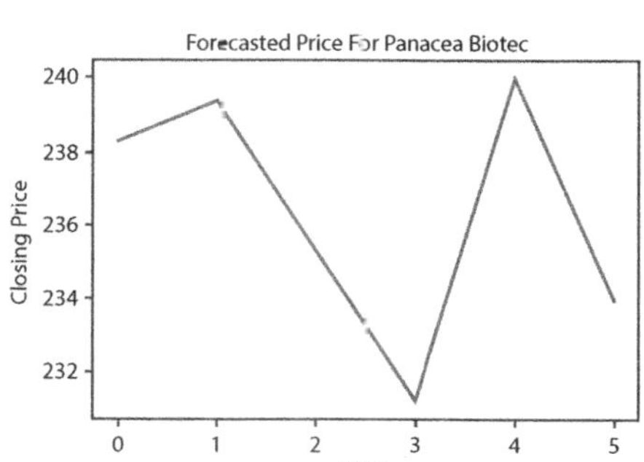

Figure 6.23 Prediction of the closing price using ARIMA.

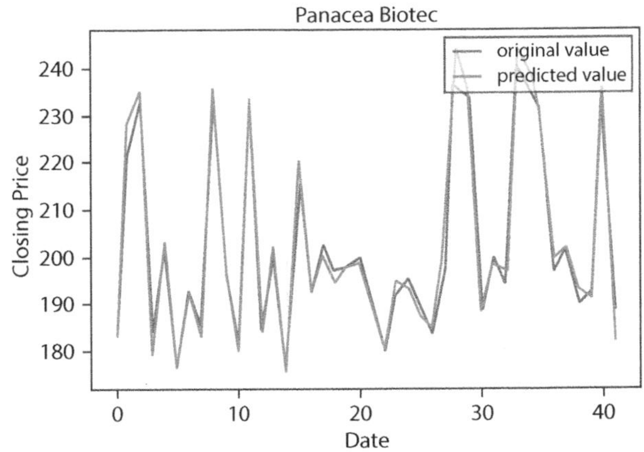

Figure 6.24 Prediction of the closing price using LSTM.

6.4.8 Poly Medicure Ltd

Poly Medicure Limited is one of the prominent medical device companies and exporters in India. Poly Medicure Limited has been recognized by the Department of Pharmaceuticals Ministry of Chemicals and Fertilizers, Government of India as "The Medical Devices Company of the Year 2018". They have been additionally perceived as the Largest Exporter of Medical Devices from India for 6 years in a row[10].

Figures 6.25, 6.26, and 6.27 depict the prediction of the closing price using LR, ARIMA, and LSTM, respectively.

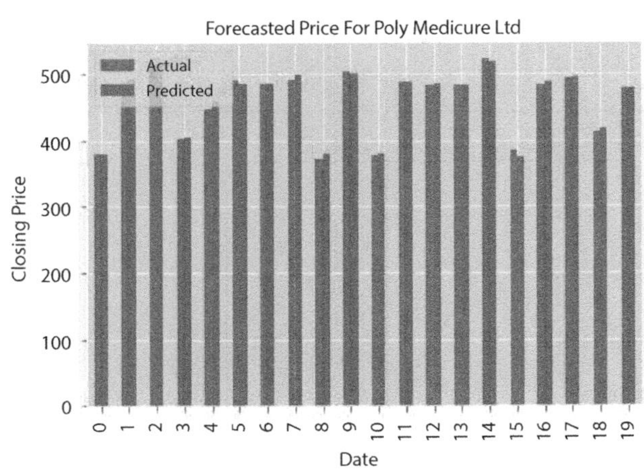

Figure 6.25 Prediction of the closing price using LR.

```
                              ARIMA Model Results
==============================================================================
Dep. Variable:                    D.y   No. Observations:                  126
Model:                 ARIMA(0, 1, 0)   Log Likelihood                 280.163
Method:                           css   S.D. of innovations              0.026
Date:                Sun, 10 Jan 2021   AIC                           -556.326
Time:                        20:20:55   BIC                           -550.654
Sample:                             1   HQIC                          -554.022

==============================================================================
                 coef    std err          z      P>|z|      [0.025      0.975]
------------------------------------------------------------------------------
const          0.0031      0.002      1.309      0.190      -0.002       0.008
==============================================================================
```

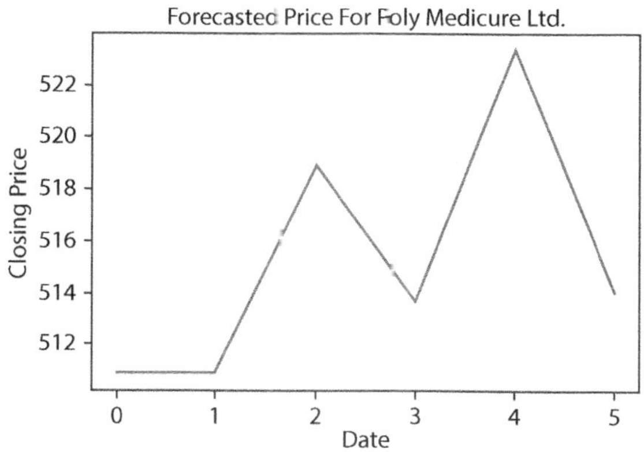

Figure 6.26 Prediction of the closing price using ARIMA.

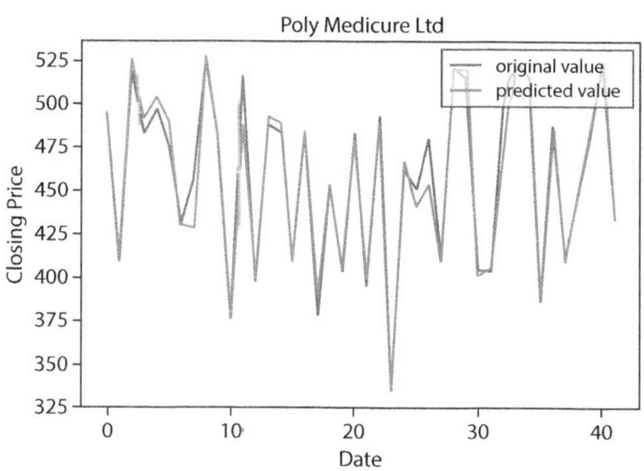

Figure 6.27 Prediction of the closing price using LSTM.

6.4.9 Thyrocare Technologies Limited

Thyrocare is India's first completely mechanized symptomatic research center with an emphasis on giving quality at reasonable expenses to laboratories and medical clinics in India and different nations. Thyrocare has utilized the high-level unrivaled innovation, equipped for chipping away at its extraordinary volumes but then conveying quality and exactness to its clients[11].

Figures 6.28, 6.29 and 6.30 depict the prediction of the closing price using LR, ARIMA, and LSTM, respectively.

6.4.10 Zydus Wellness Ltd

Zydus Wellness is also known as Carnation Nutra-Analogue Foods. Zydus wellness is a consumer merchandise company headquartered in Ahmedabad, Gujarat. It produces nourishment and skincare goods. The company is the subordinate of Cadila Healthcare pharmaceutical enterprise[12].

Figures 6.31, 6.32 and 6.33 depict the prediction of the closing price using LR, ARIMA, and LSTM respectively.

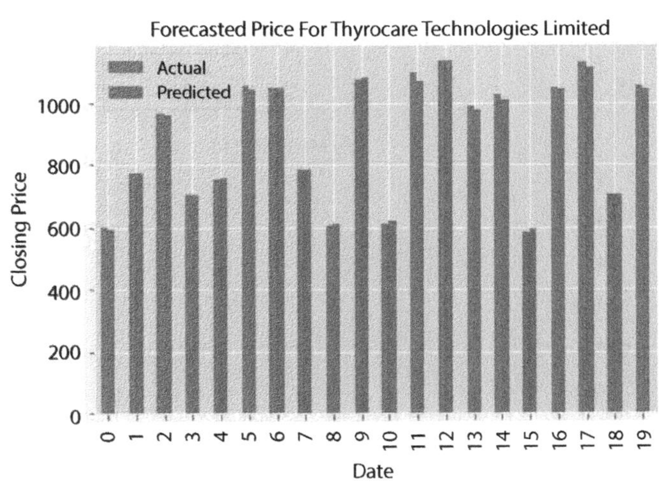

Figure 6.28 Prediction of the closing price using LR.

ARIMA Model Results

```
==============================================================================
Dep. Variable:                    D.y   No. Observations:                  126
Model:                 ARIMA(0, 1, 0)   Log Likelihood                 232.818
Method:                           css   S.D. of innovations              0.038
Date:                Sun, 10 Jan 2021   AIC                           -461.637
Time:                        20:24:29   BIC                           -455.964
Sample:                             1   HQIC                          -459.332

==============================================================================
                 coef    std err          z      P>|z|      [0.025      0.975]
------------------------------------------------------------------------------
const          0.0048      0.003      1.419      0.156      -0.002       0.011
==============================================================================
```

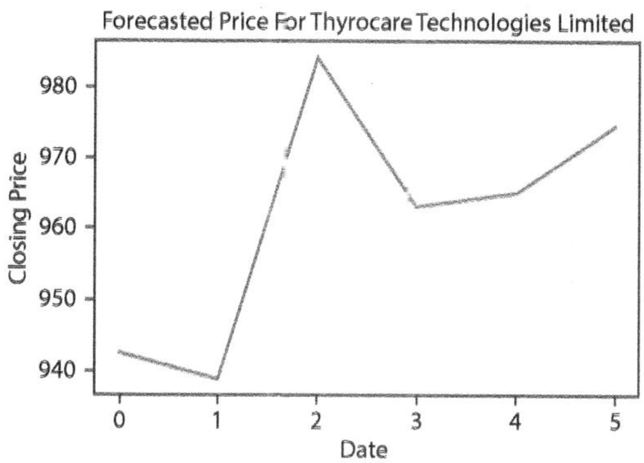

Figure 6.29 Prediction of the closing price using ARIMA.

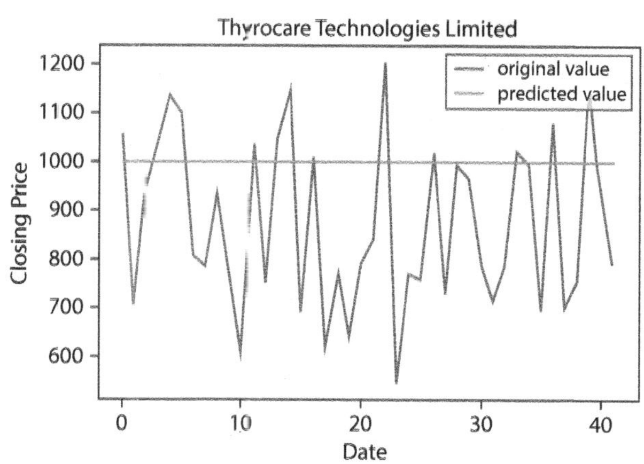

Figure 6.30 Prediction of the closing price using LSTM.

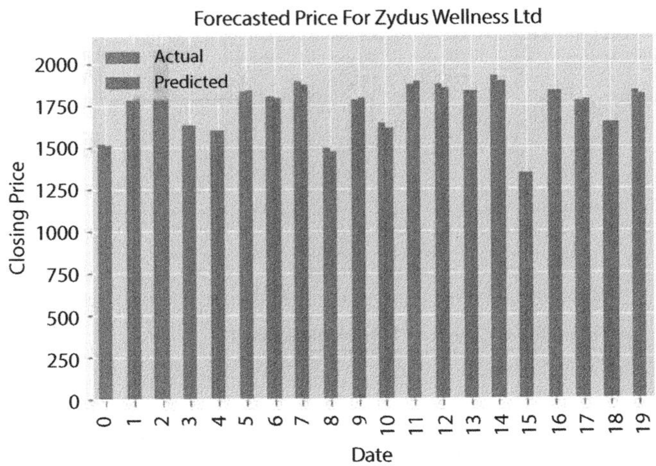

Figure 6.31 Prediction of the closing price using LR.

```
                              ARIMA Model Results
==============================================================================
Dep. Variable:                    D.y   No. Observations:                  126
Model:                 ARIMA(0, 1, 0)   Log Likelihood                 300.786
Method:                           css   S.D. of innovations              0.022
Date:                Sun, 10 Jan 2021   AIC                           -597.572
Time:                        20:26:57   BIC                           -591.900
Sample:                             1   HQIC                          -595.268

==============================================================================
                 coef    std err          z      P>|z|      [0.025      0.975]
------------------------------------------------------------------------------
const          0.0033      0.002      1.660      0.097      -0.001       0.007
==============================================================================
```

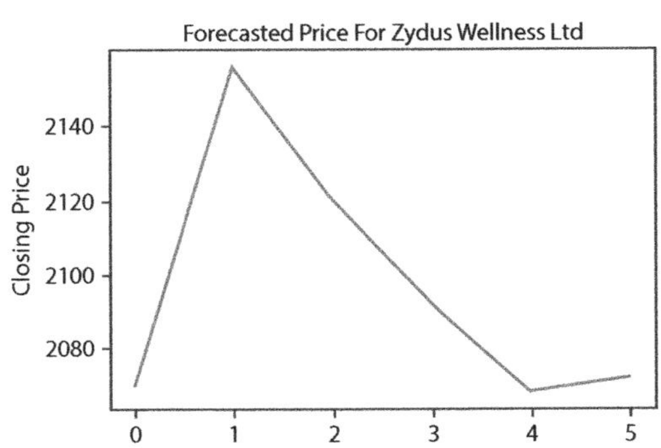

Figure 6.32 Prediction of the closing price using ARIMA.

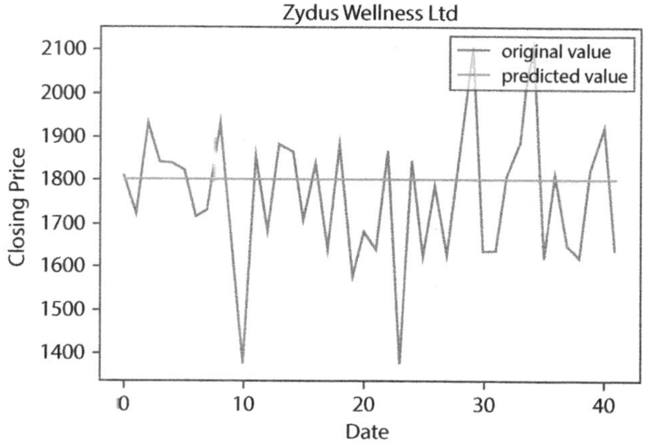

Figure 6.33 Prediction of the closing price using LSTM.

6.5 Comparisons of Healthcare Stock Prediction Framework

The healthcare stock prediction framework comprises LR, ARIMA, and LSTM. The framework models are compared with MSE and RMSE values. Table 6.1 depicts the comparison of LR, ARIMA, and LSTM with corresponding MSE and RMSE of the healthcare stocks of Apollo Hospitals Enterprise Limited, Cadila Healthcare Ltd., Dr. Reddy's Laboratories, Fortis Healthcare Limited, Max Healthcare Institute Limited, Opto Circuits Limited, Panacea Biotec, Poly Medicure Ltd., Thyrocare Technologies Limited, and Zydus Wellness Ltd.

According to Table 6.1, the LR, LSTM, and ARIMA models were predicted more accurate in positive stocks like Apollo Hospitals Enterprise Limited, Cadila Healthcare Ltd., Dr. Reddy's Laboratories, Fortis Healthcare Limited, Max Healthcare Institute Limited, Panacea Biotec, Poly Medicure Ltd., Thyrocare Technologies Limited, and Zydus Wellness Ltd. The negative stock is Opto Circuits Limited, which is a medical device industry that was worst affected the export of medical devices and critical raw materials are disturbed due to the restriction on movement, travel, and transport.

Table 6.1 Comparison of LR, ARIMA, and LSTM of MSE and RMSE.

		LR		ARIMA		LSTM	
S. no.	**Healthcare stocks**	**MSE**	**RMSE**	**MSE**	**RMSE**	**MSE**	**RMSE**
1	Apollo Hospitals Enterprise Limited	731.13	27.03	2675.81	51.72	120077.9	346.52
2	Cadila Healthcare Ltd	13.56	3.68	71.95	8.48	21.49	4.64
3	Dr. Reddy's Laboratories	3203.46	56.59	4244.89	65.15	204528.32	452.25
4	Fortis Healthcare Limited	2.38	1.54	14.65	3.827	1.93	1.39
5	Max Healthcare Institute Limited	2.10	1.45	22.06	4.69	6.92	2.63
6	Opto Circuits Limited	0.025	0.158	51.60	7.18	0.02	0.15
7	Panacea Biotec	13.22	3.636	101.05	10.05	9.89	3.14
8	Poly Medicure Ltd	24.32	4.93	81.85	9.04	103.04	10.15
9	Thyrocare Technologies Limited	108.05	10.39	598.80	24.47	45263.95	212.75
10	Zydus Wellness Ltd	241.66	15.54	1081.44	32.885	25310.09	159.09

6.6 Conclusion and Future Enhancement

Healthcare industry is massively essential to all the people around the world. Healthcare industry requires continuous innovation under increased regulation. It is scattered with risks and challenges. The proposed framework is used to predict the values of the healthcare-related stocks using a LR in SML model, ARIMA in TSF, and LSTM in CDNN. To conclude, results shown are strongly correlated with the actual and predicted stock price. Investors can predict and make their decision whether to buy, sell, or hold the stock. The performance of LR has minimized MSE and RMSE than LSTM and ARIMA. Based on the value of MSE and RMSE, LR provides minimum error compared with LSTM and ARIMA.

In the era of BI and analytics, the attractiveness of stock market trading is evolving incredible field to boost the data analytics researchers. Future research includes financial and technical domains, algorithmic trading, and reinforcement so that investors can find optimal stock prediction. Stocks are fluctuating rapidly based on the events that occurred in the world. These research works are based on the generic stock index. As a future direction, this research would like to include various factors like news, sentiments of the markets, the psychology of the investors, supply and demand, and results of the companies to predict with a high degree of accurateness.

References

1. Meher, B.K., Hawaldar, I.T., Spulbar, C., Birau, R., "Forecasting stock market prices using mixed ARIMA model: a case study of Indian pharmaceutical companies". *Invest. Manage. Financial Innov.*, 18, 1, pp 42–54, 2021.
2. Rehman, A., Naz, S., Razzak, M.I., "Leveraging big data analytics in healthcare enhancement: trends, challenges and opportunities". *Multimedia Systems*, pp. 1–33, 2021, doi 10.1007/s00530-020-00736-8.
3. Zeng, Z. and Khushi, M., "Wavelet Denoising and Attention-based RNN-ARIMA Model to Predict Forex Price". *Comput. Engineering, Finance, Sci. (cs. CE), Comput. Vis. Pattern Recognit. (cs. CV). IJCNN 2020*, 16th Aug 2020.
4. Khan, M.A., Saqib, S., Alyas, T., Rehman, A.U., Saeed, Y., Zeb, A., Zareei, M., "Effective Demand Forecasting Model Using Business Intelligence Empowered With Machine Learning". *IEEE Access*, 8, 116013–116023, 2020.
5. Nabipour, M., Nayyeri, P., Jabani, H., Mosavi, A., Salwana, E., Shahab, S., "Deep Learning for Stock Market Prediction". *Entropy*, 22, 840, 30 July 2020.

6. Ding, G. and Qin, L., "Study on the prediction of stock price based on the associated network model of LSTM". *Int. J. Mach. Learn. Cybern.*, 11, 1307–1317, 2020.
7. Kumaresan, K. and Ganeshkumar, P., "Software reliability prediction model with realistic assumption using time series (S)ARIMA model". *J. Ambient Intell. Humaniz. Comput.*, 11, 5561–5568, 2020.
8. Challa, M.L., Malepati, V., Kolusu, S.N.R., "S&P BSE Sensex and S&P BSE IT return forecasting using ARIMA". *Financial Innov.*, 6, Article number: 47, pp 1–19, 2020, https://doi.org/10.1186/s40854-020-00201-5.
9. Bukhari, A.H., Raja, M.A.Z., Sulaiman, M., Islam, S., Shoaib, M., Kumam, P., "Fractional Neuro-Sequential ARFIMA-LSTM for Financial Market Forecasting". *IEEE Access*, 8, 202071327, April 6, 2020.
10. Kulshreshtha, S. and Vijayalakshmi, A., "An ARIMA- LSTM Hybrid Model for Stock Market Prediction Using Live Data". *J. Eng. Sci. Technol. Rev.*, 13, 4, 117–123, 2020.
11. Gandhmal, D.P. and Kumar, K., "Systematic analysis and review of stock market prediction techniques". *Comput. Sci. Rev.*, 34, 100190, 2019.
12. Richards, G.S., Yeoh, G., Chong, A., Popovic, A., "Business Intelligence Effectiveness and Corporate Performance Management: An Empirical Analysis". *J. Comput. Inf. Syst.*, 59, 2, 188–196, January 2019.
13. Vijh, M., Chandola, D., Tikkiwal, V.A., Kumar, A., "Stock Closing Price Prediction using Machine Learning Techniques". *International Conference on Computational Intelligence and Data Science (ICCIDS 2019), Proc. Comput. Sci.*, 167, 599–606, 2020.
14. Umer Ghani, M., Awais, M., Muzammul, M., "Stock Market Prediction Using Machine Learning(ML)Algorithms". *ADCAIJ*, Regular Issue, Vol. 8, 4, 97–116,019.
15. Saranya, A. and Anandan, Dr. R., "Stock Market Prediction Using Machine Learning Algorithms". *Int. J. Recent Technol. Eng. (IJRTE)*, ISSN: 2277-3878, 8, 2S4, 280–283, July 2019.
16. Agrawal, S., Thakkar, D., Soni, D., Bhimani, K., Patel, Dr. C., "Stock Market Prediction using Machine Learning Techniques". *Int. J. Sci. Res. Comput. Sci. Eng. Inf. Technol. (IJSRCSEIT)*, Volume 5, 2, 1099–1103, March-April 2019.
17. Nti, I.K., Adekoya, A.F. & Weyori, B.A., A systematic review of fundamental and technical analysis of stock market predictions. *Artif. Intell. Rev.*, 53, 3007–3057, 2020. https://doi.org/10.1007/s10462-019-09754-z
18. Hiransha, M., Gopalakrishnan, E.A., Menonab, V.K., Soman, K.P., "NSE Stock Market Prediction Using Deep-Learning Models". *International Conference on Computational Intelligence and Data Science (ICCIDS 2018), Proc. Comput. Sci.*, 132, 1351–1362, 2018.
19. Bujari, A., Furini, M., Laina, N., "On Using Cashtags to Predict companies stock trends". *14th IEEE Annual Consumer Communications & Networking Conference(CCNC)*, 2017.

20. Olsen, D. H., & Dupin-Bryant, P. A., Combining Business Intelligence And Stock Market Data: A Primer For Data Analytics And Business Intelligence. *Journal of Applied Business Research (JABR)*, 29, 1, 139–156, 2012. https://doi.org/10.19030/jabr.v29i1.7562

Books

21. Alnoukari, M.,"From-Business-Intelligence-to-Big-Data -The-Power-of-Analytics", In book: *Integration Challenges for Analytics, Business Intelligence, and Data Mining*. pp. 44–62, IGI Global, November 2020, DOI: 10.4018/978-1-7998-5781-5.ch003.
22. Barai, A.K., Jain, P., Kumar, T., "NSE Stock Prediction: The Deep Learning Way", in: *Soft Computing: Theories and Applications, Part of the Advances in Intelligent Systems and Computing book series*, AISC, vol. 1154, pp. 783–791, Springer Singapore 29th, June 2020.
23. Richards, G., Yeoh, W., Chong, A.Y.L., Popovič, A., "Business intelligence effectiveness and corporate performance management An empirical analysis". *J. Comput. Inf. Syst.*, 59 2, pp 188–196, 31 July 2017.

Web Citation

1. https://www.wikipedia.org/wiki/Healthcare_industry/
2. https://www.digitalocean.com/community/tutorials/an-introduction-to-machine-learning/
3. https://www.apollohospitals.com/
4. https://zyduscadila.com/
5. https://www.drreddys.com/
6. https://www.fortishealthcare.com/
7. https://www.maxhealthcare.in/
8. http://www.optoindia.com/
9. https://www.panaceabiotec.com/
10. https://www.polymedicure.com/
11. https://www.thyrocare.com/
12. https://www.zyduswellness.in/

7

Internet of Things and Big Data Analytics for Smart Healthcare

Sathish Kumar K.[1*], Om Prakash P.G.[2], Alangudi Balaji N.[3] and Robertas Damaševičius[4]

[1]Department of Computer Science and Engineering, Koneru Lakshmaiah Education Foundation, Vaddeswaram, Guntur District, Andhra Pradesh, India
[2]Department of Computational Intelligence, SRM Institute of Science and Technology, Kattankulathur, Chengalpattu District, Tamil Nadu, India
[3]School of Computing Science and Engineering, VIT Bhopal University, Kotrikalan, Sehore District, Madhyapradesh, India
[4]Department of Software Engineering, Kaunas University of Technology, Kaunas, Lithuania

Abstract
With the impact of an array of chronic diseases on the social and economic regime of the people, everyone has started to give importance to healthcare. The monetary demands of hospitals and healthcare corporate have become severe apprehension to those who seek medicinal aids for their ailments. The lifestyles, food habits, and poor stress management usher from the young to the old invariably toward chronic heart diseases pave a path to increase the in-patient count of the hospitals. Medical system capability to continuously and constantly monitor without any inconspicuousness to beacon the signals disseminated from the various electronic devices attached to the patient in incentive care is the crucial demand to the healthcare. These high-sensitive sensor-enabled devices are effortlessly integrated with the advancement in nanoelectronics, the Internet of Things (IoT), and large datasets. IoT acts as the network infrastructure for these health systems to flood the electronic signals via diversified wireless communication technologies such as Bluetooth, Zigbee, and 4G. The physiological parameters measured from the devices associated with the patient's body are transmitted via wireless medium to reach doctors, medical consultants, and caregivers diversified over a wide geographical area. This enables to give immediate treatment, assistance, and counseling to the patients in need. Since these

**Corresponding author*: ksathish1980@gmail.com

A. Jaya, K. Kalaiselvi, Dinesh Goyal and Dhiya Al-Jumeily (eds.) Handbook of Intelligent Healthcare Analytics: Knowledge Engineering with Big Data Analytics, (147–166) © 2022 Scrivener Publishing LLC

electronic devices constantly beep out abundant signals, a vast and variety of data from veracious sources are stored and processed for fast decision making in big data with intelligent storage technology. Nanoelectronics, IoT, and big data have become inevitable technologies of medical advancements nowadays.

Keywords: Internet of Things (IoT), big data, IoT security, healthcare systems

7.1 Introduction

Big data is a container for large volumes and a variety of data. Big data stores data in any form as text, image, audio, and video as unstructured data. The healthcare systems rely on big data to store, analyze, and make critical decisions to provide better medical treatment to patients. Big data helps the healthcare systems to store and analyze the past medical history of the patients. The healthcare systems ripe good monetary benefit from the intelligent analysis as it reduces the time and expert advice involved in analyzing patient data.

Today, the medical field has witnessed tremendous advancements, thanks to the modern technological enhancements in handling electronic medical records, sensors in medical equipment, and modern handheld devices for better communication. The amount of patient-related data has increased dramatically because of the growth of the medical industry. Having an IT wing in every major hospital and healthcare system is inevitable today.

They use extensive computing and decision-making to infer important patterns or information from voluminous medicinal data. Data analysts analyze the data of patients having chronic diseases collected from the various data sources to come up with new information or pattern in the data. The treatment report of the patients, the laboratory images, clinical data, biometric information, hospital administration, and pharmacy subscription are some samples of data sources to the big data.

These sources generate on average 1.5 Mb of data per second per human being. This monumental 2.5 quintillions (2.5 followed by 18 zeros) of data lures researchers to develop healthcare systems over a network of smart devices and sensors, the Internet of Things (IoT) [13]. The IoT for big data is more data-centric. It collects huge diversified multisource data from various devices and locations and stores them in sophisticated data centers.

These data centers have the comprehensive infrastructure to support IoT to store and retrieve data with ease. The health information–centric IoT demands the uninterrupted provision of data and high bandwidth for faster dissemination of data [15]. Hence, the data centers are provided with a high speed and intelligent allocation of resources supporting enhanced data analysis. Prediction of a definite result from the set of data is vital in curing a patient. The unambiguousness of the health data prevents predicting the future of a patient. Hence, big data in medicine is inherently unambiguous with high intensive data. The existing collection models such as EMR and EHR for handling medical and health records are inferior in handling heavy data storage and analysis. Big data uses prediction modeling with the help of preparing data models of the retrieved data.

The data streamed into the big data infrastructure is preprocessed to clean the data by removing unwanted or unrequited pieces of data. This preprocessing is crucial for achieving higher accuracy in the prediction. The preprocessed data is then converted into a form suitable for further processing. This is then followed by a survey in the medical domain with big data as a foundation. Then, in the big data framework, an architecture is proposed for the medicinal domain. In all these steps, the handling of petabytes of data is a great challenge to the researchers.

7.2 Literature Survey

This section narrates the literature review of the IoT, big data analytics with reference to the medical field.

Shah *et al.* [1] suggest that the challenges faced in medical big data are storage, retrieval of the information within a stipulated time, and getting the solution at the least cost. Conventional decision-making algorithms in the healthcare systems are time inefficient, did not complete within an expected time, and suffered from dropping of some data that leads to inefficient and inaccurate information. Hence, the researchers' community has been involved in developing various healthcare systems for each one of certain diseases which provide solutions based on the data uploaded over the internet.

Diana *et al.* [2] suggest a solution to identify Obtrusive Sleep Apnea (OSA), a kind of sleep disorder. This OSA disorder will affect the quality of life due and the consequences of OSA include decrement in psychomotor performance, behavior, and personality disorders. Their proposed system enables the physician to give treatment to the elder people by continuous monitoring of sleep environment, the status of sleep, psychometric parameters, and physical activities. Their system architecture involves a two-stage process: one is pre-processing and another is predicting process. Their method utilizes big data techniques in predicting the OSA disorder with improved efficiency of up to 93.3%.

Songyun *et al.* [3] propose IoT centred toward information to support various applications such as smart health, smart city, and intelligent transportation. They mainly focus on smart health and quality of Service in recovery from data loss and robustness of the system. They employ rarity-aware data recovery algorithm to evaluate the replica distribution in data and the service requirement. The replica distribution will remove the duplicate entries that will provide quality of service and provide system robustness.

Gaspard *et al.* [4] utilize convolution neural networks (CNNs) with optimization techniques, to provide effective analysis on ECG, MRI, etc. The challenging task in developing a model in data mining is no single algorithm can provide perfect solutions to health data. Hence, they apply many machine learning algorithms to improve the quality of the analysis.

Gopinadh Sasubilli *et al.* [5] gather data from different resources and pre-process them by data mining techniques. The data mining techniques is forwarded to the framework for further implementation. Initially, researcher gathered the medical information from big data, and it is applied in different platform for prediction accuracy; to maintain prediction accuracy, the data must be genuine.

Smys *et al.* [6], in their research work, integrate big data and IoT for monitoring patient health in rural areas; particularly, the researcher concentrates to analyze the patients affected by cardiac disease. In their approach, they gather the information generated by sensors from patients in rural areas, transmit the data over the internet, and further applied big data techniques. The collected information is stored in cloud storage and further refined using machine learning algorithms to analyze and diagnose the seriousness of the disease. The big data together with IoT in integration gather 500 observation details to monitor patient health from that prediction are improved with less latency time.

Malti Bansal *et al.* [7] point out in their ECG monitoring system that the patient death rate is increased due to chronic heart diseases. So, the patient has to be more aware of his chronic heart disease. In their suggested system, the patient is advised to use hand portable devices instead to use the big machines that provide the interface between the doctor and the patient, which will save time for both doctor and patient. When the patient is in a discomfort situation and the doctor is unable to reach the patient on time due to traffic or unavoidable circumstances, with the help of IoT and big data, the doctor can attend patient from his place. The ECG monitoring system helps to monitor the remote patient with reduced infrastructure cost, which is an improvement in the medical field.

The following section explores the applications of IoT and big data analytics in various fields such as prediction of diabetes using smart devices, classification of learners based on the big data analytics, and their security issues.

7.3 Smart Healthcare Using Internet of Things and Big Data Analytics

A healthcare system that allows the patients and doctors to interact with each other and reports can be exchanged remotely gathered and analyzed from patients' daily activities via the IoT. There are a lot of applications of IoT in medical field.

7.3.1 Smart Diabetes Prediction

One of the key roles played by the IoT is monitoring diabetes of patients. The root cause of many chronic physical complications like heart failure, heart attack, memory loss, loss of sight, impotence, and failures of vital organs is a metabolic disorder—diabetes. A hormone called insulin is critical for maintaining sugar levels in the blood. Diabetes affects the flow of insulin to cause blurring in the vision, rapid weight loss, frequent urination, excessive thirst, numbness, and tingling in the foot.

Deepti *et al.* proposed employed various machine learning techniques on healthcare big data to predict diabetes. They used the WEKA tool with support vector machine (SVM) and random forest algorithm to derive a classifier model for the prediction of diabetes [8]. Vijaykumar *et al.* used the random forest algorithm with machine learning to measure the diabetes level of the patients with more accuracy [9].

The early prediction of diabetes is crucial to saving the organs of the human body and leading a healthy lifestyle. Diabetes affects irrespective of age from the children and the adults. Patients with diabetes and people prone to the risk of diabetes have to monitor their sugar levels clinically often. Due to the COVID-19 pandemic, the patients with diabetes and others confined themselves within their homes [14]. The diagnosis of the sugar levels at the designated laboratories is difficult for them to reach. They demand regular checkups and uninterrupted medical assistance from professionals to lead a quality life. A remote monitoring system to measure blood sugar and pressure levels can offer the hindered services to those at home. The details of the motor movements and food habits of the patients get stored on smart devices. When a predictive analysis analyzes the data collected from the medical equipment and smart devices, it is possible to say a person is affected or to be affected by diabetes.

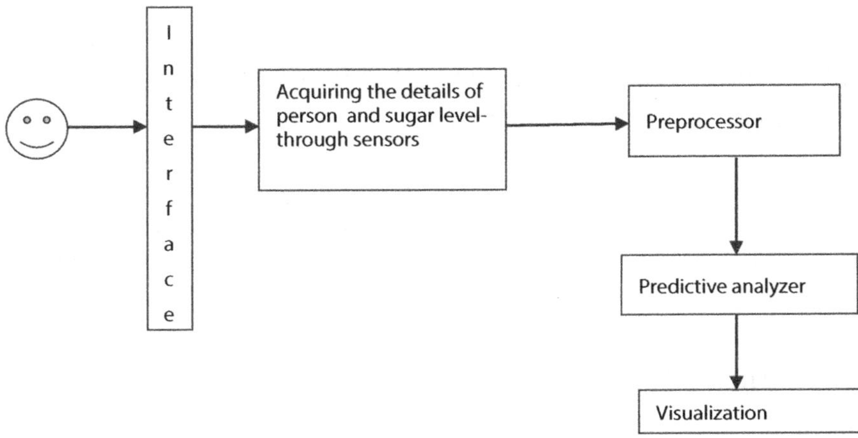

Figure 7.1 Block diagram for smart diabetes prediction.

Figure 7.1 represents the block diagram for smart diabetes prediction. In that, Every day readings collected from various smart devices get stored in a database. Then, they are preprocessed by the analyzer module. The analyzer module parses the resultant values and gives them to the predictive analyzer. The predictive analyzer attaches to the various parameters a differential weightage. The algorithm then further processes the data based on different measures to predict the diabetes status of a person.

Figure 7.2 represents the decision tree diagram for attribute age. It is categorized as young age, middle age, and adult. All types of age are terminated in to normal, overweight and obese.

The attribute food in the decision tree in Figure 7.3 is categorized into carbohydrate, protein, and fat. The food items under each such category

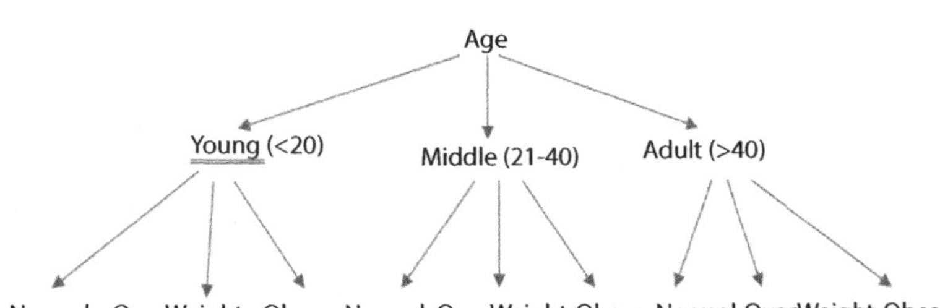

Figure 7.2 Decision tree diagram for attribute age.

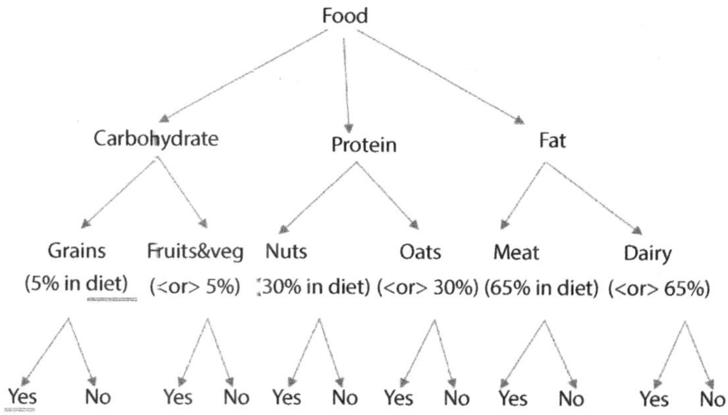

Figure 7.3 Categorized into carbohydrate, protein, and fat.

are shown in their next level. The food items are marked as either yes or no based on their percentage of inclusion in everyday food consumption in the final level of the decision tree.

The confusion matrix helps in identifying where the prediction model is confused during prediction analysis. The confusion matrix exhibits the count of incorrect and correct predictions made for each classification. The pie chart in Figure 7.4 shows the percentages of each category of persons identified as patients, non-patients, and vulnerable to diabetes from analyzed values.

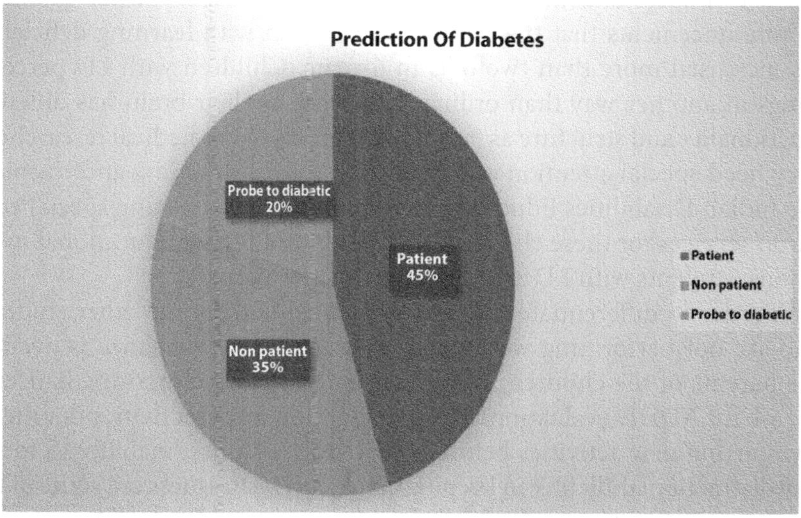

Figure 7.4 Percentages of each category of persons identified from analyzed values.

From the above research work, the smart diabetic monitoring system claims its role in healthcare systems to ensure a healthy lifestyle of the people.

7.3.2 Smart ADHD Prediction

Learning disability (LD) is one of the disorders, which, if not diagnosed, is very hard to identify as these people exhibit some level of cognitive ability on average. LD will not let the children concentrate on studies or do the things requiring concentration if not assisted or motivated by others. Children with LD should not be considered as children with other disorders like autism, challenged to hear or see, and behavioral disorders, as it is highly complex to differentiate LD from others as its characteristics vary from the rest. LD is not pertaining only to children; it also affects people of all ages ranging from toddlers to adults. The educational curriculum and its instructional system have to be improved with remedial measures along with common education for the children with LD to benefit [11].

In a group of 20 persons, nearly three to four persons are affected with LD disorders. At least one among them will have both LD and ADHD. ADHD is another attention disorder, namely, attention deficit hyperactivity disorder. For every two boys with LD, there is one girl affected by it when comparing children of both boys and girls less than 18 years old. Boys are threefolds more vulnerable than girls to LD. LD and ADHD affect nearly 15 children for every 100 children around the world [12].

The chambers of commerce for India (ASSOCHAM) warned in one of its announcements that the number of children with learning deficiency has increased more than twofolds in 6 years. Children with LD perceive things in another way than ordinary children as their brain has different functionality and structure as per the statements from medical researchers. They need special attention and different mode of teaching and learning. The Indian Disabilities Education Act enforces this by having special education services for these children. In schools and higher educational institutions, students with LD are identified as slow learners.

In order to differentiate the students with LD from the other students who are not performing well special assessment performance is needed. The parents of the children, their teachers, family doctor, or pediatrician can ask for ADHD evaluation. Lacking attention, hyperactivity, poor memory, starting new activities before completing old ones, inability to focus, easy distraction, difficulty in keeping calm, and restlessness are some of the symptoms of LD that can be found in the children with ADHD.

Since many of the symptoms of ADHD are in common with the symptoms shown by general health problems. Hence, diagnosing ADHD in its preliminary stage is a highly difficult one. The therapist or doctor has to undergo a series of diagnoses or tests over a person's symptoms and behavior. For some people who show slow progression in the symptoms, it takes more time to diagnose that they have learning problems. This demands a mechanism to predict accurately if a child is suffering from LD or ADHD. The binary classification in data analysis which will have either true or false as a result of the given problem can be used for that

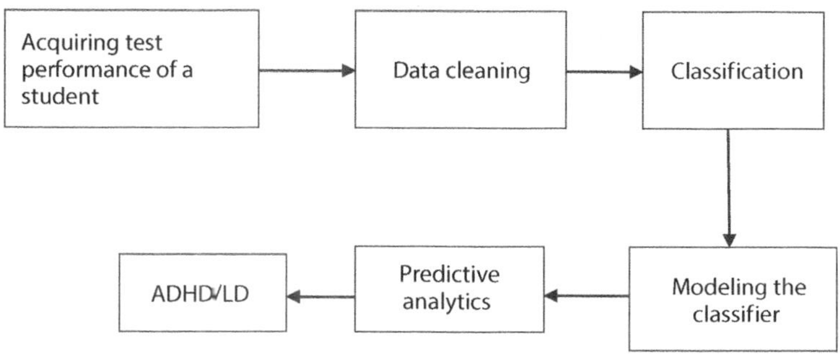

Figure 7.5 Conceptual diagram for prediction of ADHD/LD.

Figure 7.5 Indicates the conceptual diagram for prediction of ADHD/LD. The prediction of LD/ADHD among a class of students is done through the system shown in the conceptual diagram above. The system consists of four phases: data acquisition phase, data cleansing phase, training phase, and prediction phase. In the data acquisition phase, the data is collected from the continuous assessment test performance of the students. During the cleansing phase, repeated and unnecessary data and noise data are removed from the collected data. Then, the cleaned data is then partitioned as training data and test data A mathematical model is prepared and is trained using the training data as input instances. Then, the model is tested using the test data. If the test results are not fruitful, then the model is trained with adjusting the parameters. The model testing and tuning are repeated until the model produces satisfactory results.

The results of various tests performed over the students and the results are specified in a decision tree as given below. A tree traversal through the graph will result in one of the classification of the students as C1, C2, C3, and C4, as given in the Table 7 1.

In the decision tree shown in Figure 7.6 for the classification of learners, C1 represents fast learners who can study as normal; if they show poor performance in learning new skills, then they can be retrained.

Table 7.1 Results of learners.

S. no.	No. of students	Classifier	Recommendation
1.	7	C1	Repeat Training
2.	13	C2	Read the material and solve the exercises
3.	17	C3	Use visualization techniques to learn and solve the exercises
4.	3	C4	Need extensive training through video along with tutor or assistance

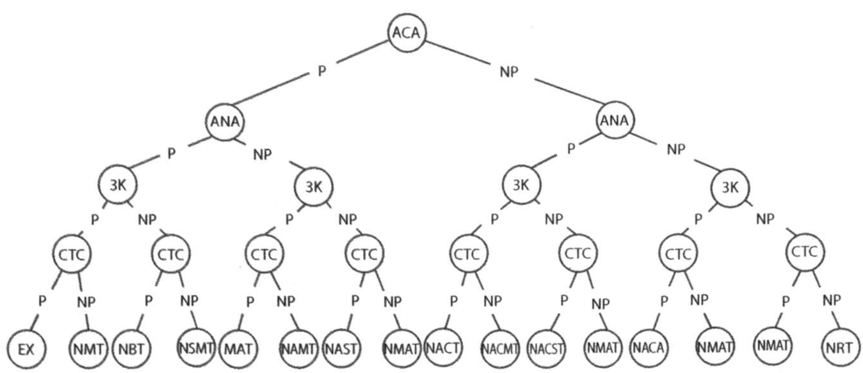

Figure 7.6 Decision tree for classification of learners.

C2 represents the average learners who need comprehensive training in all the aspects of learning. C3 and C4 are lagging in performance in the academics. They have to undergo repeated trainings on the specified skills using various visual teaching aids until the required level of mastery is achieved. A total number of 40 students from Diksha ADHD Training Center are tested using this method, and the results are tabulated in Table 7.1.

Figure 7.7 depicts the outcome of classification of a set of 40 learners. From the group of 40 students, those who have deficiency in learning are accurately diagnosed and classified exactly. The reason behind the false positives and false negatives could be the candidates might not be comfortable with the interfaces in answering a digital questionnaire. The questionnaire will get improved through accompanying with physiological experts in preparing the questions.

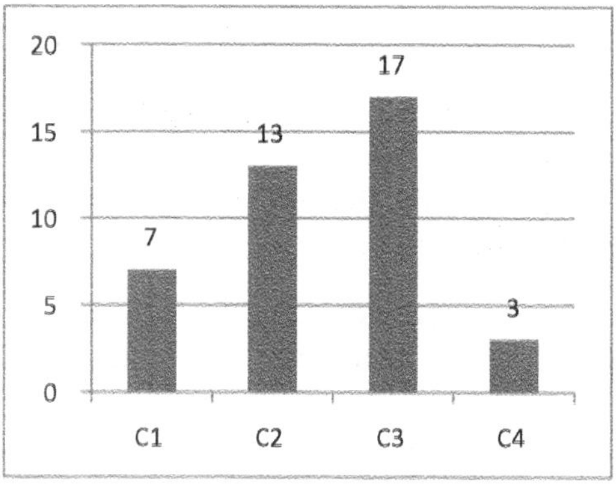

Figure 7.7 Classification of learners.

Smart Healthcare for Heart Prediction in Cardiovascular

The IoT integrated with big data plays an important role in predicting the disease with the help of medical experts, based on that doctor diagnosis the patient. The medical experts classify the different combinations of medical test results and correlate with actual symptoms of the patients; based on the analysis, the system will provide the solution from that prediction accuracy is improved.

Initially, the medical records of the patients are made available in digital form. By maintaining patient information in electronic format as several benefits, whenever medical experts need of patient information, they retrieve the data on time through rapid retrieval. By transformation of electronic to digital.

The World Health Organization submitted the statistics report in the year 2012; the 35% of adults in worldwide has increased blood pressure that leads to stroke and heart disease. Heart disease is the major cause in all over the world. The cardiovascular disease will affect the heart and blood vessels through which blood is flowing through the body. The diagnosis of heart disease is very difficult and complicated; the diagnosis of heart disease is accurate and efficient based on doctors' experience and knowledge. Due to lack of experience and knowledge, this leads to unwanted test results and medical cost will increase.

The researcher KENNEDY implements the prediction accuracy of heart disease by Naïve Bayes algorithm, and the system is having more parameters to identify the heart attack based on their patient's weight, age, doctor's consultant, and medical experts. The prediction system will identify the different levels of risk in heart attack such as normal, high, or low based on the doctor prescription that leads to predicted result.

Naïve Bayes classifier uses probability theory to classify the attributes. This classifier works by Bayes Theorem as P(A|B) = P(B|A) P(A) / P(B). Let B be a set of n number of attributes B = $\{b_1, b_2, b_3,\ldots,b_n\}$. In Bayes theorem, B is said to be evidence and A as hypothesis. We can find A if B has occurred. We can find P (A|B), the probability that the hypothesis A can occur if the evidence B has occurred.

Algorithm used	Accuracy	Time taken
Naïve Bayes	52.33%	609 m
Decision Tree	52%	719 m
K-NN	45.67%	1,000 m

The Bayesian classifier discovers and predicts the heart disease based on the historical records of the patient. The advantage of Bayes theorem is to determine the probability of diagnosis is correct, based on the observations.

Figure 7.8 describes the existing model of heart disease using naïve Bayes classifier; the CSV file is taken from UCI heart attack dataset from that pre-processing. The pre-processing contains the removal of unwanted entries from the dataset. The classifier of the heart disease is based on their patient's weight, age, doctor's consultant, and medical experts; from the above parameters, the system identifies the patient history from the system that will evaluate the results. The medical experts give suggestion to the patients based on their patient results.

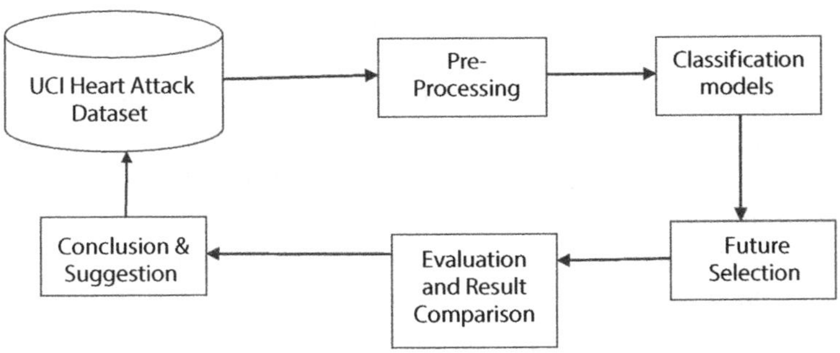

Figure 7.8 Heart disease using naïve bayes classifier.

7.4 Security for Internet of Things

There are lot of applications IoT in various medical applications. The following section discusses the security for IoT.

In the epoch of internet-based environment, secured data transfer is vital between mobile handheld devices and internet services. As internet environment is a network of networks, it is more susceptible to attacks, and therefore, it is necessitated a mechanism to authenticate the data transmission in these environments. The security protocols and the cryptographic algorithms used to offer the authentication frameworks impose heavy requirement on computing power which in turn requires more electrical energy to operate. Hence, it is required to have secured transmission between resource constrained environments and internet services with fast in time and less in energy consumption. The proposed method will provide the guarantee to optimal solution and secured data transmission.

In conventional ECC algorithm, the scalar point multiplication plays the major role and took more time than the other procedures in elliptic curve cryptography algorithm. To increase the efficiency of point multiplication, expedite point multiplication method is proposed. Expedite point multiplication–based ECC algorithm is introduced in the as research work to reduce the battery power consumption during secured data transmission. The battery consumption is examined using binary method, non-adjacent form method, and expedite point multiplication method [10].

7.4.1 K(Binary) ECC FSM

Figure 7.9 Illustrates about the ECC k(binary) FSM. ECC k-binary is utilized to control the process RSAM as well as inverse with FSM. RSAM process would perform multiplication on elliptic curves k * P. Inversion is needed to restore the value of x as well as y to affine coordinates by Itoh and Tsujii approaches. The process of multiplication kP carried out over 158 iterations in Lopez-Dahab Projective coordinates. Here, every iteration is carried out the doubling process when the addition is carried out while the key bit "1". The inversion process is carried out subsequent to iteration 158. The ECC multiplication starts by triggering the reset function (reset = 0, start = 1) subsequent to the function start (reset = 1, start = 0) to begin the data input x1, y1, a6, and k. Subsequently, the data input process is accomplished; RASM function is facilitated to carry out the operation point doubling as well as addition according to key bits.

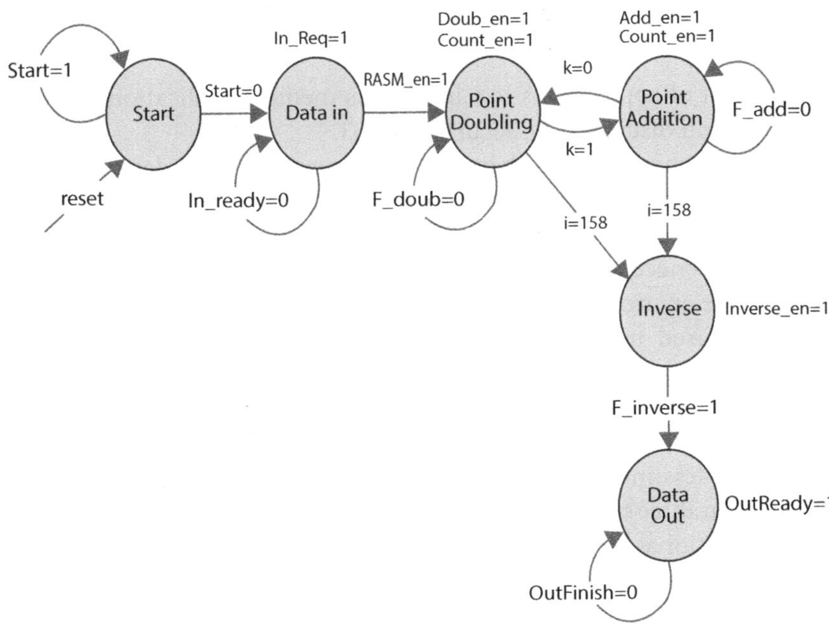

Figure 7.9 ECC k(binary) FSM.

7.4.2 NAF Method

This technique would alter the value of k to decrease the point addition. In the process of Q = 15 * P, it is carried out with Q = (16 − 1) * P = (16 − 1) * P = (2P) 2 * 2 * 2 − P, when it is matched up with Q = 15 * P = P + 2 (P + 2 (P + 2P)), the completed turn out to be fewer process as well as lets the processor to perform quicker. NAF algorithm is in this manner:

Input: positive integer $k = (k_{t-1}, \ldots, k_1, k_0)$
Set i = 0
while k is greater than or equal to 1
  if k is odd, $k_i = 2 - (k \bmod 4)$ and $k = k - k_i$;
  else $k_i = 0$;
  k is equal to k/2;
output k
Example: 10045 * P = (10011100111101)*P = (10100-101000-101)*P

In this algorithm, when a large number is represented in the binary form, if the sequence of bits in the binary number contains k successive 1s, then the last 1 in the sequence will become −1. The previous bits of

−1 up to a bit with a value 0 are substituted with 1s. The subsequent bit after the −1 turns into a 1. After that, the multiplication technique NAF (k) from the MSB to the least significant bit (LSB) is carried out in this manner.

```
Input: K = (k_{t-1}, ..., k_1, k_0)
Output: k * P
1. Q = ∞
2. For i = t − 1 to 0
       Q = 2Q;
       If k_i = 1, then Q = Q + P;
       If k_i = −1, then Q = Q − P;
3. Output Q
```

7.4.3 K-NAF Multiplication Architecture

The structural design of Elliptic Curve Cryptography based processor is illustrated. This structure includes all the fundamental elliptic curve arithmetic, point addition, point scalar multiplication, and point doubling over the group G (2^m). A normal 158 bit key security level is utilized by this cryptographic structure.

7.4.4 K(NAF) ECC FSM

Figure 7.10 shows about k-NAF ECC processor. The ECC k(NAF) includes both k-NAF and FSM. FSM k (NAF) has six states. The primary state is the initalizaiton of k (NAF). The state will move to st1 to identify 2-bit LSB k as long as NAF_en = 1. When k [1:0].

The ECC k(NAF) begins with computing k-NAF with FSM. There are six states in FSM K (NAF) method. The first state is the initialization of k (NAF). A 2-bit LSB gets identified when NAF_en = 1; at this time, the state moves to st1. At the time of k[1:0] not equal to 2'b11, the next state is st4. When k[1:0] is equal to 0, the next state is st2 for k_i =1. When k_i is set to 0, the state moves to st3. It will be in states st3 and st4, while done_count is equal to 0. If done_count is 1, then the state moves to st5 and increases the value of k (NAF) and the sign.

The process of producing k (NAF) is iterated for 158 times and every iteration produces exactly 1 bit identified in the last 2 bits of LSB value k. In every iteration, k one bit values shifted to the right and this will affect the MSB to be shifted by 1 When the last couple of bits is "11", then, after that,

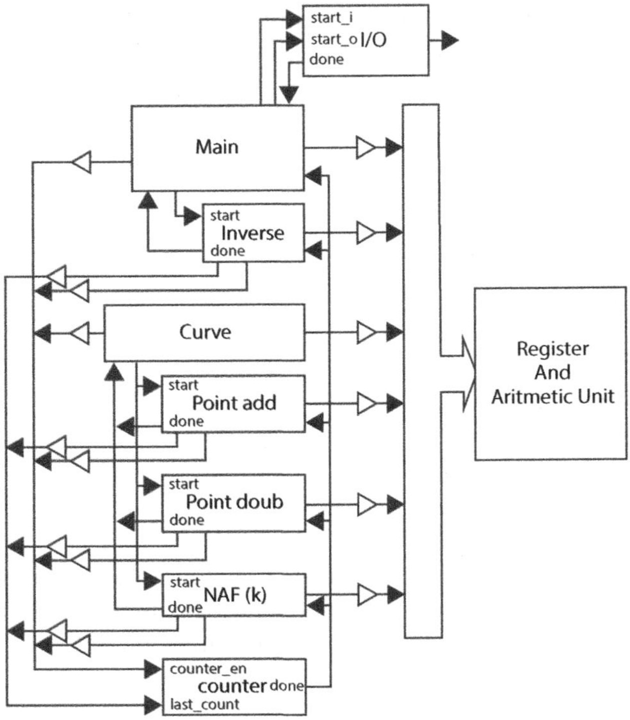

Figure 7.10 k-NAF ECC processor.

k (NAF) $_i$ = 1 and sign = 1. These steps are carried out in the state st1. If k (NAF) i = ki or k value, sign is equal to 0, and the last two bits are not equal to "11", then it is in state st4. As long as k[0] =1 after the shifting from state st1 to the state st2, the state will remain at state st2. At this state, k (NAF) I = 0 and the sign = 0. With k[0], = 0, there is a transition to the state st3 with k (NAF) i = 1 and the sign = 0. Now, the done_counter will reach 0 and the process will come back to the state st1.

Figure 7.11 represents the k-NAF FSM model. This process initially starts from the state start. When reset is set to 1 and the input transitions from high to low, the state shifts to the state data_in to record data in a register for the duration of In_ready = 0. The state will transfer to k (NAF) at this In_ready = 0. When both NAF_en and count_en are set to 1, the state will increase k (NAF) and the sgin bit 158. At that point, if f_NAF is set to 1, then the state starts the point doubling to initiate the elliptic curve multiplication k * P. The subtraction state is reached when the sign is set to 1.

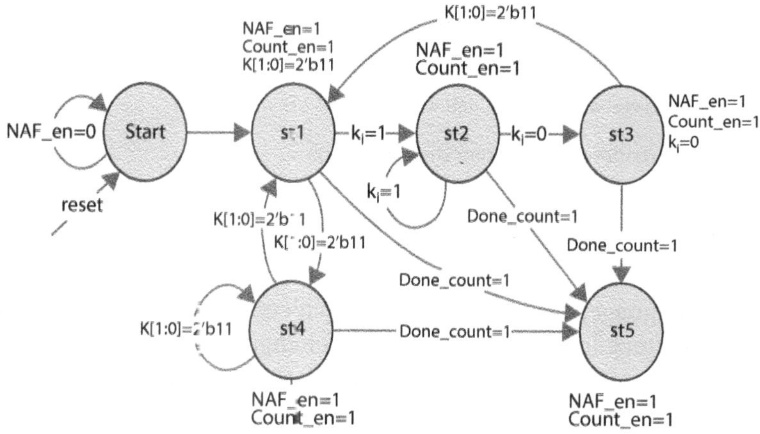

Figure 7.11 k-NAF FSM.

Figure 7.12 represents the k-NAF ECC FSM model. If the sign is set to 0, then the state reaches point addition. When i is not equal to 158, then the state returns to the point doubling, and when i is equal to 158, then the state changes to inverse state to return the outcome of multiplying the elliptic curve to organize affine. Where the elliptic curve multiplication process carried out at LD, Projective organizes to make simpler the process of the arithmetic processor ECC. Subsequently, f_inverse = 1, at that point, the state would be in state data that is out to set the data readout from 158 bits to 32 bits relate to the input port ECC processor output.

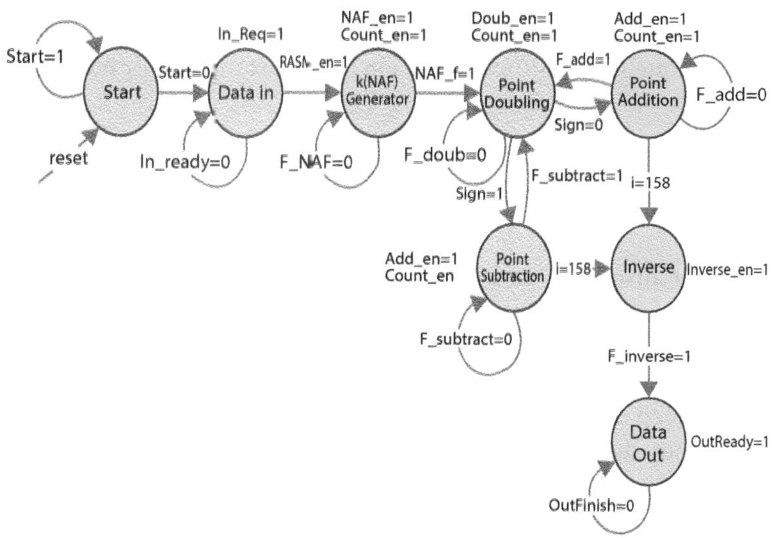

Figure 7.12 k-NAF ECC FSM.

```
Intent battery_Intent = this.getApplicationContext ().registerReceiver (null,
newIntentFilter (Intent.ACTION_BATTERY_CHANGED));

Startinglevel=battery_Intent.getIntExtra ("level", -1);

Endlevel=battery_Intent.getIntExtra ("scale", -1);
```

Figure 7.13 Battery charge level measurement in Java application using system property.

In java language, the battery charge is calculated by using Java system property "battery level" (Property name changes for diverse phones). In Figure 7.13, a snippet of code returns the current battery level from 0 to 100.

The performance evaluation of expedite point multiplication method will reveal that this algorithm consumes less computational time and extends the battery life. The time taken for data transfer and energy consumption will be compared to each of the methods. The results will infer that the proposed point multiplication algorithm can save the battery life when compared with other point multiplication methods.

All the point multiplication algorithms will be implemented in java enabled emulation software. The emulations will be conducted on android mobile phone, and the data will be tabulated. The simulation results will show that the proposed point multiplication algorithms provide improved performance in terms of less computation time and minimized energy usage. In future, the secured data can be transferred via unsecured medium using proposed ECC-based point multiplication algorithms. The measurement technique is used to calculate energy consumption in mobile handheld devices.

7.5 Conclusion

This chapter discussed various healthcare applications such as diabetes prediction, heart disease prediction, and ADHD prediction using big data analytics. The prediction supports the practitioners to provide immediate treatment, assistance, and counseling to the patients in need. IoT devices constantly beep out abundant signals, which produce the voluminous and heterogeneous data, and its security aspects have been discussed. There is lot of scope for improvement by applying IoT in various healthcare domains such as cardiology, cancer in various parts, and lungs infection prediction, which help us for COVID-19 prediction, too.

References

1. Nazir, S. *et al.*, "A Comprehensive Analysis of Healthcare Big Data Management, Analytics and Scientific Programming". IEEE Access, 95714–95733, June 3, 2020.
2. Yacchirema, D. *et al.*, "A Smart System for Sleep Monitoring by Integrating IoT With Big Data Analytics". IEEE Access, 6, 35988–36001, 2018. Digital Object Identifier 10.1109/ACCESS.2018.2849822.
3. Wang, S. *et al.*, "Active Data Replica Recovery for Quality-Assurance Big Data Analysis in IC-IoT". *IEEE Access*, 7, 106997–107005, 2019.
4. Harerimana, G. *et al.*, "Health Big Data Analytics: A Technology Survey". *IEEE. Translations content Min.*, 7, 65661–65678, 2018.
5. Sasubilli, G. *et al.*, "Machine Learning and Big Data Implementation on Health Care data". *Proceedings of the International Conference on Intelligent Computing and Control Systems (ICICCS 2020)*, pp 859–864.
6. Smys, S. *et al.*, "Internet of Things and Big Data Analytics for Healthcare with Cloud Computing". *J. Inf. Technol. Digital World*, 01, 01, 9–18, September 2019.
7. Bansal, M. *et al.*, "IoT & Big Data in Smart Healthcare (ECG Monitoring)", in: *International Conference on Machine Learning, Big Data, Cloud and Parallel Computing*, pp. 390–396, 2019.
8. Sisodia, D. and Sisodia, D.S., "Prediction of Diabetes using Classification Algorithms". *Proc. Comput. Sci.*, 132, 1578–1585, 2018.
9. Swapna, G., Vinayakumar R., Soman, K.P., "Diabetes detection using deep learning algorithms". *ICT Express*, 4, 4, 243–246, 2018.
10. S. Kumar, K. and Sukumar, R., "Achieving energy effificiency using novel scalar multiplication based ECC for android devices in Internet of Things environments". *Cluster Comput.*, 22, S12021–S12028, 2019. https://doi.org/10.1007/s10586-017-1542-8.
11. Jaya, A., "A Generic Framework for Intelligent tutor using ontology", in: *Proceedings of 2014 International Conference on Artificial Intelligence & Manufacturing Engineering (ICAIME 2014)*, Dubai, pp. 16–23, December 25-26, 2014.
12. Jaya, A. and Uma, G.V., "Role of Ontology in Case-Based Reasoning (CBR) for Diagnosing Diabetes". *IUP J. Inf. Technol.*, 2009.
13. Vanagas, G., Engelbrecht, R., Damaševičius, R., Suomi, R., Solanas, A., EHealth solutions for the integrated healthcare. *J. Healthc. Eng.*, 3846892, 1–3, 2018.
14. Kumar, V., Singh, D., Kaur, M., Damaševičius, R., Overview of current state of research on the application of artificial intelligence techniques for COVID-19. *PeerJ Comput. Sci.*, 7, 1–34, 2021.
15. Maskeliunas, R., Damaševicius, R., Segal, S., A review of internet of things technologies for ambient assisted living environments. *Future Internet*, 11, 12, 1–23, 2019.

8

Knowledge-Driven and Intelligent Computing in Healthcare

R. Mervin[1]*, Dinesh Mavalaru[1] and Tintu Thomas[2]

[1]Department of Information Technology, Saudi Electronic University, Riyadh, Saudi Arabia
[2]Department of Public Health, Saudi Electronic University, Riyadh, Saudi Arabia

Abstract

Nowadays, the health recommendation system (HRS) is one of the significant platformsfor healthcare services and it plays a great role in the decision-making process in the area of the healthcare sector. The contribution of several factors in healthcare systems, for instance, diverse specialists and embedded devices, leads IT-based healthcare system to be more complicated and competitive. Also, the utilization of several programming languages, platforms, and the information management system is directed to the constraints in the perfect swap over, incorporation, and reprocess of data through various systems.

The research work focuses on the importance of ontology in the field of HRS. In this system, the patient benefits when the medical practitioner is exposed to diverse viewpoints with varied backgrounds of the disease and provides the best recommendation for the patient's health issue. This will give the best knowledge about the disease and the remedy for the patient's medical issues. The proposed system brings a new system, where the patients can search their query to the system when they want to collaborate with expert's knowledge and find the remedy for their issue. This system consists of three phases: query preprocessing, feature selection, and recommendation. Ontology plays a vital role in checking the semantic nature of the query posted by the patient and will give the most relevant solution to the patient.

Keywords: Ontology, feature selection, health recommendation system, expert's knowledge, pre-processing, decision support system and healthcare system

**Corresponding author*: m.mary@seu.edu.sa

A. Jaya, K. Kalaiselvi, Dinesh Goyal and Dhiya Al-Jumeily (eds.) Handbook of Intelligent Healthcare Analytics: Knowledge Engineering with Big Data Analytics, (167–188) © 2022 Scrivener Publishing LLC

8.1 Introduction

The healthcare system is one of an integral part of patient information and support to the development of insights and the likelihood of diseases. The intelligent system must be capable to predict the patient health condition with the utilization of available information of patients. Electronically medical records were useful to make information regarding healthcare, which is available to clinicians and different research scholars in various medical contexts and applications [6]. However, the enormous amounts of accessible information also have a big challenge for the healthcare professional to identify the proper quality data from the large health information system. In such a situation, a consistent and efficient recommendation system is very important to get more accurate data for clinicians and healthcare researchers.

A health recommender system is a specialized health-based recommendation system; more particularly, it is a decision-making system that recommends appropriate healthcare data toward healthcare professionals and patient. The personalized recommendation is one of the effective methods to increase patient safety by evading severe drug reactions. The main aim of a health recommendation system (HRS) is to support its user with the most significant medical information of the patients linked with the patient health records (PHR) [10, 11]. For an effective integration into any health-associated information system, it is very essential to consider the system context level of HRS. There will be at least two different scenarios that can be defined based on the expert level of the PHR user. In the first case, the end user will be a healthcare professional, and the second one, the patient will be the end user.

The HSR can be used by the healthcare professional for extracting additional information for a certain case and the case associated data enhancement helpful for the doctors with the method of clinical diagnosis as the recent result can be utilized for the decision support. In the case of patient or non–healthcare professional, end user directly interacts with HSR without support from any healthcare professional. In such situations, the identifications of top-ranking file or media data of a patient are helpful for getting more confidence in terms of health information. So that, the risk of non-relevant and misleading information can be reduced with the help of evidence-based much quality health-related information to end users [12–15].

Expert's knowledge is implemented with the help of ontology. Ontology learning is the semantic way of knowledge representation that depends on transforming unstructured data sources into structured data. It is more related to the field of computer and information sciences. An ontology defines the semantic way of knowledge representation, in which the knowledge maybe shared explicitly [16, 17]. During this description, "shared" implies that the knowledge defined by ontology is often received by people; "explicit" gives the relationships between the defined concepts.

Nowadays, the selection of appropriate information is a bit challenging and time-overwhelming process. So, the recommendation system that assists to personalize the data has been selected for the particular purpose [18–20]. The system recommender has several types, and method is available to identify a recommendation that includes ontology and machine learning. Recommendation systems are being used vastly in different areas of services, such as financial investments, commerce, and medical applications [31]. Latest sign of progress in web related recommendation systems were enriched the information semantically, which permits the mining of a huge volume of information through ontology. Applicability of these methods was allowed to prioritize and group the data based on the numerous properties and rules define into ontologies [32]. Recommender systems can be grouped into three different types, which include collaborative, context-based, and hybrid systems [34].

The remaining part of the research work is structured as follows: Section 8.2 deals with literature survey. Section 8.3 describes the framework for HRS. Section 8.4 produces the experimental results. As a final point, Section 8.5 concludes theresearch by giving brief glimpses into the future enhancements in this research area.

8.1.1 Basics of Health Recommendation System

Currently, a tremendous measure of clinical information dissipated across various locales on the Internet obstructs clients from finding supportive data for their prosperity improvement. Plus, the over-burden of clinical data (e.g., on medications, clinical trials, and therapy recommendations) has acquired numerous challenges to clinical experts settling on patient-situated choices [39]. These issues raise the need to apply recommender frameworks in the medical service spaces to help both end-clients and clinical experts and make more productive and exact well-being–related choices. In this article, we give an orderly outline of existing exploration on medical service recommender frameworks [51].

8.1.2 Basics of Ontology

Ontology learning is a semantic way of knowledge representation that converts the unstructured data sources into structured data. It provides the semantic meaning of classes and gives the relationship between the attributes of the objects, which acts as the knowledge base for the domain. Ontology is the field of identifying the relationship of the database in the entire domain. A few scientists are moving in the direction of creating ontology-based models for health recommendation system. Ontology helps to find the relationships between various classes and its individuals of the selected domain. There are several uses of ontology-based systems related to information retrieval, artificial intelligence, machine learning applications, and others [3].

The purpose of ontology is a formal way of knowledge representation of classes called concepts, data, and object properties of every concept related to their attributes. In the era of artificial intelligence, ontology becomes the unavoidable knowledge repository in which a group of classes and their attributes along with their relationship create a knowledge repository. Many researchers started developing standardized ontologies in the field of medicine and used them to share and annotate information. The types of ontologies are as follows.

Top-level ontologies or upper-level ontologies: These kinds of ontologies describe general concepts common to most domains [44, 45]. These ontologies can be reused based on thedifferent purpose of the research. Examples of ontologies are object, purpose, reason, time, and space.

Domain ontologies: These ontologies are developed for a specific closed domain. For example, BASEBALL is an example of domain ontology.

Task ontologies: It describes tasks or activities that involve some input, process, and output. An ontology is related to any specific task and also helps to make a decision after completing the task.

Application ontologies: These ontologies are developed for any application, which may include the ontologies of domain, task, and top-level ontology. Examples are banking application or healthcare application.

8.1.3 Need of Ontology in Health Recommendation System

In recent days, the necessity of ontology is unavoidable in any domain. The need for ontology is also important in the field of medicine. The reason for developing patient monitoring system ontology is listed as follows [1]:

- To represent and sharing knowledge between patient and medical practitioners.
- To retrieve information based on semantically not like syntactic search.

- To apply the reasoning mechanism to validate the correctness of the ontology.
- To define the "N" number of rules.
- To share the knowledge base and its structure of information with different people.
- To support the reusability of information.
- To create an explicit knowledge base.
- To differentiate the domain knowledge base from the operational knowledge base.
- To investigate domain knowledge base.

8.2 Literature Review

The literature review provides a detailed survey of the investigations about various studies, models, methodologies, and evaluation methods of knowledge-based health recommendation system [35]. The different aspect of HRS was discussed by referring to the various authors. The summarization lists the definitions and discussions on the basic information related to the knowledge-based health recommendation system, which is important for designing any successful system.

A recommender system can be defined in such a way that the system creates individualized recommendations in the form of processed information. It has the influences of supporting the people in a personalized way to suitable answer with different opinion [1]. The important attributes of the recommender system can be summarized in five types based on the nature of the data, user profiles, information management, and the work methodology of text mode and removal capability of irrelevant information from the input stream. One of the advantages of the recommender system is that it can be suitable for both unstructured and semi-structured data [2].

Dickson *et al.* [1] presented an alert management system that supports the event-handling process of every organization. Furkh *et al.* [2] used service-oriented architecture (SOA) for solving the problems of heterogeneity and complexity with the help of ontology and established domain ontology for management of the IT-based healthcare problem, particularly in an emergency. Lage *et al.* [3] described a framework of the recommendation system that focused to make the health events available in a personalized way and presented the system that makes use of a set of signal definitions that include predefined structured queries with parameters associated with kind of threats user interests. Finally, the system gives the rating based on recommendations. Ru-Zhi *et al.* [4] proposed the construction of ontology-based system in an emergency response plan.

Subramaniyaswamy *et al.* [7] implemented a food recommendation method that is also assessed for the healthcare recommendation system. Araujo *et al.* [8] proposed an ontology-basedmethod to build simulations for training methods in an urgent situation. Recommended systems were also used in different clinical settings, which include mostly supporting health practitioners, though some interface helps their patients. However, the latest advancement in the online recommended system was enriched by the semantic web [9].

8.2.1 Ontology in Various Domain

Both ontologies and taxonomies have a diversity of applications in the research areas, namely, information integration, natural language processing, and knowledge mining [33]. Salatino *et al.* found the computer science ontology (CSO)—a significant, dynamically created ontology applicable fields that consist of approximately 14-K areas. The Klink-2 algorithm was used on a tremendous data set of 16M research articles. CSO offers two important benefits that include a huge number of areas that not belong to other classification along with that updating that can be done spontaneously with the help of Klink-2 on the latest corpora of publications [36].

Cimiano and Völker proposed an ontology learning approach created from the scratch itself, in which the model derived is based on the textual information with the careful executionof varied linguistic and statistical approach and used POM model (probabilistic ontology model)for attaining the confidence value of learning element [37]. Buitelaar *et al.* created a rule map that makes a useful way to mine ontology from the text and incorporated validation techniques to make sure proper validation before its introduction in ontology [38]. Navigli and Velardi proposed a method called OntoLearn that helps to extract domain lexicons.

Stuart *et al.* demonstrated a repository of ontology that expresses the vocabulary of SDM, which is utilized to annotate the organization's information semantically. The ontology can the represented in the form triplet format called RDF, and most of the researchers linked RDF with JSON (JavaScript Object Notation) to produce the result [40]. Larisa *et al.* introduced a general form of ontology for a scientific experiment called ontology EXPO to meet the necessity of the experimentation. EXPO connected with the SUMO (the Suggested Upper Merged Ontology) and link to domain-oriented ontology that supports design, methodology, and output demonstration. The illustration of EXPO is in the form of W3C standard ontology language OWL-DL. Larisa *et al.* also described the applicability of EXPO in various experimental domains with the highlighted application of the two experiments that include high-energy physics and phylogenetic [41].

Regarding the application of ontology in the other branches of science, for example, health science domain-oriented information is frequently determined in the ontology knowledge base. It makes use of ontologies for preprocessing and numerous ontologies already established with a wide area of diversified domains of health sciences. Oellrich *et al.* described the characteristics biomedical ontologies, which make use of describing the phenotypes found in different model organism databases [42] and also applied it in the field of human genetics [43].

Kulmanov *et al.* used a method that collaborates with many various semantic sources into ontology and that can be made use of answers for user questions from several knowledge sources. Ontologies started to apply in the field of life sciences and especially most of the biological databases; however, the combined application of ontology and machine learning is still actively developing and having its novelty of each of the methodological usage. Kulmanov *et al.* demonstrated the application of ontology in semantic similarity measures which integrated with machine learning and explained how to improve the models with that combined application [46].

Ashburner *et al.* applied a tool in ontology called Gene Ontology (GO), which is applicable in the unification of biology. The main aim of the GO consortium is to create vibrant terms, which help to derive the knowledge of gene and protein roles in cells. The GO model is planned to implement a more flexible method, the annotation of gene and protein cells in several organisms with the help of a common vocabulary that make possible to retrieve knowledge source of genes and proteins based on their ontology [47]. The Open Biomedical Ontologies (OBO) foundry is a widespread collaborative effort to construct controlled vocabularies for collective use across various biological and medical domains.

Smart Topic Miner is another tool established in collaboration with CSO that focused on classifying editorial products based on the taxonomy of research topics (Osborne, F). Ontology Forecasting is another area of application in that the Semantic Innovation Forecast (SIF) model is applied to determine novel terms of an ontology at time t+1, with the help of information available at time t (Cano-Basave) [48]. Rabahallah *et al.* used ontology to make available semantic explanation of learner and MOOC (Massive Open Online Courses), which will be integrated into the recommendation process to get a better form of the knowledge recommendations, whereas CF (Collaborative Filtering) infers predictions and then gives therecommendation. In addition to that, hybrid approach is also used to relieve cold start problem [49].

8.2.2 Ontology in Health Recommendation System

Healthcare ontologies typically have an Application Programming Interface (API) to indeed express its process. Bio ontology is an example of API, which include many health-associated ontologies. Rivero-Rodriguez *et al.* suggested appropriate associations for a division of healthcare-related YouTube videos by determining matching medical concepts from the Medline Plus API [50]. The Salus ontology network benefits to identify construing recommendation of health-linked web information for the specific user [51]. More precisely, that can conceptualize the diverse information domains included in the recommendation system in the form of an ontology.

Akmaliah *et al.* used the Kitchenham technique to identify the most frequently applicable recommendation system with a detailed analysis of 750 previous studies [31]. Farman *et al.* used type-2 fuzzy ontology that assisted healthcare recommendation systems to powerfully observe the patient's health issues while suggesting proper diets with particular foods. Those systems alsoderive the value of patient risk factors to identify the patients' health issues with the help of wearable sensors and recommended diabetes-related advice for a smart medicine box [52].

Kandula *et al.* made use of the IR approach to recommend appropriate informative resources to diabetic patients [53]. Benjamin *et al.* illustrated medicine recommender systems, which are beneficial to the healthcare that gives the identification of the most suitable medicine for the seekers and compared them based on various features. Ali *et al.* [58] described about physician recommender, which helps to identify the right physicians based on the patient preferences, introduced a novel model named SEPHYRES (semantic physician hybrid recommender expert system) that emphasizes on the patient's medical conditions and pain description features with the utilization of evidence-based ontology, and described Harrison's Principles of Internal Medicine and the latest online as the main evidence references integrated with experts' views; finally, the matched result of that model explained it would be beneficial for the overall progressof the accuracy of recommendations [54–57].

8.3 Framework for Health Recommendation System

Figure 8.1 shows the framework diagram of the HRS, and it comprises of three phases: query preprocessing, feature selection, and recommendation. This research work explores the remedy of the patient's query, and suitable recommendation to the query is presented to the patients. The query from the patient will be taken as the input and has to be preprocessed. The features of the query have been extracted using the imputation method that will be analyzed both syntactically and semantically.

Then, the similarities of the features are extracted by the TF-IDF method. The key features of the query are extracted that will be analyzed and correlated with their answer. Then, the decision is made by the experts, and the recommendation will be given back to the patients. The HRS ontology will act as the repository of the system, and the final recommendation is stored in the knowledge base. Figure 8.2 shows the flowchart of the health recommendation system.

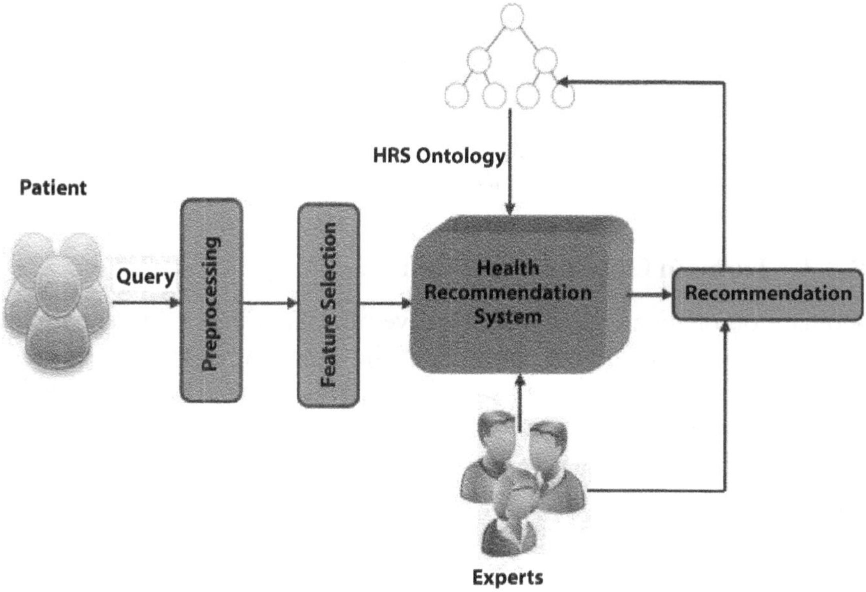

Figure 8.1 Framework of health recommendation system.

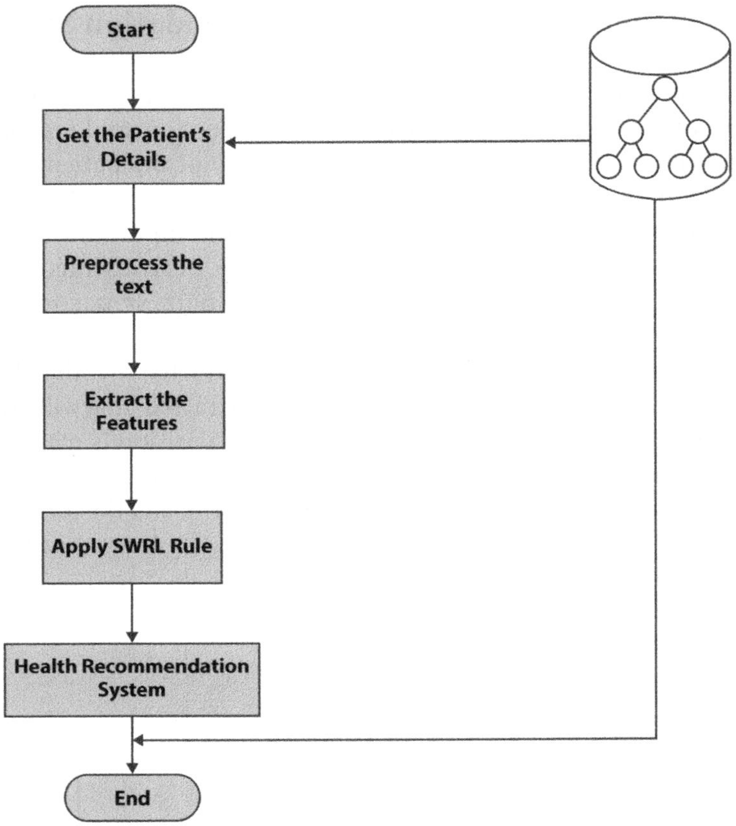

Figure 8.2 Flowchart of health recommendation system.

8.3.1 Domain Ontology Creation

Domain ontology creation is an iterative process. It process with various steps thatincludes defining concepts in the ontology, arranging the relationship between concepts in a tree structure hierarchy. Also, it is required to define the properties, instances, and their relationship related to its concepts. The expert's knowledge is involved in describing the attributes of the domain object and finding their relationship between the objects and classes. The iteration process may go as follows:

1. Select the domain and Start construction the first pass of the ontology.
2. Find the objects, classes, and their relationship to redefine the ontology.

3. If it is required to change the ontology-based on the given circumstance, then find thedifferent alternations and reconstruct the ontology.
4. Refine the process if it is required.

There are always viable alternatives for the selected domain. The existing ontology must be identified before creating the new ontology. The reusability property is always supported by the ontology for any kind of domain. Few basic rules in the patient ontology construction are described as follows:

Step 1: Find and analyze the domain and objective of the ontology
The first step is to find the domain and objective of the ontology. In this research, patientinformation is chosen as the core ontology. In the field of medicine, patient's details and their disease information is managed in the ontology. This ontology can be reused in the applications of the medical field. Naturally, the concepts describe different types of people information like their name, age, gender, address, and type of disease.

Step 2: Find the reusable ontology
Reusability of ontology is one of the best properties in any domain. It is always required to analyze the existing ontology, which is created by someone in the same domain and try to reuse the ontology in the appropriate field. For example, the Ontolingua ontology library or the DAML ontology library, and some popular domain ontologies like UNSPSC, RosettaNet, and DMOZ.

Step 3: Identify the required classes in ontology
It is necessary to find out all classes and their objects that are required to construct the domain ontology. The next two steps are closely knotted to each other, namely, developing the class hierarchy and defining the object and data properties of classes.

Step 4: Define the classes and the class hierarchy
The definition of class hierarchy depends on the top-down approach, bottom-up approach, and hybrid approach. In the top-down approach, the core class is identified first, followed by the iteration process by finding the subsequent classes. A bottom-up approach is defined as the characterization of the precise concepts with levels of hierarchy; finally, it will

reach the mergingof child classes into more specific parent class. Hybrid approach is a grouping of the top-down and bottom-up approaches. It is also necessary to check the following criteria.

- Validating the defined class hierarchy is accurate.
- Checking the position of the child in a class hierarchy.
- Checking the multiple inheritances.

Step 5: Define the properties of classes

For selected classes, the directory of properties must be defined based on the requirement of the application. The object properties can be classified as "intrinsic" properties such as gender; "extrinsic" properties such as a patient's name, and "physical and abstract parts" and "Is-a" is the relationships between instances of the class object and with other objects.

Step 6: Create instances

The next step is to build the individuals of a class object, which requires the following: 1. Select a class; 2. Create an instance of that class; and 3. Create restrictions for the class and individual.

Figure 8.2 shows the ontology for the patient's monitoring system, in which the Patient is the root or parent class; and Name, Gender, Phone_Number, and AgeGroup is the child or subclass. Again, the Gender class is classified into three subclasses: Male, Female, and Transgender.

8.3.2 Query Pre-Processing

The HRS uses the predictive mean imputation method to handle the missing and noise data. In this method, the linear equation is used to signify the relationship between the multiple input attributes and the multiple targets attributes. The missing values are considered as the dependent variable, and all other attributes are considered as the independent variable to infer the predicted missing value. Let the regression equation be Z_i, the predictors be p, and the training instances be n; then, the equation can be written as

$$Z_i = \alpha_0 + \alpha_1 x_{i1} + \alpha_2 x_{i2} + \cdots + \alpha_d x_{id} \tag{8.1}$$

for $i = 1, 2, 3, \ldots, d$.

The equation can be written in the form of a matrix. It gives

$$Z = X\alpha \tag{8.2}$$

The coefficient α can be derived explicitly by using the squared error function, and it can be written as

$$\min E(\alpha) = \frac{1}{2}(Z - X\alpha)^T (Z - X\alpha) \tag{8.3}$$

$$\frac{\partial E_a}{\partial \alpha} = X^T X\alpha - X^T Z = 0 \tag{8.4}$$

$$\alpha = (X^T X)^{-1}.\ X^T Z \tag{8.5}$$

8.3.3 Feature Selection

The important features of the recommender system can be summarized in five different types. The first one is that it can be used both unstructured and semi-structured data; the second one is that it is utilized user's health information; the third one focuses to manage great amounts of information; the fourth one is concentrated with data collected in written form [28–30]; and the last and final category aims to remove inappropriate information from the collected information. Recommendation systems allocate related tasks with information filters that help to eliminate information and decreasing overload.

The similarity between the features can be calculated using TF*IDF. The TF part indicates the number of words in the document, and IDF is the total number of words that appears in the document. The feature extraction is calculated by using the TF*IDF weight of the term based on the given formula:

TF(t) = (Number of times term "t" appears in a document)
/ (Total number of terms in the document). (8.6)

IDF(t) = loge(Total number of documents
/ Number of documents with term "t" in it). (8.7)

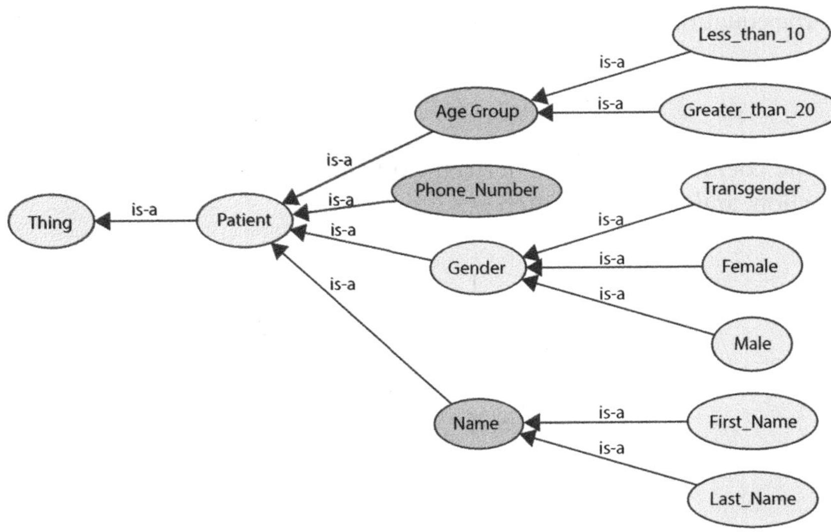

Figure 8.3 Personal information ontology.

From Equations (8.6) and (8.7), TF * IDF can be calculated as

$$TF * IDF\ (feature) = \mathrm{TF(t)*IDF(t)} \tag{8.8}$$

where *TF* is the frequency of a term in the given document, and IDF is Inverse Document Frequency. The semantic nature of the feature selection must be considered by constructing the ontology. An ontology represents the semantic way of knowledge representation. Figure 8.3 shows the personal information ontology.

8.3.4 Recommendation System

The correctness of the information is validated using the rule based information. Semantic Web Rule Language (SWRL) is used to define the entire possible rule for a given search keyword. It provides more powerful deductive reasoning capabilities, which help to infer more new information by applying the different rule on the inference engine [21–23]. An SWRL rule has two parts, namely, an antecedent part and consequent part, which represents the body and the head, respectively.

SQWRL (Semantic Query-Enhanced Web Rule Language) is predicted on the SWRL ontology query language that supports to create rules. "N" number of rules can be created for validation [9]. The correctness of the system is verified with the set of rules, which is written in SWRL. The sample rules are shown in Figure 8.4.

Rule 1: hasName(?x, ?y) -> hasFName(?x, ?y), hasLName(?x, ?y)
Rule 2: Patient (?x), hasName(?x, ?y), hasAge(?y) -> hasGender(?x, ?y)
Rule 3: Patient (?x), hasName(?x, ?y), hasSymtoms(?y) -> hasDisease(?x, ?y)Rule 4: hasDisease (?x, ?z) -> hasName(?x, ?y)
Rule 4: hasDisease (?x, ?z) -> select(?x), count(?z), orderBy Descending(?z)

Case-based reasoning is one of the reasoning methods in which the system will give a solution to a problem by reusing information and knowledge of that situation. It has four R's process to solve a problem: retrieve, reuse, revise, and retain. By applying CBR, a query can be retrieved with previous experience. Validation is used to check the correctness of an answer. A back-chaining mechanism was implemented to validate an answer [24–27]. Features retrieved from databases can be processed with different inference rules.

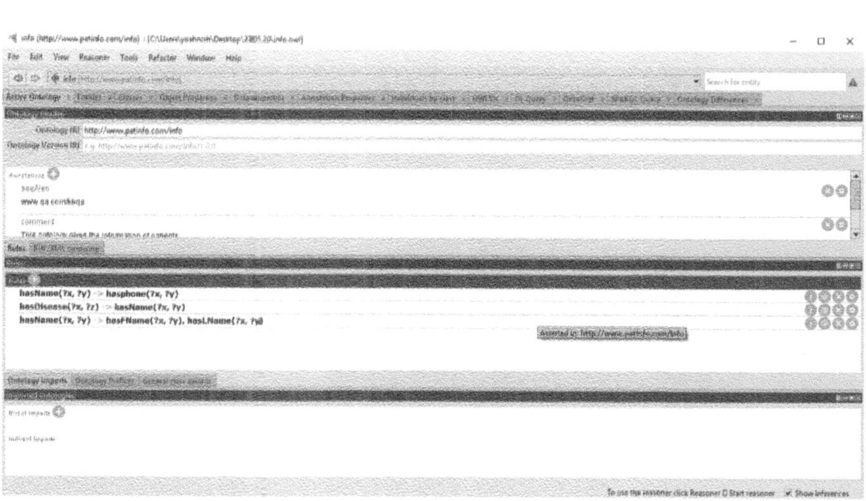

Figure 8.4 SWRL rule for the HRS.

8.4 Experimental Results

The performance evaluation of the imputation methods is verified using actual datasets of UCI. The missing value issues can be rectified to provide more accuracy on the evaluation and it must be tested with the datasets. The selected five datasets such as Iris, Ionosphere, Glass, Wine and Liver Disorder were included for comparison with the results in the work of Lau *et al.* [5]. Table 8.1 shows the dataset used in the HRS and gave the most relevant information to the user. The experiment is repeated 1500 times on the different datasets to reduce the errors. Thus, 1,767 datasets were imputed and tested in our research. It improves the accuracy, and the results show the cases of the iris dataset in Figure 8.5 and liver disorder in Figure 8.6.

Table 8.1 Dataset with cases.

Dataset	Number of cases
Iris	250
Ionosphere	500
Glass	315
Wine	278
Liver disorder	424

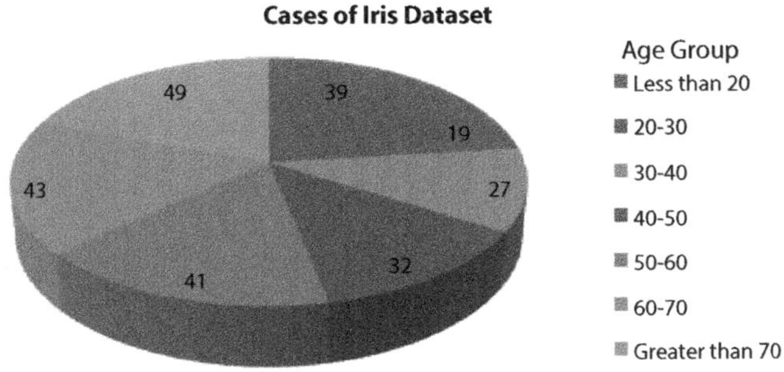

Figure 8.5 Cases of iris dataset.

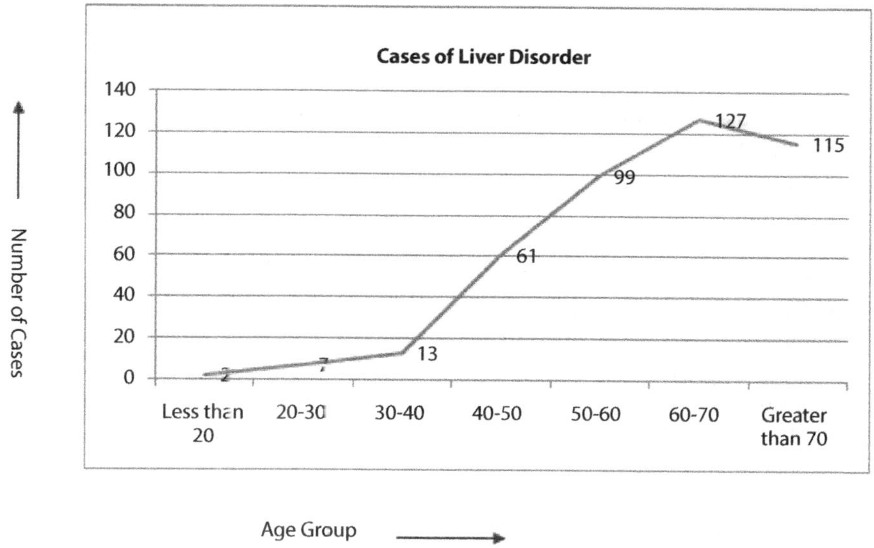

Figure 8.6 Cases of liver disorder.

8.5 Conclusion and Future Perspective

The intelligent-based human recommendation system using ontology has beenimplemented and tested with various datasets. The feature has been extorted using the imputation method, which applies the expert's knowledge. Expert's knowledge is incorporated using ontology. The semantic similarity measure helps to identify the similar feature of thehealth information of a patient. It is required to check with more features that are all not included in this research. The rules are written using SWRL and incorporated through the system. The correctness of the system is validated through the case-based reasoning method. Finally, the recommendation of the system was given to the patient. Also, chemical details of medicine characteristics must be included to our recommendation system as future enhancements, which support to get better quality of the system.

References

1. Chiu, D.K.W., Kwok, B.W.C., Wong, R.L.S., Cheung, S.C., Kafeza, E., Alert-driven E-Service Management. *Proceedings of the 37th Hawaii International Conference on System Sciences*, pp. 1–10, 2004.

2. Zeshan, F. and Mohamad, R., Medical Ontology in the Dynamic Healthcare Environment. *Proc. Comput. Sci.*, 10, 340–348, 2012.
3. Lage, R., Durao, F., Dolog, P., Stewart, A., Applicability of Recommender Systems to Medical Surveillance Systems, in: *Proceedings of the CIKM MedEX*, 2011.
4. Ru-zhi, X., Xiao-liang, D., Feng, Y., Pei-guang, L., Research on the construction method of emergency plan ontology based-on owl, in: *WISA'09 : Proceedings of the 2009 International Symposium on Web Information Systems and Applications. Nanchang*, P.R. China, pp. 019–023.
5. Lau, R.Y., Song, D., Li, Y., Cheung, T.C., Hao, J.-X., Toward a fuzzy domain ontology extraction method for adaptive e-learning. *IEEE Trans. Knowl. Data Eng.*, 21, 6, 800–813, 2009.
6. Rohrer, E., Motz, R., Díaz, A., *Ontology-Based Process for Recommending Health WebSites, I3E 2010, IFIP AICT 341*, pp. 205–214, 2010.
7. Subramaniyaswamy, V., Manogaran, G., Logesh, R., Vijayakumar, V., Chilamkurti, N., Malathi, D., Senthilselvan, N., An ontology-driven personalized food recommendation in IoT-based healthcare system. *J. Supercomput.*, 75, 3184–3216, 2019.
8. Araujo RB, R.V., Campos, M.R., Boukerche, A., Creating emergency management training simulations through ontologies integration, in: *CSEWORKSHOPS'08: Proceedings of the 2008 11th IEEE International Conference on Computational Science and Engineering - Workshops*, pp. 373–378, 20082008.
9. Bocanegra, C.L.S., Ramos, J.L.S., Rizo, C., Civit, A., Fernandez-Luque, L., HealthRecSys: A semantic content-based recommender system to complement health videos. *BMC Med. Inform. Decis. Mak.*, 17, 63, 2017.
10. Zlatanova, S., de Vries, M., van Oosterom, P., Ontology-based query of two dutch topographic data sets: an emergency response case, in: *Proceedings of the ISPRS Haifa*, pp. 193–198, 2010.
11. Adomavicius, G. and Tuzhilin, A., Toward the next generation of recommender systems: A survey of the state-of-the-art and possible extensions. *IEEE Trans. Knowl. Data Eng.*, 17, 734–749, 2005.
12. Schafer, J.B., Frankowski, D., Herlocker, J.L., Sen, S., Collaborative filtering recommender systems, in: *The AdaptiveWeb: Methods and Strategies of Web Personalization. Lect. Notes Comput. Sci.*, vol. 4321, P. Brusilovsky, A. Kobsa, W. Nejdl, (Eds.), Springer-Verlag, Berlin Heidelberg New York, 2007.
13. Resnick, P., Iacovou, N., Suchak, M., Bergstrom, P., Riedl, J., Grouplens: An openarchitecture for collaborative filtering of netnews, in: *Proceedings of the 1994 ACM conference on Computer supported cooperative work, CSCW 1994*, New York, NY, pp. 175–186, 2004.
14. Subramaniyaswamy, V., Manogaran, G., Logesh, R., Vijayakumar, V., Chilamkurti, N., Malathi, D., Senthilselvan, N., An ontology-driven

personalizedfood recommendation in IoT-based healthcare system. *J. Supercomput.*, 75, 6, 3184–3216, 2019.

15. Lampropoulos, A.S. and Tsihrintzis, G.A., *Machine learning paradigms. Applications in recommender systems*, Springer International Publishing, Switzerland, 2015.
16. Aggarwal, C.C., *Recommender systems*, Springer, Cham, 2016.
17. Espin, V., Hurtado, M.V., Noguera, M., Nutrition for Elder Care: A nutritionalsemantic recommender system for the elderly. *Expert Syst.*, 33, 2, 201–210, 2016.
18. Valdez, A.C., Ziee, M., Verbert, K., Felfernig, A., Holzinger, A., Recommendersystems for health informatics: state-of-the-art and future perspectives, in: *Machine Learning for Health Informatics*, pp. 391–414, Springer, Cham, 2016.
19. Ristoski, P., Mencia, E.L., Paulheim, H., A hybrid multi strategy recommendersystem using linked open data, in: *Semantic Web Evaluation Challenge*, pp. 150–156, Springer, Cham, 2014.
20. Achakulvisut, T., Acuna, D.E., Ruangrong, T., Kording, K., Science Concierge:A Fast Content-Based Recommendation System for Scientific Publications. *PloS One*, 11, 7, e0158423, 2016.
21. Paul, A., Ahmad, A., Rathore, M.M., Jabbar, S., Smartbuddy: defining human behaviors using bigdata analytics in social internet of things. *IEEE Wirel. Commun.*, 23, 5, 68–74, 2016.
22. Suresh, A. and Varatharajan, R., Competent resource provisioning and distribution techniques forthe cloud computing environment. *Cluster Comput.*, 22, 8, 1–8, 2017.
23. Lopez, D. and Manogaran, G., Parametric model to predict H1N1 influenza in Vellore District, TamilNadu, India, in: *Handbook of Statistics*, vol. 37, pp. 301–316, Elsevier, 2017.
24. Hawalah, A. and Fasli, M., Utilizing contextual ontological user profiles for personalized recommendations. *Expert Syst. Appl.*, 41, 4777–4797, 2014.
25. Moreno, A., Valls, A., Isern, D., Marin, L., Borràs, J., SigTur/E-destination: ontology- based personalizedrecommendation of tourism and leisure activities. *Eng. Appl. Artif. Intell.*, 26, 633–651, 2013.
26. Eysenbach, G. and Jadad, A.R., Evidence-based patient choice and consumer health informatics in the Internet age. *J. Med. Internet Res.*, 3, 2, E19, 2001 Apr-Jun.
27. Roitman, H., Yossi, M., Yevgenia, T., Yonatan, M., Increasing Patient Safety Using Explanation-driven Personalized Content Recommendation. *Proceedings of the 1st ACM International Health Informatics Symposium (IHI '10)*, ACM, New York, NY, USA, pp. 430–434, 11– 12 November 2010.
28. Agarwal, D., Chen, B., Elango, P., Ramakrishnan, R., Content recommendation on web portals. *Commun. ACM*, 56, 92–101, 2013.
29. Zhangm, Y., Contextualizing Consumer Health Information Searching: An Analysis of Questions in a Social Q&A Community. *Proceedings of the 1st*

ACM International Health Informatics Symposium (IHI '10), ACM, New York, NY, USA, pp. 210–219, 11–12 November 2010.
30. Wendel, S., Dellaert, B.G., Ronteltap, A., van Trijp, H.C., Consumers' intention to use health recommendation systems to receive personalized nutrition advice. *BMC Health Serv. Res.*, 13, 1, 126, 2013 Apr 4.
31. Akmaliah, I.F. *et al.*, Role of Ontology and Machine Learning in Recommender Systems. *2018 Electrical Power, Electronics, Communications, Controls and Informatics Seminar (EECCIS)*, 2018.
32. Yu, H.Q., Zhao, X., Deng, Z., Dong, F., Ontology driven personal health knowledge discovery, in: *Knowledge Management in Organizations*, L. Uden, M. Heričko, I.-H. Ting, (Eds.), pp. 649–63, Springer International Publishing, 2015.
33. Burke, R., Hybrid recommender systems: Survey and experiments. *User Model. User-Adapt. Interact.*, 12, 331–370, November 2002.
34. Adomavicius, G. and Tuzhilin, A., Toward the next generation of recommender systems: A survey of the state-of-the-art and possible extensions. *IEEE T. Knowl. Data Eng.*, 17, 734–49, 2005.
35. Osborne, F., Salatino, A., Birukou, A., Motta, E., Automatic Classification of Springer Nature Proceedings with Smart Topic Miner. *Semant. Web – ISWC 2016. 9982, Lect. Notes Comput. Sci.*, 383–399, 2016.
36. Salatino, A.A., Thanapalasingam, T., Mannocci, A., Birukou, A., Osborne, F., Motta, E., The computer science ontology: A comprehensive automatically-generated taxonomy of research areas. *Data Intell.*, 2, 379–416, 2020.
37. Cimiano, P. and Völker, J., Text2Onto - A Framework for Ontology Learning and Data-driven Change Discovery, in: *Proceedings of the 10th International Conference on Applications of Natural Language to Information Systems*, pp. 227–238, 2005.
38. Buitelaar, P., Olejnik, D., Sintek, M., A Protégé Plug-In for Ontology Extraction from Text Based on Linguistic Analysis, in: *Proceedings of the 1st European Semantic Web Symposium*, pp. 31–44, 2004.
39. Navigli, R. and Velardi, P., Learning Domain Ontologies from Document Warehouses and Dedicated Web Sites. *J. Comput. Linguist.*, 30, 151–179, 2004.
40. Chalk, S.J., SciData: A data model and ontology for semantic representation of scientific data. *J. Cheminform.*, 8, 54, 2016.
41. Soldatova, L.N. and King, R.D., An ontology of scientific experiments. *J. R. Soc Interface*, 3, 795–803, 2006.
42. Oellrich, A., Collier, N., Groza, T. *et al.*, The digital revolution in phenotyping. *Brief. Bioinform.*, 17, 5, 819–830, 2016.
43. Koehler, S., Carmody, L., Vasilevsky, N. *et al.*, Expansion of the Human Phenotype Ontology (HPO) knowledge base and resources. *Nucleic Acids Res.*, 47, D1, D1018–D1027, 2019.

44. Benslimane, D., Leclercq, E., Savonnet, M., Terrasse, M.-N., Yétongnon, K., On the Definition of Generic Multi-layered Ontologies for Urban Applications. *Comput. Environ. Urban Syst.*, 24, 191– 214, 2000.
45. Dumontier, M. and Villanueva-Rosales, N., Three-Layer OWL Ontology Design, in: *the Proceedings of the 2nd International Workshop on Modular Ontologies. CEUR Workshop Proceedings*, p. 315, 2007.
46. Kulmanov, M., Smaili, F.Z., Gao, X., Hoehndorf, R., Semantic similarity and machine learning with ontologies. *Brief. Bioinform.*, 1–18, 2020.
47. Ashburner, M., Ball, C.A., Blake, J.A. *et al.*, Gene ontology: tool for the unification of biology. *Nat. Genet.*, 25, 1, 25–9, 2000.
48. Thanapalasingam, T., Osborne, F., Birukou, A., Motta, E., Ontology-Based Recommendation of Editorial Products, in: *International Semantic Web Conference 2018*, Monterey, CA, USA, 2018.
49. Rabahallah, K., Mahdaoui, L., Azouaou, F., MOOCs Recommender System using Ontology and Memory-based Collaborative Filtering, in: *Proceedings of the 20th International Conference on Enterprise Information Systems (ICEIS 2018)*, pp. 635–641.
50. Rivero-Rodriguez, A., Konstantinidis, S.T., Sanchez-Bocanegra, C.L., Fernandez-Luque, L., A health information recommender system: enriching YouTube health videos with MedlinePlus information by the use of SNOMEDCT terms, in: *Proceedings of the 26th IEEE International Symposium on Computer-Based Medical Systems*, Porto, Portugal, IEEE, pp. 257–61, June. 2013.
51. Rohrer, E., Motz, R., Díaz, A., Ontology-Based Process for Recommending Health WebSites. *I3E 2010, IFIP AICT 341*, pp. 205–214, 2010.
52. Farman A., S., Razul Islam, M., Kwak, D., Khan, P., Ullah, N., Yoo, S., Kwak, K.S., Type-2 fuzzy ontology–aided recommendation systems for IoT–based healthcare. *Comput. Commun.*, 119, 138–155, April 2018.
53. Kandula, S., Curtis, D., Hill, B., Zeng-Treitler, Q., Use of topic modeling for recommending relevant education material to diabetic patients. *AMIA AnnuSymp Proc. 2011*, pp. 674– 682, 2011.
54. Wang, S., Chen, Y.L., Kuo, A.M., Chen, H., Shiu, Y.S., Design and evaluation of a cloud-based Mobile Health Information Recommendation system on wireless sensor networks. *Comput. Electr. Eng.*, 49, 221–235, 2016 Jan.
55. Jorro-Aragoneses, J.L., Ceron-Rios, G.M., Recio-García, J.A., López-Gutierrez, D.M., RecOnto: An Ontology to Model Recommender Systems and its Components, in: *2017 IEEE 29th International Conference on Tools with Artificial Intelligence (ICTAI)M*, B .Díaz-Agudo, 2017.
56. Hu, H., Elkus, A., Kerschberg, L., A Personal Health Recommender System incorporating personal health records, modular ontologies, and crowd-sourced data. *IEEE/ACM International Conference on Advances in Social Networks Analysis and Mining (ASONAM)*, 2016.

57. Stark, B., Knahl, C., Aydin, M., Elish, K., A Literature Review on Medicine Recommender Systems. *(IJACSA) Int. J. Adv. Comput. Sci. Appl.*, 10, 8, 1–8, 2019.
58. Ali, S., Mahmood, T., Mitra, A., Ali, B., Ahmad, F., SEPHYRES: A Medical Diagnosis Model based on Semantic Pseudo-Fuzzy Plan and Radar-form Interface. *Appl. Med. Inform.*, 1, 1–8, 2017.

9

Secure Healthcare Systems Based on Big Data Analytics

A. Angel Cerli[1], K. Kalaiselvi[2] and Vijayakumar Varadarajan[3]

[1]Department of Computer Science, St. Anne's Arts and Science College, Madhavaram, Chennai, Tamil Nadu, India
[2]Department of Computer Applications, School of Computing Sciences, Vels Institute of Science, Technology & Advanced Studies (Formerly Vels University), Chennai, Tamil Nadu, India
[3]School of Computer Science and Engineering, The University of New South Wales, Sydney, Australia

Abstract

Adapting the digitization of medical information, such as hospital diagnostic reports, generates vast volumes of data over a limited period in the modern healthcare environment. Data security is important for the dissemination of electronic health records (EHRs) for patient-centered study, business research, medical research, mining of health knowledge, etc. Managing large-scale data and safeguarding patient welfare have been a long-standing challenge for researchers. Blockchain technology, though, has eased some of the issues by offering a secure and distributed network. Unfortunately, existing EHR structures suffer from data distortion, delayed coordination, and efficient cooperation in data acquisition, storage, and delivery. This chapter addresses new issues regarding the protection of health records and current and potential policy in the area. This includes an overview of the blockchain network infrastructure, emerging threats, and possible scope for effective handling of current and prospective medical information. It also offers a range of blockchain implementations for potential health insurance, large data, and blockchain study.

Keywords: Blockchain, big data, data privacy, EHR security, medical big data analytics

**Corresponding author*: angelcerli@gmail.com

A. Jaya, K. Kalaiselvi, Dinesh Goyal and Dhiya Al-Jumeily (eds.) Handbook of Intelligent Healthcare Analytics: Knowledge Engineering with Big Data Analytics, (189–212) © 2022 Scrivener Publishing LLC

9.1 Introduction

The tremendous growth in data leads to the increase in big data in several fields, including advertising, industry, and banking. Health units are now also increasingly becoming a significant discipline, where there is a wide variety of potential for big data growth. Most of them are introducing simple electronic health record (EHR) programmed in NICs in the United States [1, 2]. Technology such as the Internet of Things already enables millions of people to process personal health data gradually. For example, wearable fitness trackers and mobile health applications are part of those devices [1]. In a single entity, thus, data are expanding enormously. Although there are vast numbers of such groups in the United States.

Big data creation takes place in medical units. Big data concerns certain critical health unit databases that are deemed too large, too fast, and too complicated to use for their present resources. The most important and fragile components in these datasets are health images. These photos are used by clinicians for some of the most critical measurements of the patient's health. These large databases contain both organized and unstructured data. Thus, a systematic approach to proper disease prediction requires the usage of organized and unstructured evidence from both clinical and non-clinical approaches to help identify the disorder. Big data scientists are now looking for correlations and latent trends in this critical medical data such that an improved diagnosis and treatment strategy can be established for patients by enabling a more detailed evaluation of the disease of the patient, thus saving lives and reducing the cost of medicine.

Internationally, massive health data scientists and educators utilize multidimensional health data. Similarly, insurance facilities often decline to share confidential information to their clients. The large-scale data collection and management of the patients was also a problem. The privacy and security of personal data has lately become a top concern in terms of innovation excellence. More than 112 million data papers were hacked, destroyed, or unlawful leaked according to previous analysis released in 2015 by Forbes Magazine [3]. The big data health ecosystem now plays a significant role for users, contributors, creators, and researchers. The key health results players are seen in Figure 9.1.

In order to conduct a thorough medical examination, adequate coordination and communication must be established with the patients. Often, secrecy and confidentiality instances are connected to those organizations in one manner or another. Patients are the basis of all sorts of facts in the first place. Patients use medical reports or portable devices [4] to generate

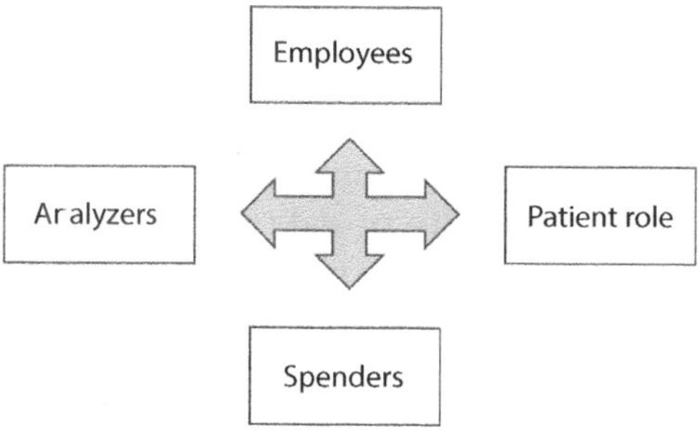

Figure 9.1 Various large data healthcare stakeholders.

this material. Then, payers are persons who support patients explicitly or indirectly by covering healthcare expenses (i.e., insurance companies, private sources, and bank loans). Moreover, patient details collect and retain the roots of medical knowledge (hospitals, clinics, medical centers, blood banks, etc.). In the end, researchers and analysts use the data generated by the schemes to render the health system more effective.

In the past, the interoperability of health data can be divided into three models for different organizations, where blockchain technology has not been available. An exchanging of data between two suppliers is feasible in the push model and the system is not open to the third party. For instance, data can be transmitted from one facility to another and data can be retrieved in the same place, even though the same data cannot be collected from another hospital, except though transported to another. End-to-end data privacy is typically not covered under this push model. One provider may informally request data from the other provider in the view model without an audit trail. For example, an orthopedic surgeon asks a cardiologist for details casually.

The display template documents can be viewed from another provider from one provider entering the text. For example, a doctor from the surgical department may analyses an x-ray obtained in the emergency room. The protection methods in this model do not rest on the customer-provider partnership, but the answer is short. The regulations that underlie the operational principle of the models alluded to above is controlled in the area by local authorities.

Through identifying data protection management elements and developing standardized and formalized data access agreements, the blockchain model adds a new layer to medical care. It is difficult to understand the identification that the employee does what and when he does the job while communicating with the EHRs, who store the information in a different workflow. The blockchain-based architecture provides a time stamp and identifier to the workflow, and copies are distributed to any of the network nodes. When a node is revamped or modified, it covers all nodes evenly and is available to anyone in the universe with connections. The model ensures the security of data is maintained without human intervention between endpoints. Any huge opportunity to revolutionize the healthcare sector by integrating the blockchain is as follows [5] (Figure 9.2).

- Decentralized storage: The blockchain holds knowledge which is available and supplied to third parties with the developer's permission. The different versions of the records are stored in many locations as a decentralized way of storing the details.
- Consent: The global algorithm of consent would govern approval, storage, and delivery. Both available stakeholders can allow data changes following an independent sanction.
- Unchanging: Such details cannot be changed. If the knowledge is inserted in a certain block, then it cannot be reversed or updated.
- Increased capacity: Blockchain is an inexpensive means of safeguarding identity protection without an agent and less regulatory problems.

The rising volume of data as well as the related study, secrecy, and safety are major issues in the field of healthcare [6]. More computing capacity is

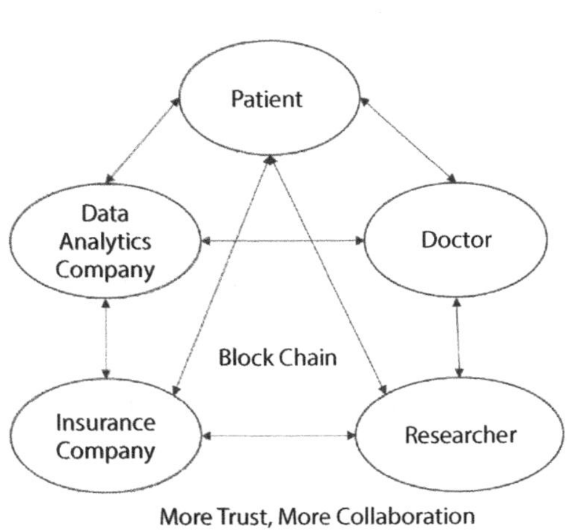

Figure 9.2 Benefits in adopting blockchain healthcare privacy information.

also required, as Pattnaik *et al.* discussed [7], with growing demand for health data mining. Many researchers have already made use of widely recognized information (PII) blockchain technologies [8, 9]. The impact of the blockchain on protection of large data in the field of healthcare is mentioned in this chapter. This section also describes the risks, ranges and the latest state of technology.

9.2 Healthcare Data

Awareness in healthcare is the data generated and obtained in the health field. This data type may mainly be described as structured, semi-structured, and unstructured data. These various data types are listed as follows [10]. Figure 9.3 shows several large data assets types.

9.2.1 Structured Data

Structured information is data which can be stored and processed as various coded values. One illustration of these is the Logical Observation Identifiers Names and Codes (LOINC) that is a database and a common tool for classifying medical laboratory observations [6]. Computer programm can quickly transfer data back and forth as health data is processed in an organized manner. This makes the acceptance process of life simpler.

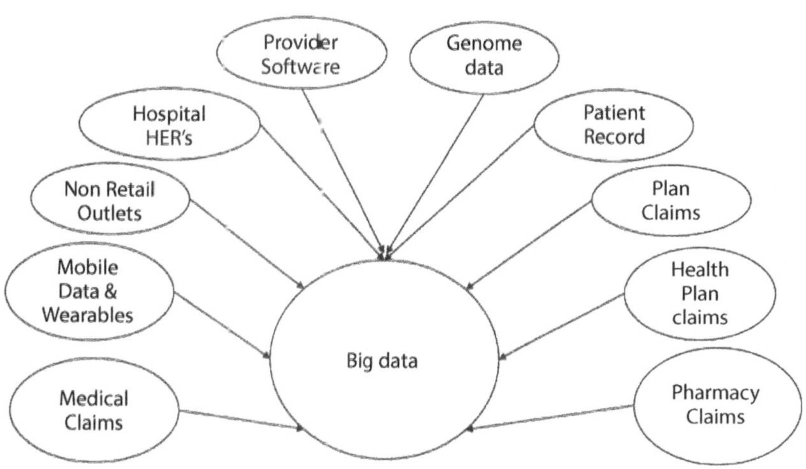

Figure 9.3 Various forms of big data tools for healthcare.

Further emphasis is assigned to structured data including patient data, clinics, procedural protocols, and medication standardized data [11].

9.2.2 Unstructured Data

Unstructured data was usually gathered and construed as free document. While human beings can easily comprehend this knowledge, it is not available on computers. Notes of development, pathological outcomes, radiology, etc., are regular instances of unstructured data [11].

9.2.3 Semi-Structured Data

Semi-structured information can be interpreted as a blend of structured and unstructured information. A semi-structured data description will be an interface for entry which could include an ordered set of data and free text.

9.2.4 Genomic Data

Genomic data are usually the details of the person's genomics and DNA. In bioinformatics, scientists have compiled, accumulated, and investigated genomes for several living organisms. This category of data is usually quite comprehensive and needs the specific tools for broad storage and bioinformatics.

9.2.5 Patient Behavior and Sentiment Data

Usually, they are collected or collected from different social networking platforms. Invaluable reviews and knowledge may be given by patients about their experiences in a hospital or clinic with physicians, nurses, and others. They can pursue support by sharing this knowledge on multiple social networking sites, patient surveys and styles of discussions. It is collected in huge amounts and useful information can be derived from it if properly evaluated using the appropriate big data analytics (BDA) framework.

9.2.6 Clinical Data and Clinical Notes

The key guide for science and medical study has already been found to be medical evidence. Even during routine medical treatment, clinical results were collected. This specific may also be established in a structured clinical trial programmed. For example, the often unstructured data are records, images, and medical or transcribed notes (about 80%).

9.2.7 Clinical Reference and Health Publication Data

Knowledge is available from different media, including scientific findings, general practice recommendations (textual reference), health articles, and clinical and medical studies services etc.

9.2.8 Administrative and External Data

The details are typically gathered from numerous external outlets, such as policy papers and associated financial statements, reporting and calendar reports, and different biometric data, including fingerprints, signatures, and iris.

9.3 Recent Works in Big Data Analytics in Healthcare Data

Several recent works are briefly presented in this section in the area of BDA for healthcare. Ojha *et al.* [12] outlined how large data processing systems, such as Hadoop, are accessible for medical data. They described how the EHRR can collect useful knowledge (EHR). For this cause, they have carried out health studies from Indore, Madhya Pradesh, India, at the largest government hospital in Central India. This hospital produces several details every day that publishers propose to process in the EHR. Per patient whose information is treated here is given a special amount. Information of any patient can be conveniently and securely obtained with this figure. This data can also be saved with the data warehouse EDH (Enterprise Data Hub). Different approaches such as sorting, clustering, and classification for data analysis can be made understandable easily on these data, which means that comprehensive patient knowledge is handled securely and effectively.

In anesthesia and surgical instruments, Simpao *et al.* [13] are focusing on large-scale data mining. Sometimes, they rely on computer analysis, simplifying logical thinking analyses streamlined by virtual artificial environments [26]. That will also contribute to serious questions regarding cognitive achievement. Strong attempts have been made through the Anesthesia Standard Institute and the Multicenter Preoperative Outcomes

Association to obtain systemic evidence of anesthesia for test results and to boasts quality. They proposed to maintain outstanding and continuous management that the correct usage of data coupled with quantitative and qualitative analysis of data at wide scale should be utilized. For example, all major implementations require clinical decision-making, quantitative risk assessment, and resource management.

Weider *et al.* [14] utilizing the Early Disease Diagnostics Large Data Driven Guidance Engine in the modern health setting. This scheme is generated using a classification algorithm, Naïve Bayes (NB). It lists patient protection problems, readback rates, maximizing treatment and adverse conditions. This algorithm operates with Apache Mahout. The thesis discusses and applies current methodologies for the usage of big data. The method suggested is very beneficial in that patients can be prescribed the required treatment after the condition is determined, and so, the typical person's life expectancy will be improved if adequate care is offered at an early point.

The BDA Platform is proposed by Chrimes *et al.* [15] for the usage of specific quantities of massive data in healthcare. HBase (NoSQL database) and Hadoop use the new high-performance computing platform (HPC) (HDFS). The database developed by NoSQL was imitated by a metadata and hospital profile framework of the Vancouver Island Health Authority. For the Hadoop and HBase worlds, a special modification was made by introducing "salt buckets" The authors propose that the current BDA data migration requirements would archive large volumes of data and reduce recovery times and connect them to novel processes and configurations.

The patient-centered, tailored treatment approach is recommended by Chawla and others [16]. The suggested framework works for physicians to classify threats to their sick patients as information-driven computational assistance. It can include early alerts of the potential risk of infection for a person, and can be turned into a dialogue between the doctor and the patient in order to help encourage the patient. The management of the wider experience of diseases of higher potential may be used to suggest previously unconsidered issues and to facilitate conversations on early testing and mitigation as well as on health-related approaches which are easier to identify and grasp.

McGregor also addressed the advantages and efficacy of large data in neonatal intensive care [17]. He argues that this would enable a wide range of lethal medical problems to be identified and dissuaded early. The capacity to handle numerous high pace evaluation data sources from various patients in several locations and in real time will both be improved greatly through medical and patient safety expertise.

Fahim *et al.* [18] proposed to create and equate relation between specific health requirements and to provide input on the person lifestyle and well-being facilities in real time, ATHENA (activity-awareness for engaged human health applications). Its purpose is to establish a framework to promote responsible citizenship by contrasting their previous actions and suggesting helpful intervention. Through the usage of an ML (machine learning) algorithm and leveraging cloud technology, the suggested design handles sensory data inside smart devices to minimize costs. Wide data architecture here is used for comprehensive sensory data storage and fast recommendation retrieval.

9.4 Healthcare Big Data

Big data is such a term as to incomplete and hybrid current data handling applications. The purpose and measurement of mass data processing or broad data is specified by 3Vs (volume, velocity, and variety) "Volume" implies the size and significance of the data. Finally, "speed" means a mix of different types of data. The scope of the processing of data is called "variety" and all the elements concerned in the healthcare system have been told by the advent of modern healthcare industry. Several diagnostic reports are accessible at different sites, from regular text, health photographs and sound files, x-rays and MRI's, magnetic resonance imaging, medical conversations, various IoT mapping devices, and medical trackers. There are also standards for specifications.

Extracting the EHR pattern will enable clinicians to estimate future troubles. Health and general knowledge are strongly related. For instance, a text-mining approach [19] investigates the correlation between Parkinson's disease and gait and the relation between Parkinson's disease and healthy older grown-ups utilizing pattern-based ML technologies. Several papers dealt with the need for the recognition of EHR for millions of medical knowledge fields.

The data from the EMR is a compilation of information from the start of the disease control service gathered by medical organizations. This is a summary of time data reported by hospitals, as seen in Figure 9.4.

IoT is widely seen on cell telephones, wearables, microphones, CCTV, environmental monitors, smartwatches, etc. In the facilities, information is gathered to track the state of several patient elements

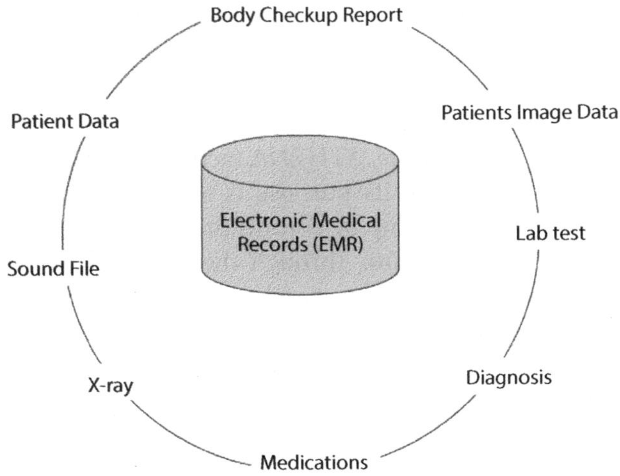

Figure 9.4 Electronic medical record (EMR).

or functions. Wearable devices, for example, have been used to analyses the PD data and to design a diagnostic system for monitoring PD patients [20].

9.5 Privacy of Healthcare Big Data

This research mentions three widely employed protection techniques: administrative, physical, and technological protections. Figure 9.5 demonstrates these methods.

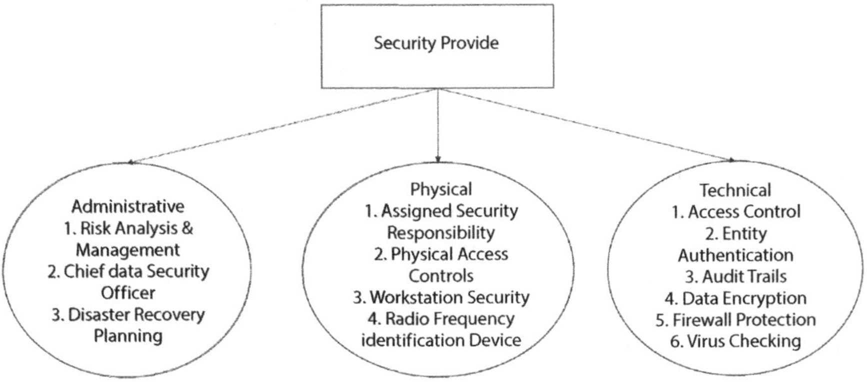

Figure 9.5 Different forms of strategies for security.

Digital ID: A digital identity is linked to an individual or company's online identification. Any details or attributes applicable to the business shall include the digital identification (individual, organization, or electronic device). The digital ID information enable a person to be assessed and authenticated on the internet. Any legal and personal details from any online transaction can be created by a digital identity (online identity).

Personally, identifiable information (PII): Any information or details that may belong to a single individual or person is referenced as PII or personal data. Today, worldwide appreciation exists for sensitive personal details (SPI) or personal information (PD) or personal data (PI). The PII is often referred to as any specification that is used to describe personal identification. All data that helps to de-anonymize unidentified data is often called the PII. Digital identification as a way of understanding customer protection was developed by numerous studies. The PII is specified as the distinguishing number [21] by "a subset of individual attribute attributes dividing this entity from another class". McCallister *et al.* [22] of the National Technical Information Service (NTIS) describes "data that have been maintained by each organization, including any information that may be used to differentiate between them".

Potential personally identifiable data (PPID): The studies above have shown that personal identification or PID is not only a number or individual but may be generated by a mixture of partial assets and persons with an entire PII or identity. "A subset of full identity attribute values in which the whole identity is a mixture of all of its attribute values is defined as a partial identity concept or a probability of personal identifying details (PPID)".

Protected health information (PHI): This applies to scientific findings individually reported by every clinic, exercise planner, or healthcare professional (i.e., clinic, doctor, pathology, government health department, and the insurance company). All knowledge about a person's physical or mental well-being in regard to a person's past, current, and potential is completely known as a PHI. Typically, PHI data may be stored or submitted in any medium, language, text, or electronic mail. HIPAA developed the first PHI programmed in 1996 (HealthCare Portability and Transparency Act). Four of the more vulnerable cases of PHI are as follows:

- Health data: First visit to a doctor, registration number, medical bed number, medical identification number, hospital address, etc

- Photos: Citizens' pictures and their attraction topics. It contains all type of radiographic photographs, medical videos, case documents and MRIs.
- Biometric data: Biometric records are specifically meant to protect the person's biological knowledge. This knowledge involves precise body indicators, fingerprints, retinal pigments, weight, origin of blood, speech sounds, DNA, race, body colors, etc.
- Details on reimbursement and correspondence: Both the forms and descriptions of payment systems, the patient contact number, the email response, etc. Mehmet Kayaalp [23] has acknowledged several links between personal identities. A daily number of PII and medical reports is found to be in PHI.

9.6 Privacy Right by Country and Organization

Each country has its data privacy policies and regulations. Information protection regulations are given to many countries with key variables in Table 9.1.

9.7 How Blockchain is Big Data Usable for Healthcare

9.7.1 Digital Trust

Everyone needs to have confidence in their company, from patients to physicians, from data scientists to health data providers. Blockchain technology would create this digital confidence among big data dealers in healthcare. Mattila [24] has provided for digital trust that can fulfill the three criteria of blockchain technologies: confidentiality, authentication, and traceability. In order to create morality among certain groups, an intellectual contract must be created between them for any data generation or handler wanting to share data. Who bears this responsibility usually found in large-scale health data? Blockchain technology can create a secure network in a true, physical world [25]. In the other side, several researchers have often called for a greater awareness and conformity to digital confidence in health technology.

Table 9.1 Knowledge security laws from numerous countries and organizations.

Information protection regulations of different countries and organizations		
Country	**Law**	**Key factor**
Brazil	The Constitution's Rules	Private details and private life and the reputation of people are viewed as extremely confidential and protected data.
EU	Government Regulations on Data Security	Security of confidential details from individuals during storage and retrieval.
Russia	Law on Personal Details under Federal Russian Law	Data operators must assume complete liability for any unauthorized access to details.
United Kingdom	Act on Data Privacy	In regulating the flow of personal data, the user has enormous influence. Privacy data must match the geographical boundaries of the land.
India	Act on IT	The victim should be liable for some sort of breach of personal details.
South Korea	Security of Personal Knowledge Act September 30, 2011	Citizens' privileges, desires, and integrity are taken care of. It does not specify any geographical framework.
EU	General Regulation on Data Security	Control on the security of personal records, of which the main considerations are the owner's permission.
New Zealand	Code of Secrecy of Health Records	Unique guidelines regarding the compilation, handling, and analysis of health data by medical organizations.
Commonwealth	Transition Authority for National Electronic Health	NEHTA seeks to open facets of the eHealth infrastructure to facilitate the processing and sharing methods of electronic health information.

9.7.2 Smart Data Tracking

Big data mining is becoming more complex alongside robotics, genetic algorithms, and ML. Incomplete prior health records may be translated to data and evaluated for production of information (e.g., for marketing and drug analysis). Specific data management is a key concern in terms of protection and the possibility of abuse. The blockchain ensures a multi-layer data protection framework with the assistance of decentralization and smart agreements. The information in question is better, from secure to wearable IoT sensors in which investors (doctors, consultants, government officials, patients, IoT businesses, etc.) can have links to the blockchain network as miners.

9.7.3 Ecosystem Sensible

The healthcare network is being updated by hospitals, businesses, and customers, their data suppliers, sponsors, or rivals. Intelligent environments may be utilized through blockchain applications to collect health info. Blockchain technology decreases maintenance expenses, time, and deterioration when handling and verifying medical records. Blockchain utilizes the idea of "smart contract" or "smart means of protection" [22] to render any clear and argument-free value. This stabilizes the transaction and holds things running. Best of all, blockchain technology guarantees the private security of terms and conditions between the parties.

9.7.4 Switch Digital

Kim and Laskowski [27] defined the 7.4. Switch Digital (SD) system, with all supplies being tracked by all healthcare stakeholders. The SD can monitor a meal, a medication, a wellness product, an insurance product, etc. Tian [28] and Aitken [29] also proposed a life-cycle chain delivery of food and farm goods leveraging blockchain technology. The system for secure supply chain blockchain for drugs and pharmaceuticals was created by Shae and Tsai [30]. A core advantage of the blockchain is that it offers a centralized transaction record without the individuality of the participant. Any group is notified with the usage of a shared key infrastructure blockchain (PKI). The purchase can be monitored by data or service providers inside the supply chain. This technology will check any part of the arrangement and everyone can abandon it if one of the components is lacking. Another study by Chhetri *et al.* [31] shows the reliability of the digital supply chain in the future market architecture.

9.7.5 Cybersecurity

The blockchain technology preserves information for the better. This is a tamper-proof technology for the keeping of contracts, decisions, transactions, and records. The U.S. Army used this technology as a cryptographic firewall, the Pentagon and the Washington Times say. Application blockchain provides digital information protection. However, this clustered storage strategy can minimize data mistreatment. The distribution and sharing of clinical data by blockchain technology was determined by Nugent *et al.* [32]. Due to the hash algorithm, the consent algorithm, and the immutability of logs, blockchain has been identified as a secure technology by Matanovic [33]. Maddux research [34] has found the potential of large data healthcare blockchain technologies. The study suggested that it would be more beneficial to carry and distribute data utilizing this technology. Blockchain retains some of the elements of the distribution of results, along with the authentication of information, time-testing, and identification purposes, etc., in order to improve communication between the parties (data holders and researchers). According to IBM and Ponemon, the lack of medical information costs around USD 380 per second, although the violation costs just USD 141 per second in the sector.

Health claims: Insurance money, proof of a breach of fitness, violation of the penalty, etc., can be reimbursed quickly. DE binding external parties to smart contracts by blockchain would ultimately provide victims with monetary benefits and unsustainable accessible networks.

Medication adherence: The conduct of patients has a partial effect on the reputation of the medication. Few patients today have a full knowledge of modern medicine, diagnosis, or diet. Once a specialist has fulfilled certain tasks or appeals, he or she will send incentives to blockchain technologies. The Intelligent Medical Device can work immersivity and remind patients not to do so. Blockchain may be used to create a drug-smart deal between the IoT device and the customer.

9.7.6 Sharing Interoperability and Data

Interconnectivity and data exchange between businesses are often connected to research and growth. Data include OMICS, care performance, IoT units, patient laboratory testing, and drug status. Much of this information is shared between hospitals, companies, doctors, patients, blood banks, hospitals, etc. However, there is inadequate interconnectivity between existing patient records and their interoperability. Blockchain technologies can enhance the management and governance of data and

Table 9.2 Blockchain flexibility concerning threats.

The capability of blockchain concerning challenge offered from healthcare big data		
Factors	**Healthcare big data challenge**	**Blockchain solution**
Fragmentation of details	In a fragmented, decentralized manner, data is generated. Data is created separately by patients, physicians, hospitals, therapists, and analytics.	A network can be generated between all groups through computer networks and the linked decentralized mechanism.
Simultaneous and prompt entry	Timely entry and parallel research were delayed by manual cooperation, privacy policy, and agreement.	Without any possibility of data theft, Distributed blockchain will protect concurrent research on the same data set (blockchain are immutable).
Scalability of the device	The problem of confidence (data and individuals) creates unrest between parties in the partnership.	Blockchain verifies participating nodes and relevant details, removing all threats.
Data handling Sensors	Thousands of IoT machines, which are hard to monitor and manage, are used to capture and transmit data.	Through building a private blockchain network between them, blockchain will protect IoT devices via user communication and data.
The consistency of data and access	Data handlers cannot be transparent and there must be certain delivery rules.	To specify the selected data processors and regulators of the owner, blockchain may use its three consumer scope categories (private, public, and consortium).

(*Continued*)

Table 9.2 Blockchain flexibility concerning threats. (*Continued*)

The capability of blockchain concerning challenge offered from healthcare big data		
Factors	**Healthcare big data challenge**	**Blockchain solution**
Value of data analysis	Some third parties conduct an audit of data along with all individuals, which raises costs.	The operating cost of data analysis and delivery may be minimized by blockchain intermediaries with less parallel infrastructure.
The connection from company to enterprise	To make a profit, middlemen and third parties are currently exploiting and sharing info.	To establish a territory-less accessible border for data distribution safely, blockchain will exclude any third party.

create confidence. There are several methods needed to improve networking. The same network ensures accuracy of performance, data dependence, and stakeholder perspective by consensus with both stakeholders through blockchain technology [33, 34]. Blockchain Smart Contract Plans are developed by all care professionals, insurance providers, and patients. In this Table 9.2 the blockchain flexibility concerning threats and challenges faced from healthcare big data is elaborated.

9.7.7 Improving Research and Development (R&D)

Extensive case studies, numerical health sector, and genomic data have also been reviewed. Blockchain technology can ensure that the accuracy of genomic and health information analyzers from generation to generation is tracked. Distorted and falsified biological and clinical facts endangers the integrity of the study. Joint research can acquire more interest if partners are to maintain data excellence.

9.7.8 Drugs Fighting Counterfeit

There are counterfeit and failed medications all over the world. Blockchain technologies can help secure medical practices by monitoring the development of input by end users. According to the study of the World Health Organization, blockchain technology will track the production and distribution of medication supply chains. The WHO invests USD 30 billion on the battle against counterfeit medicines.

9.7.9 Patient Mutual Participation

In order to take care of each other, the patient should join a collective community. Using blockchain technologies, the overall control structure will be retained. The related illness understanding continues to dissatisfy patients. Both knowledge and patient options may be discussed through private blockchain technologies. The patient group and the families would request the approval of the doctors for certain sensitive decisions. Medical personnel from different countries will work together to perform some research trials without any challenge to faith.

9.7.10 Internet Access by Patient to Longitudinal Data

For online medical records, blockchain offers a secure double-layer network. The biggest benefit is that the patient will do it without third-party intervention. It has three core characteristics, allowing blockchain an important platform for clinical data processing. Next, a trusted health archive can be built via blockchain. Secondly, blockchain may build a network and framework for secrecy between data holders. Ultimately, before any vote, automatic approval is always valid.

9.7.11 Data Storage into Off Related to Confidentiality and Data Scale

Off-chain data storage facilities became common because of privacy restrictions. Due to its decentralized nature, blockchain is inherently unable to hold large volumes of info. In the other side, there are, of course, statistics, health facilities, and biological facts. Data storage, supported by data hacking, was a good way to solve this problem. A comprehensive overview will be given in the next part.

9.8 Blockchain Threats and Medical Strategies Big Data Technology

Obstacles: While blockchain technology offers benefits and potentials, there are still several inherent disadvantages that pose some unique domain challenges. The four main problems facing this technology in the world of big data healthcare include storage, scalability, change, protection, and regulation.

Storage: A significant number of patient and wearable IoT devices with EMR and sensor data are generated in healthcare and diagnostic records. In the other side, the architecture blockchain supports very minimal data storage in the chain. The blockchain architecture, decentralized and hacked, is too expensive to store records. Similarly, since the sample size is larger, it is also expensive to register, handle, and recover blockchain data. This functionality demands that blockchain applications be built.

Correction: On one hand, blockchain code immutability safeguards the device; on the other hand, there is little risk of computer modification and deletion. Either a new block from all the nodes must be formed by agreement or a new chain must be built. Both approaches are expensive and impossible to procure. The technological development in blockchain will then be such that the data modification threshold is the lowest.

Validation: The scalability issue is less serious due to the modular design. But millions of private consumers include family physicians, dentists, IoT-specialist startups, rural clinics, and market testing companies. It is particularly doubtful to preserve the shared design of the blockchain. In addition, blockchain technology has improved energy processing and enables new services to be utilized by network implementations [35]. Large data scalability in healthcare should be taken seriously to popularize blockchain.

Privacy and regulations: Blockchain multiple times optimize content security. A stable, decentralized, autonomous, and unchanging infrastructure can ensure optimal content protection. Big data in healthcare was mostly about the hospital, the patient, and the patient's confidential information. A copy of the data in each node could also be dangerous. Indefinite storage of PII and EHRs is the main challenge with blockchain technologies presently in operation. This pattern is not practiced by other countries and coordinated bodies.

9.9 Conclusion and Future Research

This chapter provides a short description of blockchain healthcare prospects, challenges, and solutions. The use of systematic data collection in the health sector is rising. Medical imaging is most or partially based on medical records, since the automated phase of the manual diagnosis method includes constant review and assessment of a broad range of diagnostic evidence collected from patients with a similar condition from multiple clinics at different places. However, private blockchain safe storage is a common blockchain design solution for solving current problems. Overall, the blockchain provides a vast spectrum of medical data solutions, building on several research opportunities. This chapter overall reveals how important blockchain is for broad data protection in healthcare and provides a way to minimize long-term consequences. Supporting this chapter is understandable and feasible for the technological, strategic, economic, and regulatory effects of the protection of big data in blockchain and healthcare. The potential aim is to establish a genuinely interconnected data exchange and identity sharing infrastructure in the healthcare industry in a security way.

References

1. Zhang, R., Wang, H., Tewari, R., Schmidt, G., Kakrania, D., Big data for medical image analysis: A performance study, in: *Parallel and Distributed Processing Symposium Workshops, 2016 IEEE International; IEEE*, pp. 1660–1664, 2016.
2. Charles, D., Gabriel, M., Furukawa, M.F., Adoption of electronic health record systems among US non-federal acute care hospitals: 2008-2012, in: *ONC Data Brief*, vol. 9, pp. 1–9, 2013.

3. Munro, D., *Data Breaches in Healthcare Totaled Over 112 Million Records in 2015*, vol. 31, NY Forbes, New York, 2015.
4. Banerjee, S., Hemphill, T., Longstreet, P., Wearable devices and healthcare: data sharing and privacy. *Inf. Soc*, 34, 1, 49–57, 2018.
5. Holotescu, C., Understanding blockchain opportunities and challenges, in: *The International Scientific Conference eLearning and Software for Education*, vol. 4, Carol I National Defence University, pp. 275–283, 2018.
6. Sahi, M.A., Abbas, H., Saleem, K., Yang, X., Derhab, A., Orgun, M.A., Yaseen, A., Privacy preservation in e-healthcare environments: state of the art and future directions. *IEEE Access*, 6, 464–478, 2018.
7. Pattnaik, P.K., Rautaray, S.S., Das, H., Nayak, J., Progress in computing, analytics and networking. *Proceedings of ICCAN 2017*, vol. 710, Springer, 2018.
8. Halamka, J.D., Lippman, A., Ekblaw, A., The potential for blockchain to transform electronic health records. *Harv. Bus. Rev.*, 3, 257–263, 2017. Retrieved from: https://hbr.org/2017/03/the-potential-for-blockchain-to-transform-electronic-health-record.
9. Yue, X., Wang, H., Jin, D., Li, M., Healthcare data gateways: found healthcare intelligence on blockchain with novel privacy risk control. *J. Med. Syst.*, 40, 10, 218, 2016.
10. Health Data Knowledge. List of Different Types of Health Data-Health Data Knowledge, 2018. [online] Available from: http://www.healthdataknowledge.com/list-of-different-types-of-health-data/.
11. LOINC, 2018. Retrieved from https://en.wikipedia.org/wiki/LOINC.
12. Ojha, M. and Mathur, K., Proposed application of big data analytics in healthcare at Maharaja Yeshwantrao Hospital, in: *Big Data and Smart City (ICBDSC), 2016 3rd MEC International Conference; IEEE*, 2016March:1–7.
13. Simpao, A.F., Ahumada, L.M., Rehman, M.A., Big data and visual analytics in anaesthesia and healthcare. 115, 3, 350–356, 2015.AU: Please provide journal title.
14. Weider, D.Y., Pratiksha, C., Swati, S., Akhil, S., Sarath, M., A modeling approach to big data-based recommendation engine in modern healthcare environment, in: *Computer Software and Applications Conference (COMPSAC), 2015 IEEE 39th Annual; IEEE*, vol. 1, pp. 75–86, 2015.
15. Chrimes, D., Kuo, M.H., Moa, B., Hu, W., Towards a real-time big data analytics platform for health applications. *Int. J. Big Data Intell.*, 4, 2, 61–80, 2017.
16. Chawla, N.V. and Davis, D.A., Bringing big data to personalized healthcare: a patient-centered framework. *J. Gen. Intern. Med.*, 28, 3, 660–665, 2013.
17. McGregor, C., Big data in neonatal intensive care. *Computer*, 46, 6, 54–59, 2013.
18. Fahim, M., Idris, M., Ali, R., Nugent, C., Kang, B., Huh, E.N., Lee, S., ATHENA: a personalized platform to promote an active lifestyle and well-being based on physical, mental and social health primitives. *Sensors*, 14, 5, 9313–9329, 2014.

19. Aich, S., Sain, M., Park, J., Choi, K.W., Kim, H.C., A text mining approach to identify the relationship between gait-Parkinson's disease (PD) from PD based research articles, in: *International Conference on Inventive Computing and Informatics (ICICI); IEEE*, pp. 481–485, 2017.
20. Aich, S. and Kim, H.C., Auto detection of Parkinson's disease based on objective measurement of gait parameters using wearable sensors. *Artif. Intell.*, 117, 103–112, 2018.
21. Pfitzmann, A. and Hansen, M., A Terminology for Talking about Privacy by Data Minimization: Anonymity, Unlinkability, Undetectability, Unobservability, Pseudonymity, and Identity Management, 2010.
22. McCallister, E., Grance, T., Scarfone, K.A., *Guide to Protecting the Confidentiality of Personally Identifiable Information (PII)*, No. Special Publication (NIST SP)-800-122, 2010.
23. Kayaalp, M., Patient privacy in the era of big data. *Balkan Med. J.*, 35, 1, 8–17, 2018.
24. Mattila, J., *The Blockchain Phenomenon*, 2016 ed. Berkeley Roundtable of the International Economy, 2016.
25. Hesse, B.W., Nelson, D.E., Kreps, G.L., Croyle, R.T., Arora, N.K., Rimer, B.K., Viswanath, K., Trust and sTheces of health information: the impact of the Internet and its implications for healthcare providers: findings from the first Health Information National Trends Survey. *Arch. Intern. Med.*, 165, 22, 2618–2624, 2005.
26. O'Leary, D.E., Artificial intelligence and big data. *IEEE Intell. Syst.*, 28, 2, 96–99, 2013.
27. Kim, H.M. and Laskowski, M., Toward an ontology-driven blockchain design for supply-chain provenance. *Intell. Syst. Account. Financ. Manage.*, 25, 1, 18–27, 2018.
28. Tian, F., An agri-food supply chain traceability system for China based on RFID & blockchain technology, in: *2016 13th International Conference on Service Systems and Service Management (ICSSSM)*, IEEE, pp. 1–6, 2016.
29. Shae, Z. and Tsai, J.J., On the design of a blockchain platform for clinical trial and precision medicine, in: *2017 IEEE 37th International Conference on Distributed Computing Systems (ICDCS)*, IEEE, pp. 1972–1980, 2017.
30. Chhetri, S.R., Faezi, S., Rashid, N., Al Faruque, M.A., Manufacturing supply chain and product lifecycle security in the era of industry 4.0. *J. Hardw. Syst. Secur.*, 2, 1, 51–68, 2018.
31. Nugent, T., Upton, D., Cimpoesu, M., Improving data transparency in clinical trials using blockchain smart contracts. 5, 2541, F1000Res, 2016.
32. Matanović, A., Blockchain/cryptocurrencies and cybersecurity, threats and opportunities, in: *The 9th International Conference on Business Information Security*, pp. 11–15, 2017.
33. Maddux, D., Vice President, Kidney Disease Initiatives, Cybersecurity and Blockchain in HealthCare (Acumen Physical Solutions, 24 April 2017)

HibBucket, in: *Artificial Intelligence Blockchain Research and Development (HubBucket Healthcare Blockchain-HubBlockchain)*, 2017.

34. Onik, M.M.H., Miraz, M.H., Kim, C.S., *A recruitment and human resThece management technique using blockchain technology for industry 4.0. In: Smart Cities Symposium 2018 (SCS'18), 22–23rd April*, University of Bahrain, IET, Bahrain, pp. 11–16, 2018.
35. Al-Zaben, N., Onik, M.M.H., Yang, J., Lee, N.Y., Kim, C.S., General data protection regulation complied blockchain architecture for personally identifiable information management, in: *Proceeding of International Conference on Computing, Electronics & Communications Engineering 2018 (iCCECE '18)*, University of Essex, Southend, pp. 77–82, 2018.

10

Predictive and Descriptive Analysis for Healthcare Data

Pritam R. Ahire* **and Rohini Hanchate**†

Department of Computer Engineering, D.Y. Patil Institute of Engineering and Technology SPPU, Pune, Maharashtra, India

Abstract

In machine learning and statistical algorithm techniques, predictive analysis is used to predict the probable future that depends on chronicle data. The main aim is to predict the future by analyzing decades of data for prediction to provide the best evaluation for the future prediction to many real-time aspects. To perform statistical analysis, descriptive analysis is the first major step. To gain a better knowledge of distribution data, helps to identify outside average values or erroneous values, which enable us to classify associated variables to make statistical analysis. However, there are different types of summarizing and graphical representing tools that are available, which make confusion for data analysts to select which analysis method can be used. Data analysts performs various analyses, which is time-consuming, or skip the entire step of statistical analysis, which lead to erroneous decisions; predictive and descriptive methods for healthcare data analytics are the systematic ways of healthcare professionals in data analysis process to improve the management of healthcare system, like patient admission, diagnosis and expenditure. Data analysis for healthcare sectors are combination of real-time data along with chronological data to predict upcoming trends, which represents active insights for long-lasting growth. Prescriptive analytics is used as a backbone for the predictive model of hospitals. To predict future outcomes, predictive analytics plays a prominent role in healthcare groups, using prescriptive analytics nudges, and takes remedial action on those findings. It gives the healthcare company the power to influence the results. With prescriptive analytics, doctors can identify associated risk factors and provide solutions for the recovery of patients. Prescriptive analytics provide much improved results by using the decision-making process by eliminating unnecessary presumption.

**Corresponding author*: pritamahire33@gmail.com

†*Corresponding author*: rohini.shanchate@gmail.com

A. Jaya, K. Kalaiselvi, Dinesh Goyal and Dhiya Al-Jumeily (eds.) Handbook of Intelligent Healthcare Analytics: Knowledge Engineering with Big Data Analytics, (213–232) © 2022 Scrivener Publishing LLC

Data analytics gives status of patients admitted to the hospital and, in particular, how many got recovered in specified days in a month. Descriptive analytics process gives qualitative human readable reports of patient who are suffering or infected. To make big data into actionable descriptive form of data, descriptive analytics is performed by healthcare service providers to produce healthcare reports of occurred measures along with resources used or with new clinical diagnosis for patients. Due to this, it makes it very easy to identify diabetic patients, as per expected outcomes against government expectations, or identify areas for improvement on clinical quality measures or other aspects of care. In this chapter, pointed out that there are various healthcare sectors that are stuck in several steps of analytics, which should overcome this for future progress by investing in interoperable EHR technology and developing robust data monitoring system by using third-party entity for analytics services.

***Keywords*:** Machine learning (ML), predictive analysis (PA), descriptive analysis (DA), statistical analyses, outliers, healthcare systems, analytics, data modeling

10.1 Introduction

There are various categories of analytics, which are descriptive (i.e., identifying data and analyzing data sets), predictive (i.e., prediction on analyzed chronological data), and prescriptive (to get decision support). In a recent analysis, out of 92%, only 5% employed both analytics [1]. Figure 10.1 referred from [1] depicts the level of medical care research papers utilizing all examination type and shows that the illustrative investigation is the most ordinarily utilized in medical care (48%). The spellbinding examination was prevailing in other applications regions besides medical choice help. Among the application territories, pharmacovigilance concentrates in utilizing spellbinding examination as this application zone is centered on recognizing a relationship between antagonistic medication impacts with medicine. The prescient examination was utilized in 43% of medical research papers. Among the various applications, medical choice help had the most elevated utilization of prescient examination; the same number

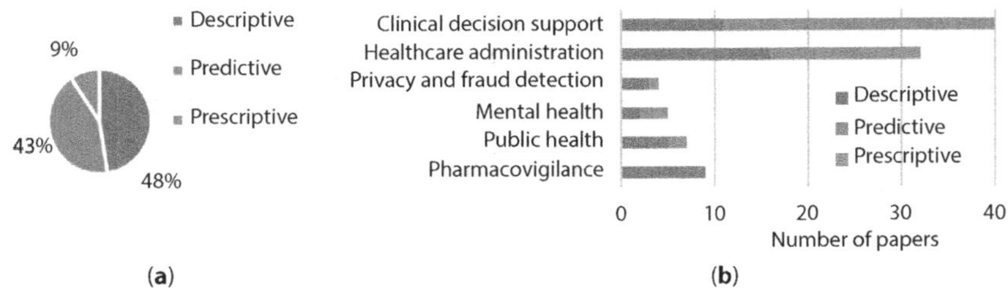

Figure 10.1 Different types of data analytics. (a) Percentage (%). (b) Types with application.

of studies here is engaged with danger and dismalness expectation of chest torment, cardiovascular failure, and different sicknesses. Conversely, utilization of prescriptive examination was extremely remarkable (just 9%); various research papers focus on population base or a specific disease scenario [2]. However, prescriptive analytics was originated in all healthcare systems. These investigations make an information data repository and a logical stage to encourage dynamic for various situations.

10.2 Motivation

Healthcare industry is undergoing major transformations; to address all challenges, a system is needed. There are three major accomplishments that are qualitative with cost-efficient and good results for *patients*. ACA aims to achieve efficient, safest, and cost-effective adoptive technology. Healthcare analytics implementation occurs at different levels, which includes tracking, resolving medical issues, data combinations, predictive modeling, and personalized modeling. Huge studies and researches have been conducted, which come up with mining huge data to analyze data, with various disputes and opportunities. In this, chapter mainly focuses on predictive and descriptive analysis models in healthcare, explains how these analytical models help to achieve efficiency and safety, and makes summarized report of viewpoints of the data analyst who set the visionaries in research areas of predictive models.

10.2.1 Healthcare Analysis

In the current era, the healthcare business strategies are moving from volume-based to value-based; to get better and efficient results, the medical workers like and nurses need to overwork. To generate outstanding outcomes and to make enormous improvements in the field of healthcare practice, certain changes in individual lifestyle and in entire system to prevent illnesses, diseases, and infections should be made. Huge amount of data is generated over the past years; managing large healthcare data is very challenging due to change in new emerging technologies, mobile apps, and simultaneously growing ratio of healthcare issues. Therefore, the healthcare data analytics tools are very important in order to mage the complex data, which improves healthcare industries and helps to opt best medical services to achieve efficient and accurate system, if proper utilization of tools is done.

With growing medical sectors producing huge amount of patient data in the like insurances, projected plan for treatments, etc., there are various dimensions of data mining healthcare applications that are addressed. However, the lack of an all-inclusive and organized description is motivated to construct a literature review. In this chapter, there is a summarized survey for big data analysis for medical healthcare analysis. To obtain optimal reports of medical healthcare, organized views are applied with analysis of meta-analyses (PRISMA) strategy, as we perform analysis of database between 2005 and 2016. There are various significant factors like medical healthcare categories and techniques, various types of data analytics, various data extractions, and various elements need to evaluated to get a systematic analysis implementation, which lead to the right direction in upcoming events as various studies mainly focus on analysis in medical data and medical admin level decision-making. By using manual, entered data is majorly considered for the approval of healthcare records in healthcare. However, data analytics depend on social activities, and other aspects are increased tremendously in past years. There is need of prescriptive analytics to perform and combine with domain expertise to enhance the research work in the area of decision-making [3].

As new technologies emerged in the era of healthcare analytics, medical clinics and centers are searching for another normalization of data procedures that will be suited to the development of data innovation and investigate an enormous measure of complex information. The proposed method is mainly focused on enhanced healthcare practice so the data can be used efficiently and smoothly for analysis purpose, handles the gaps of previous technologies and, in turn, handles challenges in big data and evades data failure, which will prompt improved care, to get cost-effective healthcare system, and helps healthcare expectation and avoidance frameworks and cost-effective. The proposed method gives promising results, which, in turn, is beneficiary to hospital managements if it is applied with time and cost-effective system; therefore, for the smooth conduction that can be conducted safely by adding true assistance to their job, which will prompt improved care, it helps illnesses expectation, defines healthcare framework, and decreases cost; there is a need for an efficient technique that provides facilitation and saves the efforts such as retrieval of old data, for effective discovery of hidden patterns, and data is kept in sorted logical form. Moreover, data analysts will be satisfied as they aim to achieve better decision-making process, which is necessary for advances in individuals' healthcare.

Analytics is one of the major approaches for the analysis of data. For the management of data which is based on oriented decision, the Centers for Medicare and Medicaid Services (CMS) [4] uses analysis to condense hospital readmission cost and to stop fraud payments. There are diverse uses of analytics—including data processing methods like mining and analytics—which help the medical doctors to predict and diagnose disease, which, in turn, results in improved quality of cost reduction. There are estimations from various studies that are made, which saves huge economies in the US; scholars have done various researches on analytical techniques to apply to different healthcare sectors.

10.2.2 Predictive Analytics

In predictive analytics (PA) process of predicting future trends, analysis of big data is performed. To serve customer as per their business requirements and to adopt their business strategies to achieve qualitative outputs, the various organization profits a benefit to detect fraud cases in inventory management [5].

PA is one of a major parts of machine learning. Regardless the various techniques, in that process, this includes rules that trains itself by having right to use to outcomes, for example, customer purchase prediction based on customer feedback.

To construct a model by performing prediction of future outcomes in the training algorithms, data such as time of day and weather are analyzed.

There are advanced analysis methods that come up with for future prediction. Predictive analysis is considered as one of the advanced analysis methods; predictive approach uses various events like ML and AI mining. Predictive approach uses data, algorithms, and ML concepts to classify the probable future prediction, which is based on chronological data. The main objective is to know what will happen in the future by providing the best assessment parameters.

10.2.3 Predictive Analytics Current Trends

PA method used the number of years and healthcare sectors turning to PA to amplify their growth, revenue, tougher economic conditions, and growing voluminous data to produce the most valuable insights to meet efficiency and usability [6]. These investigations make an information storehouse and a logical stage to encourage dynamic for various situations.

10.2.3.1 Importance of PA

Organizations focus on various PA approaches to solve the uncovered problems for new opportunities. Common uses include the following:

- Detecting fraud
- Optimal marketing promotion
- Improving event operations
- Eliminating or reducing risk

10.2.4 Descriptive Analysis

Descriptive method includes developing tables of mean, quintiles, variance, and standard deviation, and crosstabs or cross-tabulations are used to examine diverse hypotheses. Examined hypotheses are having differences among the subgroups. Numerous parameters are used to perform measures like subsets, discrimination, and inequality in the descriptive analysis. Decomposition and audit studies are used for measuring discrimination. Bundles of segregation types are needed to analyze either positive or negative outcomes; precise measures like time and space are the initial preconditions to perceive the processes.

A table of subgroup means may reveal substantial differences between subgroups, and this form of clear analysis often promotes causal deduction. When we see a profit gap, for example, we usually have to extrapolate the explanation for the existence of such instances. However, this approaches the realm of impact measurement, which necessitates the use of various techniques.

Two-way tabulation, also known as a crosstab, is a method of displaying the unit of propositions or cell proportions for two variables. For example, we can determine the proportion of the population that receives assistance by running a crosstab of education versus assistance receipt. The row proportions or fractions in each educational category that offers assistance, to see if you can find any trends, are examined. Higher education levels result in sharply lower levels of assistance. There must be a proportions column for the percentage of beneficiaries with different levels of education. We might get lucky. As a result, the proportion of recipients with a college degree is higher than the proportion of recipients who have graduated from college [6, 7].

10.2.4.1 Descriptive Statistics

Descriptive method is a process of organizing and summing data for ease of accessing data. In descriptive statistics (DS), its purpose is to explain knowledge than attempting to make inferences from sample to populations. We describe data in terms of samples. DS is not dependent on probability analysis.

10.2.4.2 Categories of Descriptive Analysis

DS provides data based on the following properties:

1. Measures of frequency: Frequency, count, percentage, and frequency of occurrence best fits for when responses defined.
2. Measures of central tendency: We calculate mode, mean, and median. The distribution is located using different points. When you want to demonstrate how an average or most frequently indicated answer works, use this.
3. Variations of differential measures: Limits, differences, and deviations. Interval is defined as spread values and range is given by high to low points. Variance is given range of scores.
4. Positions: To measure percentile and quartile. Depending on standard score, it analyzes where exactly the score falls; normalized score plus measuring positions; predictive and descriptive analyses for healthcare.

As the increased global business competitions and the need for sustainable growth for improved business decisions of organizations, PA helps to obtain insights of data. Healthcare organizations are more prone to using data analytics to ingest and find new data insights to apply. Medical and organizational problems are being tackled using innovative theoretical methods. We have summarized how PA leads to the beneficiary for different parts of the healthcare organization. Clinical research and drug development are two areas of study in the life sciences. Healthcare providers receive diagnostic assistance and clinical decision support.

- Providers of insurance: cost-cutting in healthcare and detection of fraud.
- Public health: tracks the state of healthcare and identifies disease outbreaks.
- Individuals: assist in real-time health surveillance and involvement in critical care situations.

In the descriptive analytics, there are a large number of users using different methods and software to fully comprehend what occurred in historical data and to interpret and categorize historical and organized data. Healthcare companies are increasingly focused on predictive approach, which analyzes past data by using historical data for prediction of future events and designing models using simulation and forecasts, which supports enterprise analysis. Clinical decision support systems and evidence- based medicine that examine trends, designs, deviations, peculiarities, and relationships in patients' clinical history knowledge through cultures and areas to identify any pandemic breakouts on a continuous basis are excellent examples.

Moreover, the young adults suffer from mental disorders like depression, hypertension in mid-aged persons, and asthma or respiratory diseases in pre-teens and teens (refer Figure 10.2 [8]).

Various healthcare advantages by using PA organizations depend on PA to have knowledge of effective treatments in the clinic and the ability to diagnose and personalize patient treatment efficiently and reliably; for the first time, they admitted that the patients' confidence increases in the healthcare system and the expensive readmissions decrease.

Use case scenarios:

- Critical care intervention
- Diagnostic assistance
- Clinical decision support
- Disease management
- Optimized healthcare costs
- Fraud detection and prevention
- Individual healthcare
- Readmission prevention

Majorly predictive models are categorized into two types of classification model, which predicts the members of class. For example, to classify

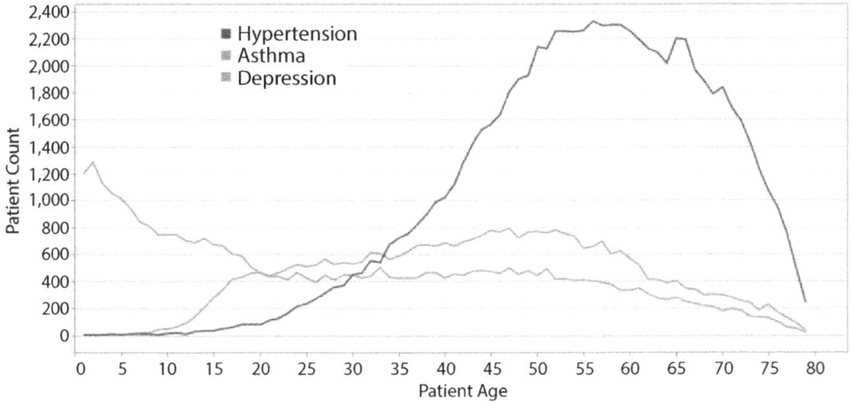

Figure 10.2 Disease categorization by age.

likelihood of person leaves, whether the person will respond to a solicitation, the results will be either 1 or 0 and the event of being 1 is targeted. Regression models forecast a numbers, for example, the calculation is done in numbers like revenue of the customer or the number of components will fail on the machine.

10.2.5 Method of Modeling

Decision trees are classification models that divide data into subsets based on input variables, allowing you to see how someone makes decisions. A decision tree is similar to a tree in that it has a root node, branches, and leaf nodes. Each internal node represents entropy and information gain factor, each branch represents the resultant entropy, and the leaf node represents the decision or class. In a decision tree, the data is examined and the variable that divides the data into logical classes is sought. The decision tree is a very useful tool that is simple to understand and can manage missing data.

One of the most widely used statistical methods is regression (both linear and logistic). Regression predicting is key dependent or independent variable in the large data set and identifies relationships among key variables to classify the key.

To model extremely complex relationships, neural networks (NNs) are preferred technology. NNs are very flexible and powerful as it represents relationship for all entities in data set. They function admirably when no numerical equation is realized that relates to the prediction of higher priority than clarification, inputs to outputs, or there is a huge training data.

10.2.6 Measures of Data Analytics

- Central Value (Tendency)
  It refers to concepts in which there is a number that summarizes all the entries of sets, and number represents a central value of the set.
- Mean
  Mean or average is referred to as the central values in data where whole data is spread out around the number, which is the only number used to estimate the values of whole data sets.
- Median
  Median is the process of dividing data into two equal halves, in which data is equally distributed. When data is arranged

in ascending or descending order, the number of data points on the right side would be the same as on the left side.

- Mode
  It is the maximum number of time that the data appears in the data set or the highest frequency of data.
- Spread/Dispersion
  Dispersion refers to the unpredictability measure within data.
- Standard deviation
  It is a measure of average variance among all available points and its mean value; the lower standard deviation value specifies that the data points are closer; otherwise, higher deviation specifies the data values that are dispersed over a large spectrum. During the collection process, samples or populations must be chosen.

  To calculate standard deviation of the population, it is given by sample standard deviation in Equation (10.1).

$$S.D = \sqrt{\frac{1}{n-1}\sum_{i=0}^{n}(x-\bar{x})^2} \tag{10.1}$$

  where x is the mean value.

- Mean Deviation/Absolute Deviation
  In the data set of values, it specifies the average of the difference between points and average of all data points of sets given in Equation (10.2).

$$M.D = \frac{1}{N}\sum_{i=0}^{n}(|x_i - \bar{x}| \tag{10.2}$$

- Variance
  The square of the average distance between each quantity and the mean is variance. That is, it is the standard deviation squared, which is given in Equation (10.3).

$$Variance = (S.D)^2 \tag{10.3}$$

- Range
  The difference between the lowest and highest data set value is defined by the range.

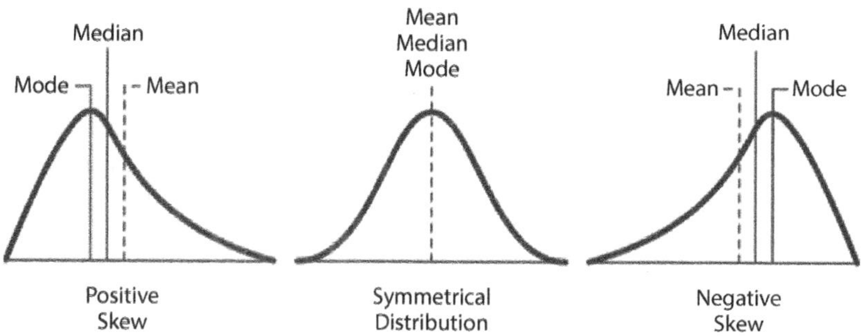

Figure 10.3 Disease categorization by age.

- Percentile
  It represents positions, and to calculate, it should be always in ascending order to calculate percentile.
- Skewness
  Skewness specifies a normal distribution, which specifies asymmetric values in the distribution; the normal distribution specifies zero skewness; the values can be zero or non-zero, not defined.

Figure 10.3 [8] shows the exact normal distribution, and the tail values represent exactly the mirror image of the other side. Negative skewness is, when a normal distribution is skewed to the left side of data points, the tails on the curve's left region are more than the tails on the right region, and the mean value is less than the mode. If the distribution is regular and if the data points are skewed to the right, then the tails on the right-hand side of the curve are longer than the tails on the left-hand side, and if the mean value is higher than the mode, then the skewness is positive.

10.2.7 Healthcare Data Analytics Platforms and Tools

When we compare traditional analytics and advanced analytics, the traditional analytics mainly focuses on data normalization; on the other hand, advanced analytics focuses on prediction of future based on analysis.

To majorly contribute for predictive methods with various advanced tools and techniques to get improved healthcare analytics [8]:

- Advanced Data Visualization (ADV): It is one of the advanced methods to visualize data since it ranges millions of data values for visualization and handles the data types; due to this, ADV differs from other standards charts; ADV is simple to use and allows researchers to conduct extensive data analysis. While retrieving medical data, ADV gives efficient and effective results of the analysis. ADV may also provide rich findings and fluid interactions to expose clinically hidden trends in data.

- Presto: In this, a large amount of data is gathered; it is based on a distributed SQL query engine. For healthcare sectors, such products play a very essential role to handle a large amount of data which is generated every milliseconds to analyze huge data; it may take many hours or days or years; by using Presto, data analysis is done in just a span of seconds.
- Hive: It is one of the programs designed to manage large amounts of data; it does not process and analyze data as quickly as Presto, but it efficiently performs all Excel tasks that do not require real-time performance; as a result, businesses can use both Presto and Hive for the best results, since Presto can access data stored on Hive.
- Vertica: It is more cost-effective architecture, which is used for a large amount of data; this is very similar to the Presto method. Also, this method reduces the operational cost and accelerates medical report analyzing patient patterns using Vertica features, which are scalable and cover all hospital data.
- Key Performance Indicators (KPI): It is a tool for assessing how well companies are carrying out their strategic goals. It improves the quality of healthcare for patients as it is used for specifying significant indicators that are to be observed and corrected and for identifying weakness. Also, it uses the electronic medical data analysis.
- Online Analytics Processing (OLAP): OLAP calculates data integrity checks, quality control, and monitoring services by using hierarchical and multidimensional structured data statistics. By providing a clearer monitoring of medical history and diagnoses, OLAP has the potential to enhance the healthcare decision-making system.
- Online Transaction Processing (OLTP): OLTP is similar to OLAP in that it is structured to manage operations such as patient registration, hospital records, and performance.
- The Hadoop Distributed File System (HDFS): Hadoop file processing mainly is used to divide large amount of data

and distributed across the systems; distributed file system reduces the data redundancy and mainly focuses on a feature built into the storage layer. It primarily focuses on patient recovery preparation, diagnosis aid, patient sign tracking, and fraud detection.

- Cassandra File System (CFS): It is one of the file systems in a distributed system like HDFS; this method performs the analytic operation with no point failure.
- MapReduce System: MapReduce technique breaks the large tasks into smaller subtasks to manage large data efficiently. MapReduce technique keeps track of all tasks that are being performed. Using MapReduce technique parallelization for many tasks is performed parallel with zero waiting time.
- Complex Event Processing (CEP): It is one of the emerging methodologies for healthcare sectors; if event occurs, then it changes the state, for instance, relating complex event of real time environment, EMR and HER processes will be improved.
- Text mining adds value to the healthcare system by allowing hospital emergency departments to review clinical data from the emergency department, record them, and respond to the call. Such issues may have an effect on both the quality and the cost of healthcare. As a result, text mining will include a treatment plan that develops certain standards and protocols to better understand the situation.
- Mahout: It is an Apache project that aims to create applications for Hadoop systems that support healthcare data analytics.
- JAQL: JAQL used to process large data set, and it is a functional query language. JAQL along with MapReduce works efficiently; it mainly focuses on translating high-level queries to low-level queries in parallel processing, which can be achieved.
- AVRO: AVRO offers data encoding and serialization as well as enhancing data structure.

10.2.8 Challenges

There are primarily three types of obstacles that must be overcome in order for health analytics to be successfully implemented and used: 1) technical

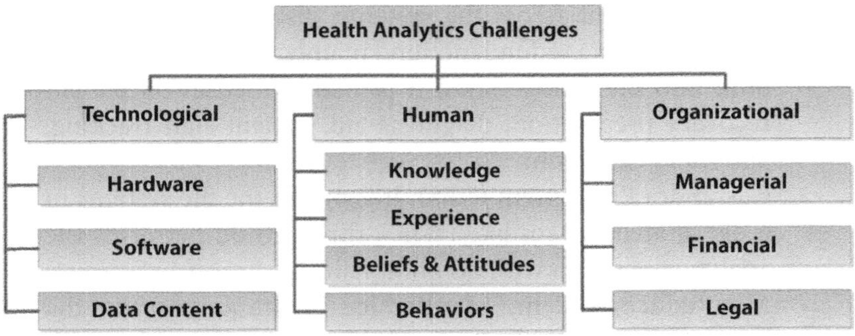

Figure 10.4 Challenges in healthcare.

difficulties involving hardware, software, and data; 2) human problems, including awareness, experiences, misconceptions, and beliefs about health analytics and end-user behavior; 3) organizational issues, such as managerial, financial, and legal impediments to optimum health analytics use. Since at least two of the 24 qualifying studies addressed each defined difficulty within the three categories, we only included 12 studies in the references. Figure 10.4 depicts all three categories and the issues that arise in healthcare analytics.

10.2.9 Issues in Predictive Healthcare Analysis

10.2.9.1 Integrating Separate Data Sources

To achieve higher qualitative analysis, PA should be dependent on the excellence and ease of accessing data. Data must be integrated to ensure an effective implementation for many sources, and they need to store and process the huge data before proceeding to PA solutions.

One of the major data resources in healthcare data is the US-based Electronic Healthcare Records, which leads to numerous data difficulties in integrating. In the market, there are a lot of different vendors, and the average hospital uses up to 16 different platforms. Moreover, the medical data are set aside separately from other departments of healthcare management [9].

10.2.9.2 Advanced Cloud Technologies

As the use of cloud storage is rising where huge data is stored on medical healthcare industries, they focus on security concerns likes privacy and security concerns [10].

As a result, healthcare data environments are often hybrid. There should be advanced tools that need to be developed PA, which is used to perform analysis of large data, which is to be carried out easily and securely.

10.2.9.3 Privacy and Security

Healthcare industries need to focus on protecting patient data along with maintaining patient privacy; organizations need to be extra careful in protecting it, as cloud service providers focus on all security majors.

All healthcare data need to store behind the firewalls and need to keep track, when the data will be accessed from the cloud.

10.2.9.4 The Fast Pace of Technology Changes

As the new emerging technology, the data are stored in the cloud need to be carefully updated. They should also be adaptable for new technology, new data sources, and systemic improvements, which are all on the horizon. A modular solution and adoptable technologies need to have for healthcare sectors [11].

10.2.10 Applications of Predictive Analysis

10.2.10.1 Improving Operational Efficiency

Currently, healthcare sectors are focusing on various industries investing in BI and data analytics software to boost performance and to get higher deliverables [12], e.g., real-time activities that can benefit from reporting to look for various functionalities to perform significant actions, if required, assigning more numbers of resources to get relevant output.

Healthcare sectors use advance tools to provide services and to study both historical and real-time patient data in order to come up with a better approach and gain a better understanding of patient real-time data and staff performance so that they can provide better service for the incoming patients; to add any new equipment is needed to deploy to get the services or the medical staff in a timely manner [13].

Due to the data analysis process, various healthcare sectors are improved to achieve an optimal staff-to-patient ratio with analysis. Using those solutions helps healthcare sectors to plan and to provide required services to the users.

10.2.10.2 Personal Medicine

In this field, it allows doctors or specialists to provide relevant treatment to the particular diseases. This is valid even though the illness is unknown at the time [14]. To give accurate modeling, PA allows healthcare sectors, but this is just the imaginary part like tip of the iceberg.

It is highly difficult for a single doctor to predict or to analyze the patient data and provide the medicine for the diseases because human bodies are very complex for different patients; it may be possible that those medicines work for the only group of patients; the PA tools helps to analyze and identify the correlations and hidden patterns for huge data sets to get efficient results and to get better treatments for the patients.

10.2.10.3 Population Health and Risk Scoring

For healthcare management, predictions and preventions go hand in hand. By generating risk scores with the aid of predictive analysis [15], healthcare sectors use PA to identify individuals with early disease prediction so that patients can prevent long-term health issues. Risk scores are evaluated on patient data and different laboratory tests to enhance risk evaluation, which helps to give the best analysis for patients.

10.2.10.4 Outbreak Prediction

Using standard prediction models like ML and AI tools are preferred at the government sectors. Predictive analytic models use both historical and real-time data and define possible development with different regions, cities, etc.

10.2.10.5 Controlling Patient Deterioration

In hospitals, patients face various problems like getting an infection of disease or sudden decrease in health condition; using predictive analysis, doctors can analyze patient condition, hospital management system can be managed well, and deterioration can be avoided [16]. To achieve effective results, using machine learning technology gives the most promising results for predicting clinical events.

10.2.10.6 Supply Chain Management

It is one of the costliest areas of healthcare, and using supply chain management, revenue can also be increased in healthcare services; SCM related to what kind of services will be provided what are essential parts of it. The purpose of supply chain insights is to evaluate patterns. of supply chain are to analyze patterns and to invest in predictive and descriptive analyses, which could lead to profits to healthcare service sectors [17].

10.2.10.7 Potential in Precision Medicine

Healthcare providers have access to millions of documents that they can use to find people who have experienced a similar reaction to a particular drug [18]. Since the data sets in question are huge, only machine learning–based PA solutions can reveal such insights. The data that is relevant to healthcare systems included gender, age, and location. Predictive analysis helps doctor to customize all medical treatments and practices, and it can lead to improved decision-making [19].

10.2.10.8 Cost Savings From Reducing Waste and Fraud

As per survey done, healthcare system costs more than $200 billion in the UK; there are various frauds and abuses per year, which includes performing various studies like fake billings and unnecessary treatments identifying [20] such issues, and medical health service providers aims to eliminate such frauds.

10.3 Conclusion

The major analysis on predictive and descriptive analysis is defined in this article with their tools that are used to perform data analysis process. Both predictive and descriptive analysis processes support decision-making for healthcare analysis, and predictive approach is to be considered more effective and gives best results for medical healthcare; it gives prominent results. Various applications of predictive analysis are defined further to improve healthcare care analysis with the various prediction approach can be used.

References

1. Bernardita Calzon Durcevic, S., *Business Intelligence*, datapine, 2020, https://www.datapine.com/blog/big-data-examples-in-healthcare/.
2. von Wedel, P. and Hagist, C., Economic Value of Data and Analytics for Healthcare Providers: Hermeneutic Systematic Literature Review. *J. Med. Internet Res.*, 22, 11, e23315, 2020.

3. Nussbaumer, P. and Burkholder, B., *The best predictive Analytics techniques*, 2020, https://blog.udacity.com/2020/09/the-best-predictive-analyticstechniques.html#:~:text=predictive%20analytics%20combines%20several%20data,%2C%20data%20mining%2C%20and%20statistics.
4. Dixon, M., Selerity-blog, 2019 46(98),*Selerity-blog*, https://selerityssas.com/blog/2019/12/12/types-of-predictiveanalytics-models-and-how-they-work/.
5. Narkhede, S., *Understanding Descriptive Statistics*, 2018, https://towardsdata-science. com/understanding-descriptive-statistics-c9c2b0641291.
6. Alkhatib, M.A., Talaei-Khoei, A., Ghapanchi, A.H., Analysis of Research in Healthcare Data Analytics, ISBN 978-0-646-95337-3 Australasian Conference on Information Systems, 2015 https://arxiv.org/ftp/arxiv/papers/1606/1606.01354.pdf.ACIS/2015/107
7. Raden, N., Amir, T.-K., Ghapanchi, A., Analysis of Research in Healthcare Data Analytics, Get analytics right from the start, 2010. file:///C:/Users/Mohammad%20al%20khatib/Downloads/X/Analytics-from-the-Start-WP.pdf. file:///C:/Users/Mohammad%20al%20khatib/Downloads/X/Analytics-from-the-Start-WP.pdf.
8. Wulff, F., Bertsimas, D., Kallus, N., From predictive to prescriptive analytics."Presto.", e-prints, 2013, https://prestodb.io/
9. Pešic, S., Stankovic, T., Jankovic, D., "Benefits of using OLAP versus RDBMS for data analyses in healthcare information systems". *Electronics*, 13, 2, 56, December, 2009.
10. Wongsuphasawat, K., Gomez, J.A.G., Plaisant, C., Wang, T.D., Shneiderman, B., Taieb-Maimon, M., "LifeFlow: visualizing an overview of event sequences". *Proc. ACM SIGCHI Conference*, ACM Press, New York, pp. 1747–1756, May 2011.
11. Khalifa, M., Macquarie University, Sydney, Australia. "*Challenges of Health Analytics Utilization: A Review of Literature, Health Informatics Vision: From Data via Information to Knowledge*. J. Mantas, (Eds.), IOS Press, 2019,
12. Roy, K., "Advances and scope in big data analytics and healthcare". *Curr. Trends Biomed. Eng. Biosci.*, 9, 2, 1–7, 2017.
13. King, I., Lyu, M., Yang, H., *"Online learning for big data analytics"*, 2013, Available: http://cci.drexel.edu/bigdata/bigdata2013/iee.bigdata.tutorial.1.slides.pdf.
14. Bertsimas, D. and Nathan Kallus , 2014 ,66(3), "From Predictive to Prescriptive Analytics",https://doi.org/10.1287/mnsc.2018.3253
15. Zhuhadar, L.P. and Thrasher, E., "Data analytics and its advantages for addressing the complexity of healthcare: A simulated zika case study example". *Appl. Sci.*, 9, 11, 1–16, 2019.
16. Deka, G.C., Big Data Predictive and Prescriptive Analytics, in: *Handbook on Big Data Pridictive & descriptive analytics for medical healthcare, 2014,* https://www.igi-global.com/chapter/big-data-predictive-and-prescriptiveanalytics/103222
17. Johansson, E., Krishnamurthy, C., Schlissberg, H.E., 2003 "Solving the solutions problem, McKinsey Q., 3, 40, 5, 116–125, 2003.

18. Shirsath, S.S. and Shubhangi, P., Disease Prediction Using Machine Learning Over Big Data. *IJIRSET*, 7, 6, 6752–6757, 2018.
19. Gopalakrishna Palem, Palem, G., The Practice of Predictive Analytics in Healthcare, 2013. https://www.researchgate.net/publication/236336250.
20. Kou, Y., Lu, C.-T., Sirwongwattana, S., Huang, Y.-P., "*Survey of fraud detection techniques*, pp. 7803–8193-9, 2014.

11

Machine and Deep Learning Algorithms for Healthcare Applications

K. France[1]*, A. Jaya[1] and Doru Tiliute[2]

[1]Department of Computer Applications, B.S. Abdur Rahman Crescent Institute of Science and Technology, Chennai, India
[2]Ştefan cel Mare University of Suceava, Strada University 13, Suceava, Romania

Abstract

In the healthcare system, medical images are playing a vital role to identify the symptoms of early diseases by using image patterns. In the past few decades, because of increasing advancements in the healthcare systems, it produces a large volume of imaging data with different modalities (MRI, fMRI, tomosynthesis, x-ray, computed tomography, etc.) and different dimensionalities like 2D, 3D, and 4D. So, there is a necessity to develop machine learning (ML) tools to manage these healthcare data. In the medical image diagnosing process, these ML tools are used to automatically identify the different disease patterns that appear from the various modalities of medical images. One of the key challenges in medical image diagnosis using ML tools is representing a medical image in the semantic space, and extracting effective features is the crucial step; it is called the semantic gap. In the past 20 years, there are a lot of enhancements in ML techniques to reduce the semantic gap in the diagnosis of medical images. Deep learning (DL) is one of the predominant techniques, which is extensively used to reduce the semantic gap in medical image diagnosis. DL is subset of Artificial Neural Network (ANN). It has many hidden layers to learn complex patterns with different stages of abstraction. Also, in the medical image diagnosing process, DL algorithms give more accurate result than a radiologist. Particularly, the Convolutional Neural Network (CNN) was developed to study medical images. This paper gives an overview of various DL algorithms used in medical image diagnosis. We aimed to provide the key reach areas, like classification of medical images, segmentation, disease localization, and image retrieval. This will help the researcher to identify the emerging trends, research obstacles, and possible future directions in medical image diagnosis.

**Corresponding author*: franceapm@gmail.com

A. Jaya, K. Kalaiselvi, Dinesh Goyal and Dhiya Al-Jumeily (eds.) Handbook of Intelligent Healthcare Analytics: Knowledge Engineering with Big Data Analytics, (233–254) © 2022 Scrivener Publishing LLC

Keywords: Deep learning, machine learning, medical image retrieval, medical image analysis, supervised learning, unsupervised learning, healthcare, disease detection

11.1 Introduction

The impact of digital technology on healthcare systems in the last two decades generates a large volume of patient data. However, there is a technology impact on all medical disciplines; the area of radiology is most affected than other disciplines because it is the only medical specialty fully dependent on technologies. Also, there are different ways the healthcare data is generated including sensors, databases, mobile applications, healthcare websites, and wearable technologies. This paradigm shifts in the radiology department; today's research is data-driven, and there are multiple solutions to a single radiological problem. These medical images are stored and analyzed by the physician to increase their clinical decision support. To analyze the huge volume of medical images and different modalities, involving tedious efforts by the medical experts and the expert analysis is subjective and may vary across different experts and leads to human error. Authors in [1] explained what are the challenges the radiologist should go through to adapt the digital evolution in the radiology department from film-based to film-less imaging analysis during their practice. Hence, it will be better if machines could involve storing and extracting medically high-risk factors by using programming and algorithms. In the radiology systems, we can reduce human error and will identify the correct diseases in the early stages, by using these machine learning algorithms, which will improve the diagnostic process and reduce the patient risk factors. With an increasing patient population in the world, there is a shortage of medical experts, which leads to focusing on improving the efficiency of clinical services using information technologies. This chapter provides an overview of emerging trends in medical image diagnosis using machine learning (ML) and deep learning (DL) algorithms.

11.2 Artificial Intelligence, Machine Learning, and Deep Learning

Computer science is a technology that used to solve more complex problems. It has many subfields called artificial intelligence (AI), ML, DL, and

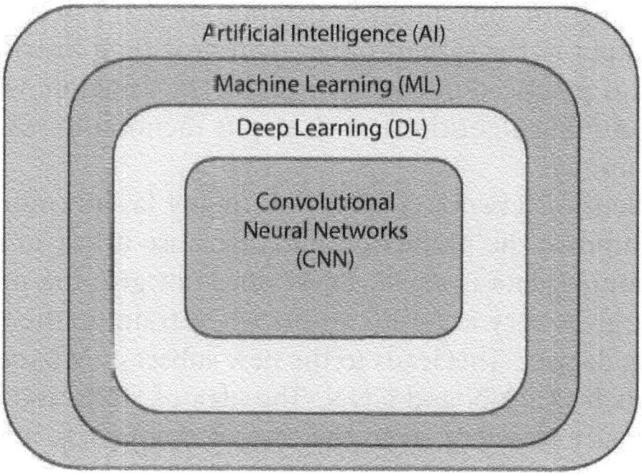

Figure 11.1 Schematic representation of computer science subfields.

Convolutional Neural Network (CNN). See the schematic representation of computer science subfields in Figure 11.1. The main aim of an AI machine is to create a system that can perform the task like human intelligence. The AI algorithms learn through experience by adjusting the new input. AI systems are used in many fields like economic, healthcare, education, and robotics. In the healthcare sector, the premature form of AI system is used to translate the physician decisions into a set of commands which systems can carry out the task. In clinical situations, it is not possible to convert the expert's instruction to convert the machine understanding language for various diseases. To overcome these difficulties, the ML techniques emerged in the 1990s. In ML algorithms, the rules are trying to learn from the examples instead of encoding by humans.

ML is a form of a subset of AI. This is done by creating computer programs that can be learned without explicitly planning to perform a specific task. The main purpose of ML is to learn through experience and achieve the task with less human assistance. The ML system learns and distinguishes the patterns from the training data given to the system during the training period. This is different to conventional computer programming algorithms, which need exact algorithms that describe individual steps that the program must execute.

The common purpose of conventional software engineering and ML technique is to solve a specific problem or set of tasks; they differ only in the way of approach to solve the problems. In a conventional software program, the output will depend on the input data. This is done logically, generally with if-else statements, Boolean operators, and while loops. ML algorithms vary from conventional operations in that it provides x as input and y as a output to the system rather than providing instructions on (f) function and is practiced to control or guess the function, while in traditional programming algorithms, each step is mention to solve a problem by the programmer.

A Deep Neural Network (DNN) have many layers (more than five), which will improve the prediction data. Because it has so many layers, when the training data is small, DNN does not give the more accurate result, and it gives very accurate results when training the model uses a large training dataset. This leads to the new subset of ML tools called DL techniques such as DNN and CNN. The idea of DNN and CNN is that rather acting as a classifier, as in the case of classic ML, it also acts as a feature extractor. DNN and CNN techniques allow for end-to-end training because they are recognized and output from the input data; these techniques are widely used in the image segmentation and classification process [2].

11.3 Machine Learning

Solving problems through computer programming methods using training data is called ML. In the past few decades, the growth of digital data generated in various fields increases rapidly. Computational techniques are needed to analyze such data especially in the field of radiology, which produces numerous clinical images every day around the world.

Several types of ML methods are used to store and investigate hidden features from the large volumes of medical image databases. The ML algorithms are largely classified as four types; see Figure 11.2.

11.3.1 Supervised Learning

The input data is also called the training dataset and has unique labeled values for all the input data. In supervised learning techniques, first algorithms are trained using input data and their ground truth labels; this training process will continue until the algorithm predicts accurate results. These supervised learning algorithms are classified as follows:

Classification: Classification techniques perceive the result based on a trained database, where the output variables are in different categories. In classification, the algorithms are trained using input data along with their

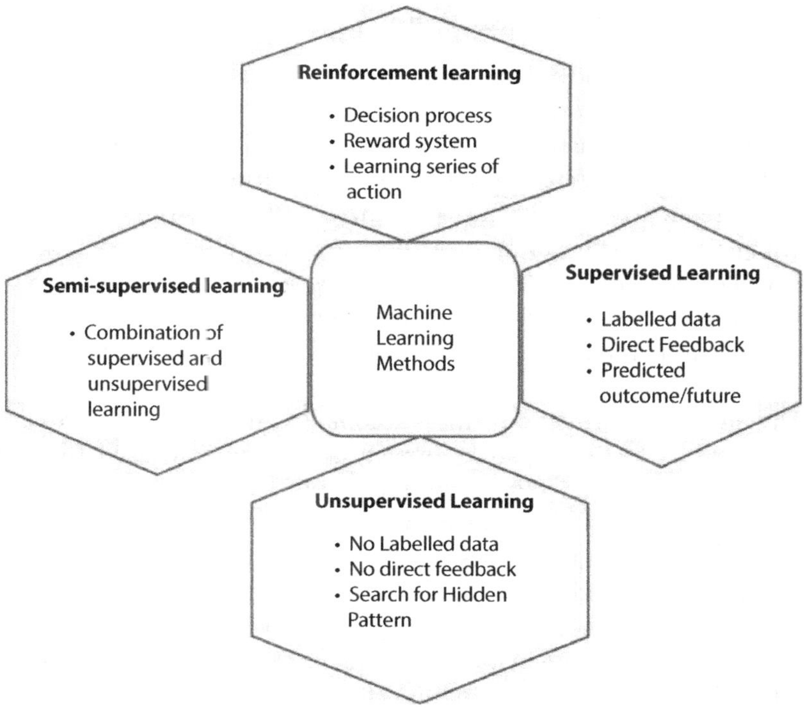

Figure 11.2 Methods of machine learning algorithms.

labeled data. Classification techniques such as logistic regression, k-nearest neighbor (KNN), Naïve Bayes, and support vector machine (SVM) use the decision boundaries to differentiate different categories of training data using their unique feature pattern. An example of real-world use of classification techniques used in the process of diagnosis is to classify a person as healthy or not based on their symptoms.

Regression: The regression is similar to the classification technique. The regression methods differ from the classification methods, only in the output values. Instead of categorizing the output data, the regression techniques will give the actual value of the output variable. Examples of regression models are linear regression, ensembles, polynomial regression, and neural networks.

Forecasting: In the forecasting method, the output result is based on past and current data. It is also called time-series prediction.

Ensemble: Models combine the different types of ML algorithms to predict the output of new input data. Sometimes, the model itself becomes the features of this method.

11.3.2 Unsupervised Learning

The system will decide by itself rather than training based on some dataset. Labeling details are not given to the training system to predict the data. This unsupervised learning method extracts the hidden pattern using features from the given data. Clustering association rule learning and dimensionality reduction methods fall into these categories. The unsupervised learning algorithms never predict the correct output to the data; instead, they will classify the data to which categories belong to.

Clustering: Clustering is the process of combining the data points that are similar in nature; also, it differs from the other cluster groups of data points.

Dimensionality reduction: This is accomplished based on feature choice and feature extraction. Feature extraction methods are used to reduce high-dimensional data into low-dimensional space. Feature extraction approaches are used for dimensionality reduction. The dimensionality reduction technique creates a set of new features which have fewer dimensions compared to the input data. This will allow for easier visualization; also, it reduced searching time and storage space. The Principal Component Analysis (PCA), Singular Value Decomposition (SVD), and Hidden Markov Models fall into this category.

11.3.3 Semi-Supervised

In semi-supervised learning methods, input data has a combination of unlabeled and labeled data. It is learning to organize the obtained sample data and make the prediction. Since the real-time data is a combination of labeled and unlabeled data, this method is widely used to analyze the real-time data.

11.3.4 Reinforcement Learning

Reinforcement learning aims to create an intelligent agent (RL-agent) that works with the dynamic problem environment. The RL-agent learns like humans through trial-and- error methods. Model feedback is provided as a reward for success and as a punishment for wrong decisions made in complex situations. RL-agent will learn by reviewing past actions that have been rewarded. Reinforcement learning is not like a supervised learning method; in the reinforcement learning method, the proper input data are not provided. In the real world, these techniques are used to create autonomous vehicles; the agent needs to consider features of the location including recent speed, road impediments, nearby traffic, and operator controls [3, 4].

11.4 Advantages of Using Deep Learning on Top of Machine Learning

Precise diagnosis of the diseases depends on the procurement of image quality and understanding of the medical images by the radiologist. Today, world medical imaging machines have updated and producing high-resolution images in different modalities like x-ray technology, computed tomography (CT), and MRI. The feature extraction technique in the image interpretation using the traditional ML algorithms fully depends on the expert's knowledge. Since the expert knowledge is subjective, the same patient image may be interpreted in different ways by the different radiologists. So, the traditional ML methods do not have consistency in disease diagnosing. The ML algorithms are advanced from the handcraft features extraction techniques to the automatic feature extraction from the images. This approach is called the DNN, which has different layers of neurons each layer fully connected to the other layers. The DNN takes multiple inputs and combines them with the weights and generates the outputs.

11.5 Deep Learning Architecture

The AI is an efficient technology utilized in the contemporary healthcare area. The advancement in medical image acquisition devises a massive number of medical images, which is produced in the field of the radiology department. With the present success of AI, the radiology department is using AI to manage and analyze these medical images. Based on the radiologist's input, AI techniques analyze and provide the details that can help the radiologist during the decision-making process. The overall neural network structure is exhibited in Figure 11.3. The AI neural network architecture consists of input and output layers and more than one hidden layer; each layer has many neurons all the neurons are interrelated to each other; every neuron combines all the inputs and uses an activation function to give output to the following layer. Hence, the addition of more hidden layers will allow a system that affects more difficult problems because it is capturing the non-linear relationships. This kind of network architecture

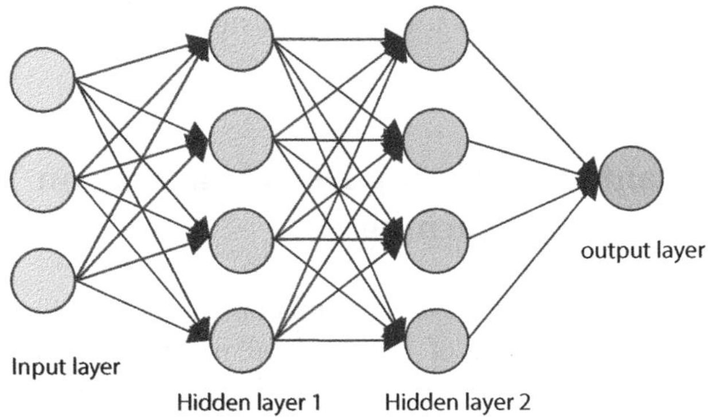

Figure 11.3 Neural network architecture.

is called a DNN; see Figure 11.4. The DNN methods are having capable of performing better than human beings, particularly in recognizing the early symptoms of various diseases pattern.

The DNN architecture has many layers and neurons to create a classified representation of the feature. When training these deep network structures use a large amount of data, it will learn all possibilities of outcome and makes intelligent predictions. Therefore, DL has a profound influence on the computer vision field and medical image analysis. Different DL methods are proposed in the last decades like CNN, DNN, DBN, dAs, DBM (DC-ELM), and RNN and their alternatives like BLSTM and MDLATM. Table 11.1 shows their advantages and disadvantages [5, 6].

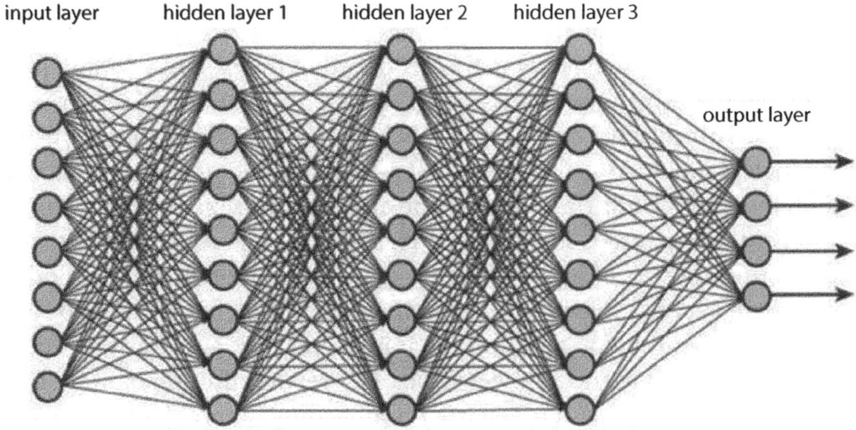

Figure 11.4 Deep learning architecture with multiple layers.

Table 11.1 Comparison of different DNN architectures.

Name of the network	Network architecture details	Advantages	Disadvantages
Deep Neural Network (DNN)	DNN has many hidden layers (more than three). It will allow the network to capture the nonlinear relationships.	It gives more accurate results.	The vanishing gradient problem affect the training of the DNN methods while doing the backpropagation.
Convolutional Neural Network (CNN)	It is performing well with the two-dimensional data.	The model performance is good; it learns rapidly.	During the training process, we need to provide the labeled data for the classification.
Recurrent Neural Network (RNN)	The RNN learns continuously and the weights are shared to the neurons.	Learn consecutive events; it has time dependencies; it has many variations of architecture.	It has the vanishing gradient problem, and it requires a large dataset to train the model.
Deep Conventional Extreme Learning Machine (DC-ELM)	The model uses the Gaussian probability function for the sampling of local connections.	It gives high accuracy and the training model is easy and very fast.	The weight initialization should be effective when training the model which has small sample data and labeled data.
Deep Belief Network (DBN)	DBN has only one direction of connection in its architecture does not have a backpropagation and is used in the machine learning model.	Since it is used in all the layers, the interpretation of the layers and increases the system accuracy.	Computational cost is high when training the model because of the initialization of weight.
Deep Autoencoder (dA)	This method is mainly used for feature extraction and dimensionality reduction using an unsupervised learning method.	Labeling data not required; it has a different architecture model	This model needs to be pre-training.

DL has a more number of hidden layers to learn high-level features in the absence of individual human involvement and gain the knowledge of the domain database. The CNN is extensively used in analyses of medical images and object detection. It is contradictory to a conventional multilayer perceptron (MLP), which has fully connected layers, to reduce the complexity of network structures; CNNs are using the convolution filters while obtaining the image information from the segmented images. The CNN consists of various components such as an input layer, an output layer, convolutional layer, pooling layer, and fully connected layers.

11.6 Medical Image Analysis using Deep Learning

Suzuki *et al.*, [7], overviewed the changes that occurred in the ML field after the DL techniques emerged in the medical imaging field. Also, they compared the variation between traditional ML with handcraft features and DL techniques. Conventional handcraft feature-based ML techniques are subjective and, for the same image, different people can select the different features based on their experience to get the accurate result. This method has given the more accurate result before the DL method involved in the medical image analysis.

The DL techniques mainly replace the conventional ML techniques by learning the features directly from the images and objects. Also, it gives more accurate results than conventional ML methods, because. They compared the two different techniques: CNN and Massive Training Artificial Neural Network (MTANN); among these two methods, the MTANN is performing better than CNN; also, it requires fewer images to train the MTANN model, whereas CNN needs a huge amount of images to train the model and it takes more time to train the model. In medical image analysis, the combination of conventional ML and DL technology is going to dominate the next few decades. Ke Yan *et al.* developed different CNNs to classify the frontal chest images whether it is normal or abnormal.

It will help to train the clinician and radiologist in terms of identifying the normal and abnormality in the chest x-ray images [8].

In [9, 10], the authors provide the current state of application of big data and DL in a various fields other than the computer vision like education, NLP, and finance; particularly, they concentrate on DL implementation in various healthcare data like radiology images and non-medical imaging data (electronic healthcare record). In [11], the authors reviewed the AI techniques used in three types of diseases: cancer, neurology, cardiology, and stroke care.

In [12–14], the authors introduced the CNN, which is highlighting the clinical point of image analysis. The benefit of using CNN in healthcare data is that the hidden pattern and features of the data are automatically identified without explicitly mentioned by the expert. These papers focused on the medical image classification, disease localization, detection of disease patterns, segmentation of images, and registration. In [15, 16], authors exposed how the new ML and DL techniques are adapted into the radiology department. Also, they highlighted challenges that occur when implementing these technologies in the regular radiology practice.

11.7 Deep Learning in Chest X-Ray Images

Since there is no acceptable public biomedical image database, it is not possible to recognize the best disease detection method among the existing algorithm, which is suitable to find the different abnormalities. In [17], Mohammad Tariqul *et al.* used Deep CNN (DCNN) to identify the different abnormalities from three different databases (chest x-ray, JSRT, and Shenzhen dataset); all the databases contain only the frontal chest x-ray images. They found that the single DCNN architecture did not suitable to find the different abnormalities from the chest x-ray. Also, the DCNN methods are extracting the new features from the images, which are not considered in conventional ML algorithms. Authors in [18] proposed CNN-ATTENTION-LSTM (CAL) to identify different lesions and classification of multiple diseases from single chest x-ray images. This method gives 85.4% of precision and recall value to identify atelectasis and infiltration diseases. They used the Conv5-layer network architecture for feature extraction purposes.

In [19], Pranav Rajpurkar *et al.* created an algorithm called CheXNet, which has 121 layers of CNN to detect pneumonia from the chest x-ray images. They used the chest x-ray14 dataset for their experimental purpose. This algorithm performs and shows better results than training-level radiologists. But this algorithm works only in the frontal view x-ray images. Also, they proved that, with the small changes in the algorithm, it can detect all 14 types of thoracic diseases. To evaluate the radiologist's performance to identify the pulmonary nodules from the chest images with and without the assistance of DCNN software, they selected 12 radiologists from four hospitals and they have randomly collected 600 lung cancer images and 200 normal images for their experimental studies. All the images are separately reviewed by the radiologist and marked the suspicious region of images. With help of DCNN methods, the radiologist has reviewed the same images again. The final results indicate that the radiologists perform well with the assistance of the DCNN method. The results show that performance was increased 5% after using the DCNN method. This shows that, when we integrate the human knowledge with the DCNN model, it gives a better result than using the algorithms alone [20].

To identify tuberculosis, they used a DL CNN and radiology augmented approach. They implemented four databases, which are collected from different education institute databases and NIH databases. DCNN method gives the more accurate result AUC of 0.99. Eui Jin *et al.* [22] studied a DCNN algorithm that is identified automatically detection of various thoracic diseases. They developed a new CNN, which consists of a sequence of a dense block with five classifiers. The first four classify is trained to classify the major diseases, and the final classifier is to find the abnormality from the images. Also, they used two types of loss function classification of image and localization of different types of diseases. The algorithm gives an AUC of 0.979 for image classification and localization of 0.923. The algorithm gives better performance than the physician.

In [23], the authors studied different CAD systems used in chest x-ray image processing techniques like segmentation, image classification, and identification of specific and multiple diseases. The traditional algorithms are identified and classify the only diseases; DL algorithms can detect the appearance of the pathologies. The handcraft features are affecting the accuracy of the classification, but the SVM and random forest algorithm are performed better than the other rule-based methods. DL methods give the more precise result by learning features directly from the images. Authors in [24] proposed an Attention-Guided Convolutional Neural Network (AG-CNN); in this algorithm, they combined both global and local features to diagnose chest pathologies. AG-CNN increases diagnosing accuracy through altering image orientation and reducing the influence of noise. They experimented on the ChestXray14 dataset. This method gives the best results compare to the other standard methods.

In [25], the authors introduced a new approach called category-wise residual attention learning to classify the multiple disease classification. This method has a two-step process; the first part can combine the image features, and the second part focuses the attention based on the important image features. It gives around 82% of accuracy in classifying multiple diseases. In [26], the authors developed an integrated approach, which increased the disease detection and identification of a disease-affected region in the images, which is used by a radiologist to understand better the disease pattern. In this CNN method, it split the images into a different patch to detect disease information. Then, extracts the features from the images to classify the images and identify the location of disease affected place. They used multiple instance learning methods to find the disease from each patch. If the image patch does not have the disease pattern, then it is labeled as the disease-free path. Using this pattern, it will find the disease location from the images.

Hyo-Eun *et al.* [27] proposed a multiple-instance learning method to identify the abnormality region, and a weakly supervised Deconvolutional Neural Network for semantic segmentation is used to reduce the false positive rate when lesion segmentation in tuberculosis detection using chest x-ray images. Many works have been done to identification and place disease from the chest images, but there is limited work that has been done in the area of automatic image annotation based on the image condition. In [28], Hoo-Chang Shin *et al.* proposed the CNN model to extract the features from the chest x-ray images and RNN used to describe the image affected area and their severity. Qingji Guan *et al.* [29] proposed the category-wise residual attention learning model to classify the multi-label chest pathology. The main aim of this work is to give more weightage to the relevant pathologies which appear in the chest x-ray images. It has two parts: the first part feature is extracted using a CNN; in the second part, attention module focuses on assigning the weights to the different features of different categories.

Zongyuan *et al.* [30] proposed a pair of different error functions that can fit into any DL model to classify the multi-label chest x-ray disease diagnosis. Generally, the chest diseases are visually similar and more than one disease can present in a single scan image. In multiple-label classification DL networks, the output layers are trained for individual diseases. Also, they did study the label association among present and the absent classes. In this work, they suggested the Multi-label SoftMax Loss (MSML) and Correlation Loss (Cor-Loss), to report the properties of various labels and visually similar data.

11.8 Machine Learning and Deep Learning in Content-Based Medical Image Retrieval

Medical images are playing a vital role in the diagnosis process. The digital technology development is evident that improved the radiology department a lot. Because of this digital revolution in the radiology department, it produced the different modalities (x-ray, MRI, and CT) and high-resolution medical images every day and improved the diagnosis accuracy. Although, the method of medical image understanding has taken place recently to get advantages from computer technology. Generally, the radiologists will look at all the images one by one and generate medical reports during the diagnosis process. However, it is very difficult to see all the medical images one by one and create the report by the radiologist. Since this process is subjective, there will be a huge difference between the radiologists. To overcome this issue, there is a necessity to create the computational tools, which are used to store and analysis these medical images, it will help to reduce the radiologist's time and increase their diagnostic accuracy. In the past three decades, there have been many researchers who developed automatic ML tools to analyze medical images and implement them in the daily clinical workflow system.

There are many CAD systems that are improved based on the new technology to assist the radiology, in the recent day; the Content-Based Image Retrieval (CBIR) is one of the emerging fields in the radiology department, which is used to manage to store and retrieve from the medical image repository. The CBIR methods help the radiologist to increase the accuracy of their decision-making process during the diagnosing time by retrieving similar images from the database based on the input image. Also, there is a problem in the CBIR system, which is to retrieve similar disease patterns from the stored database images. There is a difference between human understanding and machines learning process of images. Humans understand the images from the semantic space, but ML algorithms understand the image based on the details provided by the humans. The accuracy of retrieving images depends on the representation of images in the semantic space. It is a key challenging task to represent the images in semantic space. The common block diagram of the CBIR system is shown in Figure 11.5. References [21–35] provide more information about the various feature extraction techniques and different image features used in the CBIR systems to extract database images that are similar to the input image. The image visual features are grouped into three categories: low-, mid-, and high-level features. Although there have been many CBIR systems

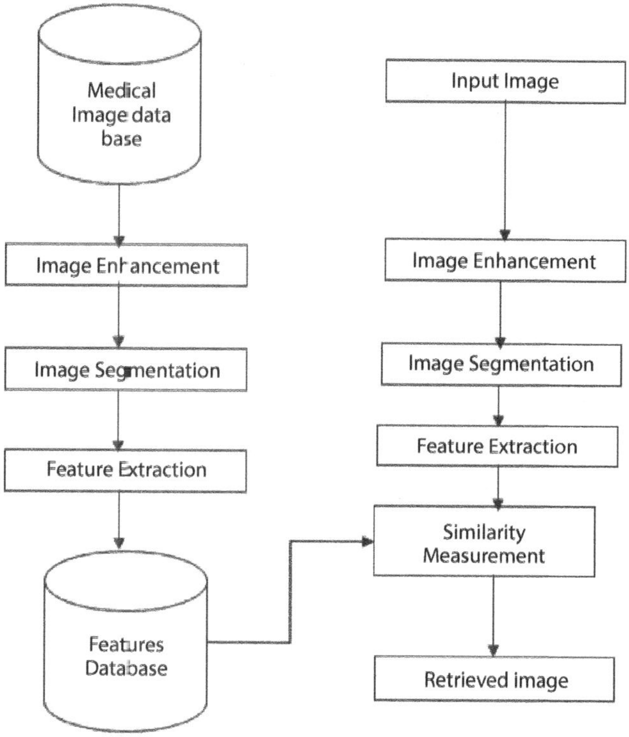

Figure 11.5 Block diagram of the CBIR system.

developed by the researchers, only a few systems are used in real-time hospital environments like high-resolution computed tomography images (ASSERT) system, Spine X-rays (MIRS), IRMA for the different images, and biology images (Bioimages).

In the CBIR system, the image representation is playing an important role, because the image representation methods directly affect the accuracy image retrieval system. The key tricky in an image retrieval system is that gap between human understanding of image and ML process it is called the semantic gap. Authors in [36] identified different gaps in the CBIR system and grouped them into four groups, as shown in Table 11.2. When using the DCNN techniques, it will excerpt the features from the database images and extract the primitive and high-level image features. Digital Imaging and Communication in Medicine (DICOM) is a common system; it is used to store and analysis of medical images. In [37], to reduce the semantic gap, the authors used the DCNN algorithm that can pre-filter the DICOM image metadata. This algorithm not only reduces

Table 11.2 Different gaps in CBIR systems [36].

	CBIR system aspect		
Gaps heading	**Slighter understanding**	**Gap title**	**Fuller realization**
Content	Quantifiable, understanding of machine	Semantic	Images are analyzed by the humans.
Input data	Features are physically selected	Feature extraction	Algorithms will select the appropriate features.
	Only the whole structure is identified	Structure	Global and local image objects are predictable.
	Image details are managed at one scale	Range of image scale	Visual detail managed at multiple scales.
	Input data not used spatial and time dimensions details.	Space + time dimension	All longitudinal and time details used to calculate the suitable features.
Performance	Not executed; theoretical only	Real time application	Feasible to executable through internet.
	System unconnected	Combination	Information is exchange with the all the system.
	Linear search used in the database	Indexing	Data organized in spatial trees. Data are clustered and maintained by correspond to hardware construction.
	No assessment	Assessment	Quantifiable valuation with original data.
Usability	Query by text only	Query	Query by text, feature of numerical description.
	Results are fine-tuned without feedback from the system	Feedback	Results include a similarity measure and help the human to identify the reason of fine-tuning.
	No modification capability is provided	Refinement	The system learns the user's preferences through feedback.

the semantic gap but also decreases the search space. It gives a precision value is 0.80.

To support the eardrum diagnosis process, authors in [38] developed the CBIR system using a digital otoscope CNN model for otoscope images, which is extract the normal, middle ear effusion, and tympanostomy tube conditions to process the images. Most of the existing methods in the CBIR system are developed based on a particular modality or particular dimensional of image data. In [39], the authors analyzed the medical CBIR in the retrieval of various collections of different modalities of 2D images. Also, authors investigated nonimage medical data instead of visual features to retrieve the database image. There is a lot of research that has been done to reduce the semantic gap that occurs a difference between image features extracted by ML algorithm and semantic perception by human beings.

In recent times, the development of DL techniques is widely used in the CBIR methods to reduce the semantic gap by extracting features directly from the images. It gives more accurate results than the traditional ML algorithm [40].

The traditional handcraft features do not perform well in the classification of a different modality of image database. To overcome this problem, in [41], the authors proposed an enhanced residual network (ResNet) to classify and retrieve various types of modalities of images with different dimensionality of images. They experimented on 12 databases and achieve the accuracy and F1 score, respectively, 81.5% and 82.42%, which are higher than the already existing methods.

11.9 Image Retrieval Performance Metrics

The CBIR image retrieval systems evaluated using precision, recall, F1 score, mean average precision (MAP) given by Equations (11.1), (11.2), and (11.3) [42].

- Retrieval time: How much time the system will take to give the output images based on the input image.
- Precision: The ratio between the how many related images from the total number of retrieved images.

$$\text{Precision} = \frac{\text{Number of correct retrieved images}}{\text{Total number of retrieved images}} \tag{11.1}$$

- Recall: The ratio between relevant output images with the number of similar images in the database.

$$\text{Recall} = \frac{\text{How many relevant images retrieved}}{\text{Overall similar images in the image database}} \tag{11.2}$$

- F1 score: The symmetrical mean of precision and recall.

$$\text{F1 Score} = \frac{2 \times \text{Precision} \times \text{Recall}}{\text{Precision} + \text{Recall}} \tag{11.3}$$

11.10 Conclusion

Different types of imaging modalities in healthcare applications (MRI, fMRI, Tomosynthesis, x-ray, CT, etc.) and different dimensionalities (like 2D, 3D, and 4D) are available. To analyze this variety of data, various ML and DL algorithms are used such as supervised, unsupervised, semi-supervised, and reinforcement learning, and CNN is used to improve the disease detection and early disease diagnosing process using various performance metrics like total time taken to retrieve the similar images from the database, recall, and precision, F1 score, and MAP. In this chapter, we reviewed various algorithms that existed: AI, ML, and DL algorithms used in medical image diagnosis, like early disease identification, localization, image classification, and CBIR. From this survey, we concluded that DL algorithms give more precise results compared to AI and ML algorithms. Only a few projects have been done in multimodality and different disease classification using DL algorithms. There is a research scope in various DL algorithms to improve their accuracy in earlier disease identification, disease localization, and content-based medical image retrieval (CBMIR) using multimodality and multidimensional image data.

References

1. Bruce, R., Siegel, E.L., Siddiqui, K., Evolution of the Digital Revolution: A Radiologist Perspective. *J. Digit. Imaging*, 16, 4, 324–330, December) 2003.
2. Ahuja, A.S., The impact of artificial intelligence in medicine on the future role of the physician. *PeerJ*, 7, e7702, 2019.

3. Sweety Bakyarani., E., Dr. Srimathi., H., Bagavandas, Dr. M., A Survey of Machine Learning Algorithms In Healthcare. *Int. J. Sci. Technol. Res.*, 8, 11, 288– 2292, November 2019.
4. Latif, J., Xiao, C., Imran, A., Tu, S., Medical Imaging using Machine Learning and Deep Learning Algorithms: A Review. *International Conference on Computing, Mathematics and Engineering Technologies – iCoMET*, 2019.
5. Razzak, M.I., Naz, S., Zaib, A., Deep Learning for Medical Image Processing: Overview, Challenges and the Future. *Classification in BioApps. Lecture Notes in Computational Vision and Biomechanics*, vol. 26. pp. 1–30, Springer, Cham, 2018. https://doi.org/10.1007/978-3-319-65981-7_12
6. Datta, S., Barua, R., Das, J., *Application of Artificial Intelligence in Modern Healthcare System*, Alginates - Recent Uses of This Natural Polymer, Intech open, December 2019. http://dx.doi.org/10.5772/intechopen.90454.
7. Suzuki, K., Overview of deep learning in medical imaging. *Radiol. Phys. Technol.*, 10, 257–273, 2017.
8. Tang, Y.-X. and Tang, Y.-E., Automated abnormality classification of chest radiographs using deep convolutional neural networks. *NJP Digit. Med.*, 3, 70, 1–8, 2020.
9. Bote-Curiel, L., Muñoz-Romero, S., Gerrero-Curieses, A., Rojo-Álvarez, J.L., Deep Learning and Big Data in Healthcare: A Double Review for Critical Beginners. *Appl. Sci.*, 9, 2331, 2019.
10. Wang, F., Wang, S., Jiang, X., Dudley, J.T., Miotto, R., Deep learning for healthcare: review, opportunities and Challenges. *Briefings Bioinf.*, 19, 6, 1236–1246, 2018.
11. Jiang, F., Jiang, Y., Zhi, H., Jiang, F. *et al.*, Artificial intelligence in healthcare: past, present and future. *Stroke Vasc. Neurol.*, 2017.
12. Ker, J., Wang, L., Rao, J., Lim, T., Deep Learning Applications in Medical Image Analysis. *IEEE Access*, 2017.
13. Lundervold, A.S. and Lundervold, A., An overview of deep learning in medical imaging focusing on MRI. *Med. Phys.*, 29, 102–127, 2019.
14. Kim, M., Yun, J., Cho, Y., Deep Learning in Medical Imaging. *Neurospine*, 16, 4, 657–668, 2019.
15. McBee, X.P. and Awan, A., *Deep Learning in Radiology*, Journal of academic radiology, March 2018. doi: 10.1016/j.acra.2018.02.018.
16. Mazurowski, M.A., Buda, M., Saha, A., Bashir, M.R., Deep learning in radiology: an overview of the concepts and a survey of the state of the art. *J. Magn. Reson. Imaging*, February 2018.
17. Islam, M.T., Aowal, Md A., Minhaz, A.T., Abnormality Detection and Localization in Chest X-Rays using Deep Convolutional Neural Networks. *International journal of Computer Vision and Pattern Recognition*, 1–16. arXiv:1705.09850v3 [cs.CV] 27 Sep 2017.
18. Xu, S., Guo, J., Zhang, G., Bie, R., Automated Detection of Multiple Lesions on Chest X-ray Images: Classification Using a Neural Network Technique

with Association-Specific Contexts. International Journal of Computer Vision and Pattern Recognition, *Appl. Sci.*, 10, 1742, 2020.
19. Rajpurkar, P., Irvin, J., Zhu, K., Yang, B., *CheXNet: Radiologist-Level Pneumonia Detection on Chest X-Rays with Deep Learning*, vol. 25, Dec 2017, arXiv:1711.05225v3 [cs.CV].
20. Sim, Y., Chung, M.D.M.J., Kotter, M.D.E., Deep Convolutional Neural Network–based Software Improves Radiologist Detection of Malignant Lung Nodules on Chest Radiographs, in: *Radiology*, vol. 00, pp. 1–11, 2019.
21. Lakhani, M.D.P. and Sundaram, M.D.B., Deep Learning at Chest Radiography: Automated Classification of Pulmonary Tuberculosis by Using Convolutional Neural Networks. *Radiology*, 284, 2, 574–582, August 2017.
22. Hwang, M.D.E.J., Park, M.S.S., Jin, MD K-N., Development and Validation of a Deep Learning–Based Automated Detection Algorithm for Major Thoracic Diseases on Chest Radiographs. *JAMA Network Open*, 2, 3, e191095, 2019, 2019.
23. Qin, C., Yao, D., Shi, Y., Song, Z., Computeraided detection in chest radiography based on artificial intelligence: a survey. *Biomed. Eng. Online*, 17, 1, 113, Pages 1-23, 2018.
24. Guan, Q., Huang, Y., Zhong, Z., Zheng, Z., Zheng, L., Yang, Y., Diagnose like a Radiologist: Attention Guided Convolutional Neural Network for Thorax Disease Classification. International Journal of Computer Vision and Pattern Recognition, (page 1-10), 30, Jan 2018. arXiv:1801.09927v1 [cs.CV].
25. Guan, Q. and Huang, Y., Multi-label chest X-ray image classification via category-wise residual attention learning. International Journal of Computer Vision and Pattern Recognition, (page 1-16), *Pattern Recognit. Lett.*, 130, 259–266, 2020.
26. Li, Z., Wang, C., Han, M., Xue, Y., Thoracic Disease Identification and Localization with Limited Supervision. *International Journal of Computer Vision and Pattern Recognition*, 20, 1–12, Jun 2018. arXiv:1711.06373v6 [cs. CV].
27. Kim, H.-E. and Hwang, S., Deconvolutional Feature Stacking for Weakly-Supervised Semantic Segmentation. 12, Mar 2016. arXiv:1602.04984v3 [cs. CV].
28. Shin, H.-C., Roberts, K., Lu, L., Demner-Fushman, D., Yao, J., Summers, R.M., Learning to read chest X-rays: recurrent neural cascade model for automated image annotation, in: *Proceedings of the IEEE conference on computer vision and pattern recognition*, pp. 2497–2506, 2016.
29. Guan, Q. and Huang, Y., Multi-label chest X-ray image classification via category-wise residual attention learning. *Pattern Recognit. Lett.*, 130, 259–266, 2020.
30. Ge, Z., Mahapatra, D., Chang, X., Chen, Z., Chi, L., Lu, H., Improving multi-label chest X-ray disease diagnosis by exploiting disease and health labels dependencies. *Multimed. Tools Appl.*, 79, 14889–14902, 2020.

31. Francea, K. and Jaya, Dr. A., Classification and Retrieval of Thoracic Diseases Using Patch- Based Visual Words: A Study on Chest X-rays. *Biomed. Phys. Eng. Express*, 2020.
32. Muller, H., Michoux, N., Bandon, D., Geissbuhler, A., A Review of Content Based Image Retrieval Systems in Medical Applications Clinical Benefits and Future Directions. *Int. J. Med. Inform.*, March 2004.
33. Akgül, C.B., Rubin, D.L., Napel, S., Beaulieu, C.F., Greenspan, H., Acar, B., Content-Based Image Retrieval in Radiology: Current Status and Future Directions. *J. Digit. Imaging*, 24, 2, 208–222, April 2010.
34. Mansourvar, M. and Ismail, M.A., Content-Based Image Retrieval (CBIR) in Medical Systems. *Int. J. Inf. Technol.*, 20, 2, pages 1–9, 2014.
35. Hwang, K.H., Lee, H., Choi, D., Medical Image Retrieval: Past and Present. *Healthc. Inform. Res.*, 18, 1, 3–9, 2012 March.
36. Thomas, M.D., Antani, S., Long, R., Ontology of Gaps in Content-Based Image Retrieval. *J. Digit. Imaging*, 22, 2, 202–215, 2009.
37. Haripriya, P. and Porkodi, R., An Insight Framework for Content based Medical Image Retrieval using Deep Convolutional Neural Network. *Int. J. Eng. Adv. Technol. (IJEAT)*, 9, 1S, 2249 - 8958, October 2019.
38. CamalanI, S., Niazi, M.K.K., Moberly, A.C., OtoMatch: Content-based eardrum image retrieval using deep learning. *PloS One*, 15, 5, 1–16, May 15, 2020, https://doi.org/10.1371/journal.pone.0232776.
39. Kumar, A., Kim, J, Cai, W., Fulham, M., Feng, D., Content-Based Medical Image Retrieval: A Survey of Applications to Multidimensional and Multimodality Data. *J. Digit. Imaging*, 26, 1025–1039, 2013.
40. Wan, J., Wang, D., Hoi, S.C.H., Wu, P., Zhu, J., Deep learning for content-based image retrieval: A comprehensive study. *ACM International Conference on Multimedia*, November 3-7, 2014.
41. Effective Diagnosis and Treatment through Content-Based Medical Image Retrieval (CBMIR) by Using Artificial Intelligence. *J. Clin. Med.*, 8, 462, 2019.
42. Kurian, P. and Jeyakumar, V., *Multimodality medical image retrieval using convolutional neural network Department of Biomedical Engineering*, SSN College of Engineering, Chennai, India, pp. 53-95, January 2020, http://dx.doi.org/10.1016/B978-0-12-819061-6.00003-3

12

Artificial Intelligence in Healthcare Data Science with Knowledge Engineering

S. Asha[1]*, Kanchana Devi V.[2] and G. Sahaja Vaishnavi[2]

[1]School of Computer Science and Engineering & Centre for Cyber Physical Systems, Vellore Institute of Technology, Chennai, India
[2]School of Computer Science and Engineering, Vellore Institute of Technology, Chennai, India

Abstract

Prospective research in Artificial Intelligence (AI) in various domains has changed engineering around the world. It is difficult to build clinical decision models due to lack of large datasets. Knowledge engineering with AI techniques provides a solution to health services by improvising the safety and quality at low cost. Combining the power of machine learning and deep learning offers a strong base for clinical decision support system. With increasing number of patients worldwide, there is a severe setback in healthcare system to fight against control of COVID-19, a pandemic and a threat to lives of human beings. The whole world is looking for a novel technology to fight against this pandemic. This research focuses on use of various AI technologies to analyze and predict COVID-19 patients by collecting real-time data from various age groups, gender, and health conditions. Hence, in this chapter, in application of AI concepts to predict the future health predictions of COVID-19–recovered patients, use of data science to obtain statistics of COVID-19, and to provide a personalized care for each and every COVID-19 patient, knowledge engineering concepts are applied. Also, this chapter focuses on possible health issues met by COVID-19–improved patients. By collecting real-time data through social media, in person and surveys, data is analyzed and implemented to predict future health issues that may be confronted by the COVID-19–recovered patients. Existing machine learning and statistical analysis methods are applied to predict future health problems that a COVID-19 patient. Results obtained are promising with real-time statistics of COVID-19–recovered patients.

**Corresponding author*: asha.s@vit.ac.in

A. Jaya, K. Kalaiselvi, Dinesh Goyal and Dhiya Al-Jumeily (eds.) Handbook of Intelligent Healthcare Analytics: Knowledge Engineering with Big Data Analytics, (255–284) © 2022 Scrivener Publishing LLC

Keywords: Global pandemic, COVID-19, recovery patient assessment, artificial intelligence, machine learning, statistical analysis, data, knowledge engineering, post recovery

12.1 Introduction

The novel coronavirus disease is the outbreak first observed in 2019 at Wuhan in province of Hubei in China. It is an unpredicted epidemic and has caused much fear among people around the world [2]. Basically, corona is based on RNA virus that is in corona viridae family, which can infect both mammals and birds. They mainly infect the respiratory track from mild to lethal. There are no drugs that can help in prevention or cure. The corona cases were observed in more than 218 countries (April 24, 2020); thus, The World Health Organization (WHO) has declared COVID-19 as a global pandemic disease on March 11, 2020. As of November 11, 2020, the total affected cases are 51,621,135 of which 1,274,923 are deaths and 33,638,712 recovery cases of corona. In the international level, there is a frequent increase in the number of patients who have recovered from COVID-19 remain with the indications of the infection, even though they are tested negative of the infection. As the whole humankind is fighting this epidemic, the interesting part is to overcome the corona sequelae with symptoms like body pain, fatigue, and, in some people, lung disease. It is important to have a health follow-up with these recovered patients and conduct a thorough diagnostic and appropriate assessment of their physical, mental, and social behavior. Some of the diagnosed patients from different ages and countries are known to be experiencing certain mental and physical health problems.

Many zoonotic coronaviruses spread the disease from animals to humans. It is a serious disease with a high infectious rate and a high mortality rate worldwide. Earlier, after the outbreak of a Severe Acute Respiratory Syndrome (SARS), many patients lost their life and many recovered from that disease [24]. But still, there was a doubt that whether the patients who returned home after treatment from this disease are completely recovered or whether there is any chance that the patient will be infected again by this deadly virus. The same question arises now; the answer is yes. Many researchers monitored the health condition of the COVID-19–improved patients by checking them on a regular basis. Based on the research, it is found that there is a possibility of the active alveolitis that developed during the protected comeback to the antigen COVID-19 can cause pulmonary fibrosis in few patients after treatment [12].

As COVID-19 infection cases continue to soar worldwide, the number of patients recovering from COVID-19 is again visiting the hospitals complaining of heart issues, mental distress, and many more. It is clear evidence that COVID-19 is simply not just a cold or flu-like infection. It poses risk for people across all age groups, in different degrees, and affects some more gravely than the others. On its impact on one's quality of life, we do know whether the deadly respiratory virus can have longer-lasting side effects. For many viral load may have depleted, but some typical symptoms, such as cough, itching throat, and fatigue may continue to persist for weeks after combatting the illness. Hence, it is even more crucial that it is vital to predict the possibility of future health hazards of the patients recovered from COVID-19 and that motivates to collect some real-time data and work on it.

Reports [1] have suggested that COVID-19 largely infects the people having previous medical illnesses associated with lungs, kidneys, heart, and GI tract. In the lungs, the production of cytokine that mediates the virus in alveolar macrophages damages the lungs by causing lung fibrosis. Active nucleotide-binding domain (NOD) receptor protein 3 (NLRP3) signaling may lead to pyro ptosis and cell death in kidneys. Also, this COVID-19 infection can break down the necrotic lipid spine that causes blood clots to lead to myocardial infarction in the heart. The olfactory entry bulb in the brain infected with SARS-CoV-2 may leads to anorexia nervosa (inability to smell the fragrance) and causes non-neurodegenerative diseases in patients diagnosed with COVID-19. Tears in the eyes cause the spread of the disease through nasolacrimal system to other parts of the human system. Also, patients diagnosed with COVID-19 may experience depression as a result of this disease and may be confused about the acceptance of their community people after infection [1].

Coronavirus is the newfangled pandemic that is hitting the entire world. Patients in international level are facing various symptoms. The severity in these symptoms leads to death, specially the elderly [4]. The epidemiological dataset has been developed by researchers with the help of different reports obtained from real-time cases in order to represent the various attributes that has been given as the important factors for prediction toward recovery [4]. Further, analysis with other available machine learning techniques incorporated with artificial neural network algorithms and regression model were tested and compared using the results of SVM. As a conclusion, the patients who does not recover had severe fever along with cough, overall tiredness, and mostly ill. Understanding the various factors controlling the patients' mental and physical health after recovering from COVID-19 and predicting their health using various machine learning algorithms is the ultimate aim of this research study. To predict the after effects in health of recovered COVID-19 patients, various machine learning methods are applied on factors such as age, previous health issues faced, and severity of COVID-19 of patients and to predict the possibilities of future health issues in corona recovered patients. Even after getting recovered from COVID-19, many of them are not capable of leading a normal life, as the symptoms prevail for a long time. So, it is important to understand for coming up with a drug which is capable of making it less "long COVID" as said by doctors. Figures 12.1 to 12.3 give the comparative study of number of positive cases affected by COVID-19 in various countries, number of deaths due to COVID-19 in various countries, and number of recoveries country-wise as of January 2021.

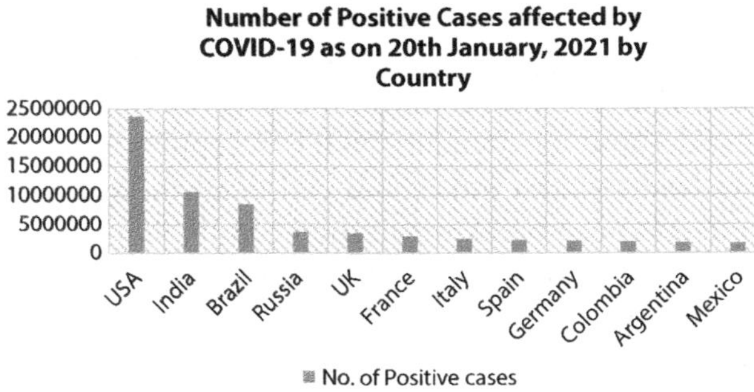

Figure 12.1 Comparative study of number of positive COVID-19 cases in various countries.

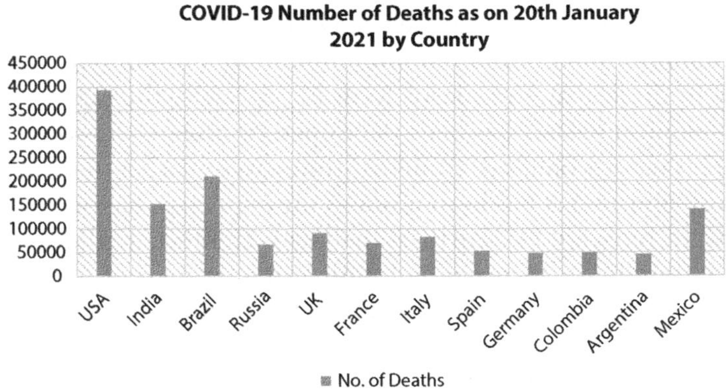

Figure 12.2 Comparison of number of COVID-19 deaths in various countries.

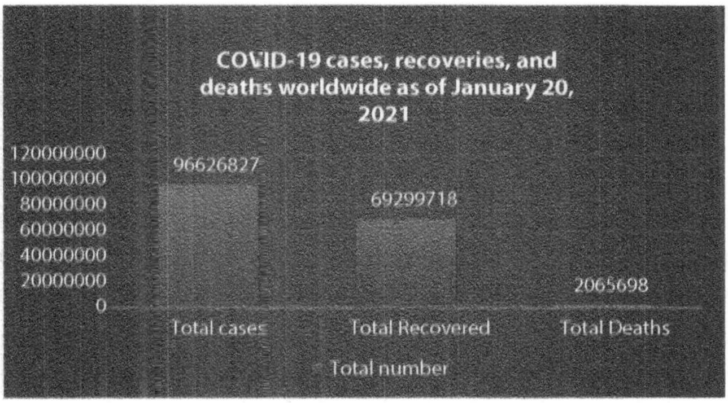

Figure 12.3 COVID-19 statistics worldwide based on total cases, recovered, deaths.

The real challenge is to understand all the factors contributing for patient getting affected and factors that may cause symptoms to prolong. Gathering the data and preprocessing is also one of the problems and understanding in which way we can collect the data so that there will be less noise in data and can be easily preprocessed.

- Data collection
- Importing data
- Understanding of data and dependencies using various plotting techniques
- Pre-processing of data
- Applying regression model
- Removing class dependences using sampling.
- Obtaining performance analysis

This research focuses on the post health issues that arise in recovered COVID-19 patients. The sectional organization of this paper is as follows: Section 12.2 describes the related background in literature review. Section 12.3 gives the concept of AI in healthcare system, Section 12.4 describes the COVID-19 data science and knowledge engineering in healthcare, and Section 12.5 gives the proposed implementation methodology to predict health issues in recovered COVID-19 patients. Sections 12.6 discusses the arrived results and future scope.

12.2 Literature Review

Artificial Intelligence (AI) is an innovative technology along with data science and machine learning algorithms to automate and make decisions promptly with less human intervention. AI has been used in the healthcare industry for the past few decades. AI techniques can be used to track, monitor, and predict the current and future health conditions of COVID-19–affected patients. AI can also be used for categorizing the symptoms and diagnosis of COVID-19. Generally, AI algorithms are applied in mobile application to identify the general public violating the rules to be followed such as social distancing, wearing mask, and mass gathering of people, during this pandemic in virus containment zones. Apart from these, AI techniques are used for online consultation of patients with specialist doctors; use of robots for distributing hand sanitizer and disinfecting in public places. AI can also be used in drug industries for production purposes. As COVID-19–infected patients' number is increasing rapidly, it becomes difficult for doctors to examine the huge number of medical reports and to complete the diagnosis within a short period of time [35]. Time plays a dynamic role in the case of COVID-19 patients. In order to speed up the process involved in diagnosis, AI technologies can be used. Nowadays, healthcare professionals use the concept of deep learning and machine learning algorithms to predict the possibilities of occurrence of any disease from the medical images. With the availability of many intelligence tools, machine learning algorithms can be extensively used in any clinical process such as testing, diagnosis, prediction, prevention, and drug manufacturing. Complete health condition about a COVID-19 patient can be maintained and a data model can be created for the same. This will help in giving a personalized care for each and every patient. To fight against this COVID-19 pandemic, it is required to acquire the up to date information about the spread of this disease and also the statistics of the number of persons affected, number of patients recovered, and number of deaths happened on an hourly basis. The collection and maintenance of data and using it for further processing, prediction, and cure is the need of the hour.

In December 2019, an increasing number of pneumonia cases were reported in Wuhan, China. The cause of this rapid outbreak was determined to be the novel coronavirus, called COVID-19. The WHO, on March 11, 2020, declared COVID-19 as a global epidemic with high rates of infection and death [1]. In early August, the number of COVID-19 cases worldwide increased to 20.4M and confirmed the death of 445K and the recovery of 12.6M with a 61.7% recovery rate, which has shown that the death rate varies from province to province due to demographics, age distribution, and health infrastructure. But, many new patients diagnosed have been recorded showing some after effects on patients. There exists a mutable extent in affected patients, and also radiological changes also exist [25, 26]. The main goal of COVID-19 is the respiratory system. Also, it affects the other organs such as gastrointestinal, endocrine, and cardiovascular systems [27, 28]. It is clear that COVID-19 is a disease that affects multiple organs in a human

COVID-19 mainly disturbs older people, undernourished and those who are having previous health problems toward lungs, kidneys, heart, and GI tract. Therefore, it is highly important to recognize the probable side effects of patients diagnosed with COVID-19 and the possible health issues that they may experience with other serious diseases. After recovery, there may be a significant number of COVID-19 patients, who may experience a disability in their general health condition. COVID-19 patients can show contagions through the GI tract, kidneys, heart, brain, eyes, and lungs. Severe pneumonia can spread to lungs through the air sacs and may result in Acute Respiratory Distress Syndrome (ARDS) [1]. According to [23], it is evident that infants infected with COVID-19 had encephalopathy. About 80% of children who are less than 3 months old, tested positive nasal swabs of SARS-CoV-2, had axial hypotonia, drowsiness, or moaning.

In this research, a study on how the various parts of the body are affected by COVID-19 infection is carried out. The virus passes through an olfactory bulb that triggers a provocative response that leads to a cytokine storm. It is stated that COVID-19 patients are exposed to neurological symptoms and headache. This causes emotional disturbances. In a latest research report [1], a 50-year-old woman from USA was spotted with COVID-19, also indicated with severe neurological symptoms. The woman was affected with acute necrotizing encephalopathy. In another research [1], a 74-year-old man was tested positive of COVID-19. He already had a chronic obstructive pulmonary disease and Parkinson's disease. He was also displayed neurological signs such as difficulty in speaking and confiscations [1]. The virus enters the heart through the bloodstream, which causes the formation of necrotic lipid. Myocardial infarction is caused due to blood clots by these lipid ruptures. Numerous reports from different parts of the world and countries like China and the USA also confirmed that COVID-19 causes heart failure [1]. In the kidneys, the virus enters toward tubules that are near the podocytes through ACE2 receptors. Later infiltration, it causes the insertion of an extra matrix of cells causing fibrosis, diuresis, and proliferation of kidney cells leading to severe kidney damage [1]. The coronavirus also spread through the tears and connect to the nasolacrimal system to other organs of the body that can cause conjunctivitis in the eyes, chemosis, inflammation of the conjunctiva, and epiphora/overflow of tears on to the face [1].

In Italy, a study in [20] was conducted in the month of April and May 2020. In that study, they judged that the symptoms in 143 patients were persistent, even after their discharge from the hospital after they are tested negative for COVID-19. In another study in [21], 55 patients after their recovery from corona and 90 days after their discharge from hospital were considered for research. The pulmonary function and related physiological features of 39 corona patients were found to have various degrees of radiological abnormalities. At the time of admission to the hospital, the presence of CT abnormalities was associated with blood urea nitrogen concentration [21].

Kupffer cells are activates in the liver by the coronavirus. This triggers the activation of hepatic stellate cells and hepatocytes, leading to apoptosis-induced damage and fibrosis. Also, the virus causes damages to the lung walls, thus causing the formation of debris. Lungs gets damaged and compressed because of these accumulated debris on the walls of alveolar cells. Also, in COVID-19, shortness of breath is a common symptom. In a follow-up study, it was reported that most people suffer from pulmonary fibrosis [6]. Children seem to be more immune from the serious problems of COVID-19. Nowadays, news reports say that a rare and potentially fatal inflammatory disease that is associated along with coronavirus troubles a small number of children. Cases with multi-system inflammatory syndrome (MIS-C), earlier called as Pediatric Multi-symptom Inflammatory Syndrome (PMIS), are reported in countries such as USA and Europe. Certain children suffered organ failures and two deaths have been reported in New York (in least).

Table 12.1 gives the complete characteristics details of COVID-19 in various organs. The aim of the system is to understand and find a pattern on how COVID-19–recovered patients are suffering even after the recovery. Understanding and identifying the key factors that cause the recovered patient to go through mental trauma or getting affected by permanent organ damage. The next step is gathering and processing of data and using a machine learning approach to analyze the data.

It is believed that patients with persistent symptoms should check with their doctor after 2 months after their recovery from COVID-19 for an evaluation, with a chest computed tomographic scan to examine intensely the relationship between COVID-19 and the chronic inflammatory symptoms. Based on these reports, follow-up treatments with specialists are carried out [11]. It is presented in many researches that the most common

Table 12.1 Characteristics details of COVID-19 in various organs [2].

Organs	Site of symptom	Manifestation	Confirmative test	Viral Nucleic Acid (VNA) presence	ACE-2 receptor's presence	Remarks
Brain	Olfactory bulb	Hyposmia, hypogeusia, hypopsia, encephalitis, and headache	CT Scan, MRI Scan, UPSIT (University of Pennsylvania Smell Identification Test)	Nasopharyngeal swab	–	SARS-COVID is attached to an olfactory bulb that leads to the odor of malfunction and a cytokine storm in the thalamus
Eye	Conjunctiva	Conjunctivitis, chemosis, swelling of conjunctiva, epiphora, or tears flowing onto the face	Slit lamp diagnostics	–	Retina and Retinal pigment epithelium	SARS-CoV-2 could shed tears in the form and could affect the disease to different organs
Kidney		AKI: Acute Kidney Injury	Creatinine blood test	Urine (uncommon)	Proximal convoluted tubules Podocytes	Inflammasome method; moderate damage done in bulk

(*Continued*)

Table 12.1 Characteristics details of COVID -19 in various organs [2]. (*Continued*)

Organs	Site of symptom	Manifestation	Confirmative test	Viral Nucleic Acid (VNA) presence	ACE-2 receptor's presence	Remarks
Lung	Pulmonary lobe	Bilateral pneumonia, and unilateral pneumonia	CT scan	Lung sputum	Transient secretory cells	Damage to walls and lubricating cells by SARS-Covid; inflammasome passage mediated lung injury
	Pulmonary nodules	Ground glass opacities, and irregular lesions	-	-	-	-
	Bronchovascular bundles	Thickening	-	-	-	-
	Alvcolus	Shortness of breath ARDS	Physical symptom	-	-	-
Heart	Myocardium and cardiac muscle cell	Cardiac failure or myocarditis, myocarial infraction	Increased levels of troponin protein	-	Viable myocardium	Inflammation and storm associated with myocardial infarction or myocarditis.

(*Continued*)

Table 12.1 Characteristics details of COVID -19 in various organs [2]. *(Continued)*

Organs	Site of symptom	Manifestation	Confirmative test	Viral Nucleic Acid (VNA) presence	ACE-2 receptor's presence	Remarks
Gastro-intestinal system	Stomach and digestive organs	Anorexia, vomiting, nausea, and abdominal pain	Physical symptoms	Nasopharynx throat, stool	Oesophagus, ileum and colon	GI symptoms can occur in the first phase of COVID-19
	GI tract	GB: Gastrointestinal-Bleeding	Endoscopy, colonoscopy	-	-	-
	Liver	Liver injury	AST, ALT	-	Cholangiocytes	Possible drug convinced liver injury - Liver fibrosis using AP-1 or the inflammasome method

abnormality is the functionality of the lung in corona recovered patients are weakening of diffusion capacity, with the harshness of the disease [15, 16]. Both reduced alveolar volume and KCO donate to the pathogenesis of impaired diffusion capacity [17]. After 3 months from discharge, certain abnormalities noticed in pulmonary function remain in the patients were observed in 25.45% of the cohort which was less compared to abnormal pulmonary function in COVID-19 patients after discharge [13]. Lung function abnormalities were sensed in 14 patients out of 55 patients, and the measurement of D-dimer levels at admission may be useful in prediction of impaired diffusion defect [14]. Several mechanisms of lung injury in COVID-19 have been defined, with both immune-mediated and viral mechanisms in [18].

12.3 AI in Healthcare

Authors in [29] presented the complication and increase of day-to-day data in healthcare applications demands the usage of AI within these fields. Numerous kinds of AI-based algorithms are now being used by customers, suppliers, and enterprises. The basic classifications of applications implicate analysis and recommendations of the appropriate treatments, meeting patient and observance, and other managerial activities. There are certain occasions where AI performs healthcare actions comparatively better than human being; some practical factors avert automation of healthcare specialized jobs for a significant period of time.

Authors in [30] have discussed on "Explainable AI in Healthcare (XAI)" where AI is an empowering technology capable of predicting the existence of health conditions in users by capturing and analyzing their health data when incorporated into healthcare applications and wearable smart devices. The incorporation of AI and wearable smart devices has a wide range of potential in healthcare applications but there is a challenge in the black box operation of decisions made by AI methodologies which leads to lack of liability and trust in the decision-making. Explainable AI (XAI) is a common field in which several techniques are developed in order to elucidate predictions which are made with the help of AI systems. In this paper, AI is assessed as a system that is commonly used for the purpose of calculation and other analysis in gathered healthcare data. The methodologies presented in this paper provide liability, tracing the results, and improvement in various model in healthcare data and application.

In paper [31], the authors have presented a different perspective of AI as "Artificial intelligence in healthcare: past, present, and future", in which AI act as an imitator of human cognitive functionalities. This AI brings a shift in modeling toward healthcare applications. The authors have surveyed the latest state of art toward AI in healthcare applications. This AI is capable of applying in both structured and unstructured healthcare data. Current AI methodologies include ML models for applying in structured healthcare data, some of the AI-based algorithms commonly used are Support Vector Machine (SVM), Neural Network (NN), Deep Learning (DL), and Natural Language Processing (NLP) for other unstructured healthcare data. Commonly, AI is used for healthcare applications for diseases such as cardiology disease, cancer, and neural disease. This paper focus more on stroke in more details, in the three different views as for identifying and analyzing it earlier, the way for treating it, along with predicting is results, for valuation purpose.

In paper [40], the power of AI is used to improve the quality of clinical data and improvise it to human understandable form and provide patient-specific recommendations. In this paper, authors used a class of AI techniques that uses knowledge engineering of human expertise in decision-making and plan about the healthcare of the patients and about the other reasoning skills in an instinctive through a formal language known as "PROforma". This PROforma models are implemented with a help of a system in order to produce patient-specific recommendations, to elucidate the details for supportive evidence on demand. In paper [41], case studies of real-time data from COVID-19 were collected, preprocessed, and analyzed the COVID-19–affected patient with different attributes like age, gender, and symptoms, that provide the knowledge about the characteristics of this epidemic. In paper [42], there is no supporting model for clinical data decision-making, provided suggestion related to AI and big data usage to build decision models with the use of analytic tools, and showed that there is a good evidence that these models have practical value and user appeal.

In paper [38], the various AI techniques used in collecting data, segmentation of data, classification of data and diagnosis of COVID-19 were reviewed. The data used in this paper are biomedical data. The integration of AI techniques with X-ray images were mainly focused in this paper. Chest X-ray and CT scan plays a dynamic role in the diagnosis of COVID-19. Hence, it is highly important to be focused on a contactless and automated method for identifying the COVID-19–infected patients. Healthcare specialists are particularly exposed to high risk in first level of occupational viral acquaintance. Imaging specialists and technicians are provided with high priority toward spread of infection, such that any primary contact along with the virus is under control. This chapter presents safe, precise, and effective imaging solutions toward COVID-19 applications.

In [49], authors have discussed about development and validation of a deep learning methodology for detection of diabetic retinopathy in retinal fundus photographs. The features of the algorithm for the detection of referable diabetic retinopathy (RDR), described as sensible and poor: (i) diabetic retinopathy (DR), (ii) referable diabetic macular edema (RDME), or commonly both, the decision has been made with the help of ophthalmologist panel. Later, the correctness of the algorithm is evaluated based on the two parameters toward the dataset: the first one through specificity and the second one through sensitivity. The authors in [50] have introduced classification of dermatologist-level in skin cancer using deep neural network. Skin cancer considered to be the common human spite, which is mainly "visual diagnosis", which always begins with screening techniques takes place in clinical level and the second stage of testing includes (i) dermoscopic analysis, (ii) biopsy, and (iii) histopathological examination. First of all, an automated classification procedure upon skin lesions using images takes place, which is an important action/activity toward inconsistency in the presence of skin lesions. Deep convolutional neural networks (CNNs) project a new future for highly varying activities toward many different categories. In this paper, the authors have incorporated the CNN concept for better results. The use of CNN algorithm accomplishes the performance on similarity with all other tested experts toward various actions; this testing proves that AI is highly capable of such skin cancer classification with a proper level of competence when compared to other dermatologists results. Equipped with deep neural networks modeling, mobile devices can hypothetically cover the usage of dermatologists other than clinic.

12.4 Data Science and Knowledge Engineering for COVID-19

Data science plays a vital role in the statistical analyses of COVID-19. It helps in focusing on the analysis of large sets of COVID-19 data. Together with knowledge engineering, it helps in building an expert data model for this pandemic outbreak. Many researchers shared their views about the relationship between data science and COVID-19. In paper [32], the researchers in Peru from two private universities designed a tool for knowledge management to analyze the educational reforms before and after recovery from COVID-19. They identified the following main aspects:

i. Innovative new methodologies for sharing of knowledge have become popular, and the researchers introduced new platforms for sharing of knowledge.
ii. Various virtual modes of communications have come into usage throughout the world for online classes, video conferencing, virtual meetings, etc.
iii. There is a great demand for online and virtual conferences, workshops, and webinars.
iv. Anywhere, anytime, has become a trend.
v. Less human investment in most of the educational institutions.
vi. The main factor is the combination of innovative skills and knowledge in the learning process.

According to the authors of paper [32], it is evident that, before COVID-19, in these institutions, educational institution process is satisfactory. During COVID-19, it affected a lot. In paper [33], the researchers collected various heterogeneous data and developed a model for precise impediment and control of these deadly virus COVID-19–affected patients. However, their model is not successful as this pandemic is not showing a regular pattern. But, the support toward further promotions of certain intelligent prevention methodologies and control toward major epidemic outbreaks and helps in improvising the accuracy in detecting epidemic transmission pathway and avoiding it from further disease spreading. In paper [34], the researchers formulated certain COVID-19–related questionnaires and tried to find data related to that. They identified the rapid increase in the COVID-19–related news. In addition, the findings show that there is an increase in COVID-19–related news. In paper [36], authors focused on identifying the number of infected cases of COVID19 on a weekly basis using various technologies. Using a numerical approach with eight different independent parameters, they are able to achieve the desired results. Based on these results, a predictive model was developed.

In [37], COVID-19 data, its impact on the research, and the various methods such as deep learning methodology and machine learning approaches used to predict the nature of the data were presented. Various intelligent diagnostic tools used for prediction of COVID-19 were investigated in this paper. The challenges were in collecting huge volumes data for predicting and forecasting using machine learning and deep learning methodologies. The real-time dataset is also poorly quantified and training this unrepresentative data will result in poor outcome. Online data, clinical data, and biomedical data were used in the investigation in this paper. From this paper [37], it is has been observed that many researches have been carried out related to COVID-19 and the majority is classification, then detection and forecasting. Due to the lack of real-time dataset, only less research is carried out in prediction and forecasting of COVID. But, the outcome of this paper is not considered to be accurate as authors have not included the impact toward other factors like hypertension, age, diabetes, chronic liver, gender, and kidney disease. But, in our research study, we have included these factors.

Monitoring the health of patients through mobile devices is discussed in paper [39]. Several AI-based algorithms and big data frameworks, data sources, techniques used and application areas, specific challenges in m-health, and various models used were also explored in this paper. Table 12.2 gives the various AI methods used in healthcare sector.

12.5 Proposed Architecture and Its Implementation

Figure 12.4 shows the architecture of the proposed methodology. Here, the COVID-19–recovered patient fills his/her medical details like age, recovery time, preciously treated dieses, long-term dieses, and bad habits. The user imports and pre-processes the data in order to remove noises in data, so does the feature extraction and model fixing. It splits the data into training and testing model. Here, we give more fractions of data in training the model as it increases the precision. From the valuation of correlation matrix and accuracy score, we know the precision of the model.

12.5.1 Implementation

12.5.1.1 Data Collection

The data is collected using Google forms and also collected from rural COVID-19 officers.

12.5.1.2 Understanding Class and Dependencies

After importing the data using pandas to understand the data, various functions are used such as unique () function that gives unique elements

Table 12.2 AI methods and big data for health care sector.

Framework	System	Technique	Application domain
Cognitive machine learning (CML) algorithm [43]	Cognitive tool for computing	Machine learning algorithm in association with memory classification	Machine learning algorithm is used to normalize the data obtained from I Echocardiography
CNN: Convolutional neural network [44]	Natural language processing along with AI is used.	chest X-ray reports are examined that will assists and alert the physicians for anti-infective therapy	For effective diagnosis of diseases, AI and NLP are combined
MLA: Machine learning algorithms [45]	SVM: Support vector machine	Analyzing of data and then classifying a multidimensional data	To differentiate between normal heart and infected heart
Smart healthcare framework [46]	Datasets from the radiology department and from different sources	Pattern recognition techniques	Smart healthcare
Smart framework for healthcare system [47]	Data from various sources such as social media, biometric information, monitoring data, diagnosis report data and HER data	Smart healthcare service provider	Smart infrastructure services
NFS-HIS: Novel framework for secured HIS [48]	Electronic-based health records	All kind of security and control techniques	Secure healthcare system

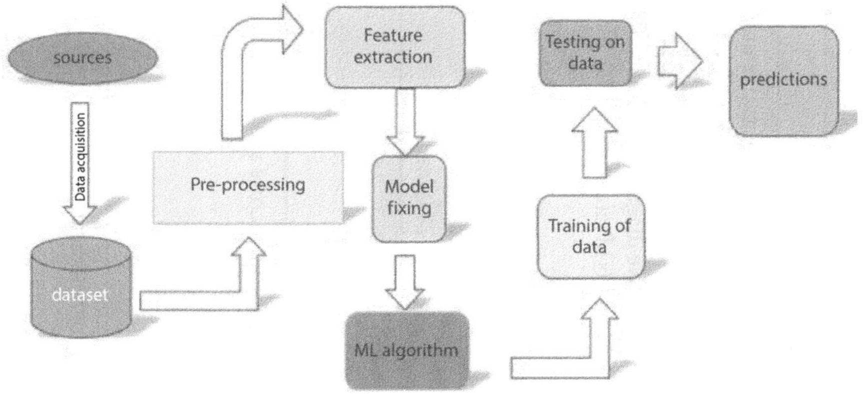

Figure 12.4 Architecture of the proposed methodology.

in a column. seaborn.countplot () function is used to understand which unique value of column has how many records. Seaborn.countplot() with hue argument gives one column values against other. plot.hist() method is used for columns having numerical as values.

12.5.1.3 Pre-Processing

In pre-processing of data, data processing is one of the important stages involved in data process involved for mining [10]. The term "garbage in, garbage out" applies especially in data mining process and ML projects [8]. Data collection methods are usually ordered by discrepancies, leading to cross over boundary values. For example, income of an individual may be a negative value, dubious data combinations may be involved like gender (male) and pregnant status (may be yes), and, sometimes, missing values is also possible. Analyzing such kind of data that is not carefully tested for such specified problems can mislead to obvious results. Consequently, representation and data quality check must be done before applying any analysis. Isnull() method is used to find if a cell in a row or a column is null. seaborn.heatmap() is used to pictorially see the null values. drop () is used to remove unwanted rows or columns whose presences affects the system negatively. In this research project, Blood_group column is not considered and planned to include it in the future work. Label encoding is used for encoding the columns which are categorical.

12.5.1.4 Sampling

As the "damaged" column that is class label column is multi-label column, this system undergoes class imbalance problem. In order to overcome this problem, SMOTE () function is used, which helps in oversampling. Oversampling involves replicating the minority class and add them to the training dataset so that the proposed model does not show favoritism to any class that has a greater number of occurrences.

12.5.1.5 Model Fixing

Model fitting is a measure of is a measure of how a machine learning model performs well the same data as the one it is trained in. A well-designed model produces accurate results. The embedded model is very similar to the data. The model included below does not fit well enough. In this research, there is no such problem. Figure 12.5 gives the description of the flow of the proposed methodology.

12.5.1.6 Analysis of Real-Time Datasets

The real-time data collected from 694 men and women in three different age groups, namely, 18–25, 25–50, and 50 and above, were analyzed. Also, the data is collected related to their food habits, whether they have

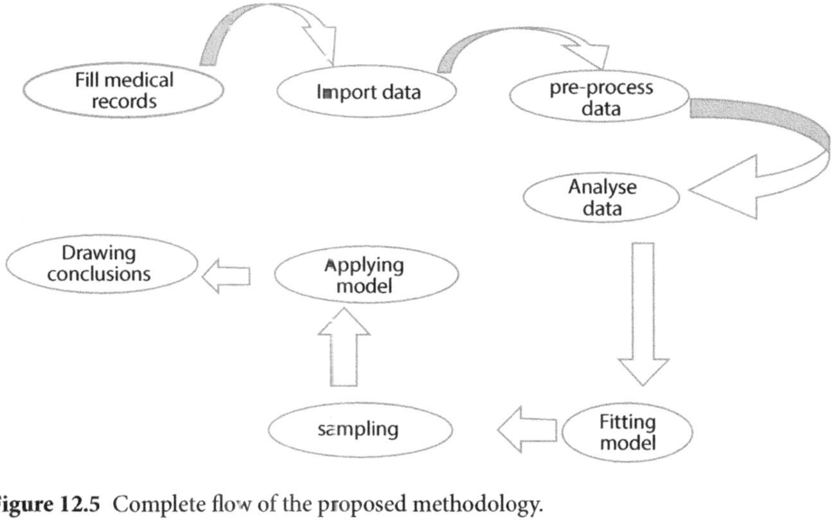

Figure 12.5 Complete flow of the proposed methodology.

premedical issues like blood pressure, heart disease, lung disease, kidney disease, etc., and also from healthy patients. The data is classified into male and female categories, and the data is analyzed. Figure 12.6 shows the result analysis of the observed data from male patients. It is observed from the figure that the post-medical problems of COVID-19–recovered patients in the age group 50 and above are considerably high even though the pre-health conditions are less in that group. Also, it is observed that the young people in the age group are more prone to health issues when compared to the middle age group 25–50. Similarly, Figure 12.7 shows the result analysis of female patients. It is noticed that the young age group 18–25 is affected high when compared to other age groups. This shows that the food habits play a vital role in COVID-19–recovered patients. Most of the people in the age group of 50 and above is affected by severe heart problems and lung disease though they do not have prior symptoms of these diseases. From Figure 12.8, it is evident that a COVID-19 patient, though recovered from COVID-19, may be affected by severe health issues after recovery.

The same data has undergone further analysis process using various MLA in order to predict the possible health problems of the COVID-19–recovered patients. Depending upon the collected data, on various parameters, using machine learning algorithm, it is predicted the possibility of health issues and lungs are the most affected part after recovery from

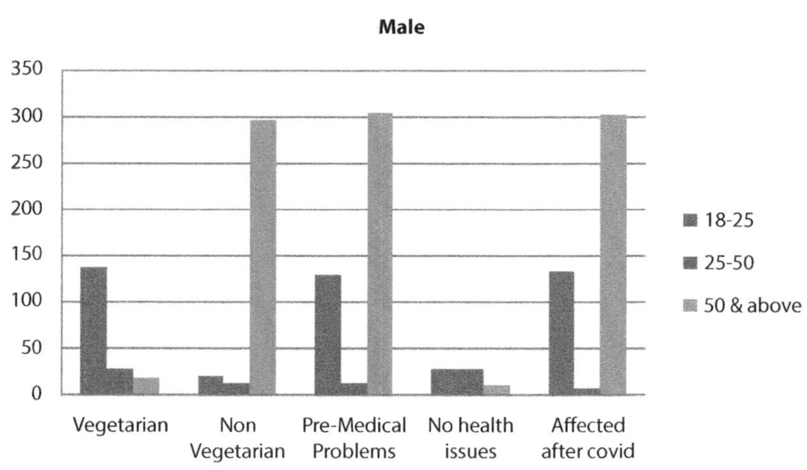

Figure 12.6 Statistics of COVID-19 recovered patients (male).

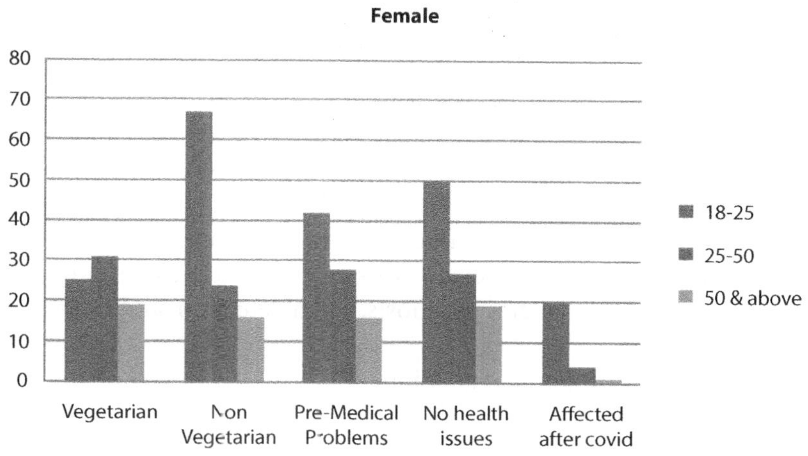

Figure 12.7 Statistics of COVID-19 recovered patients (female).

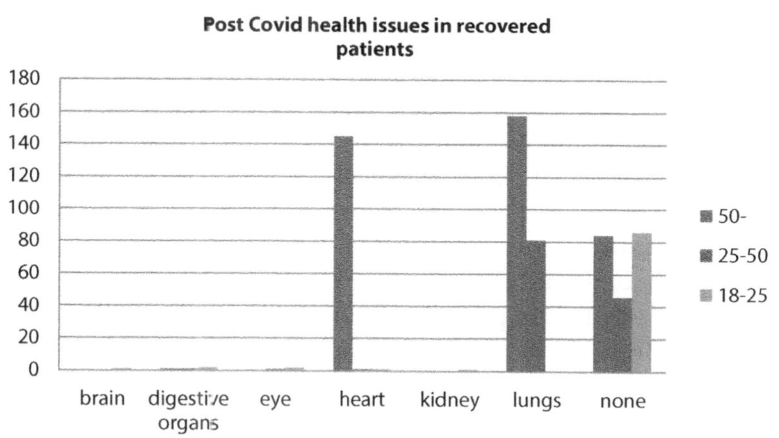

Figure 12.8 Analysis of real time data collected.

COVID-19. It has been observed that 40% of patients with corona advance to ARDS and 20% of such ARDS cases are highly severe [19]. The existence of post-corona fibrosis becomes deceiving toward time, but initial level analysis from various patients infected with corona on discharge from hospital endorses that more than normal one-third of disease recovered patients involved in the development of "Fibrotic Abnormalities".

12.5.1.7 *Machine Learning Algorithms*

i. Logistic regression
   It is a supervised learning algorithm for classification which is used to predict the probability of a variety of objectives. The type of variable dependency is clear, meaning that there will be only two probable categories [3]. Statistically, the systematic model predicts P (Y = 1) as an X function. This is one of the modest machine learning algorithms that can be implemented for an extensive range of investigative problems such as diabetes prognosis, spam detection, and cancer detection.
ii. Binary or Binomial regression
   In binomial regression, there will be only two options available. That is the dependent variable will have only two values, either 0 or 1. That is, these variables can indicate wither win or lose, yes or no and success or failure [3].
iii. Ordinal Regression
   In this type of classification, there may be three or more types of dependent variables that may be ordered to have a measurement value [3]. In this case, the dependent variable may represent poor or good, very poor, or very good.
iv. Support Vector Machine (SVM)
   SVM is generally a computationally complex kernel-dependent tool mainly used for data regression process and classification process. Compared to MLA, this SVM focuses on performance toward generalization [3]. Thus, SVM has reached popularity toward performance involved in various real-world applications such as image processing toward applications of healthcare. If sigmoid function acts as an activation functions to update weights, then the SVM will be considered as ANN. Nevertheless, in place of using multi-layers for solving nonlinear problem, SVM manages the features toward higher dimensions which are divided using hyperplane. The ultimate goal here is to maximize the hyperplane since a greater number of features are extracted for gaining accuracy in classification process. Equation (12.1) shows the SVM equation. Consider a series of data points in the equation $D = [x_i, d_i]i\ n$ in which n represents data size, d_i provides the target value, and x_i shows space vector input [5].

The SVM evaluates the function using Equations (12.1) and (12.2)

$$f(x) = \omega\ \varphi(x) + b \tag{12.1}$$

$$RSVMs\ (C) = 1/\ 2\ \omega 2 + C(\ 1/n)\Sigma L(x_i, d_i) \tag{12.2}$$

n, i=1 here, $\varphi(x)$: space feature
b: Scalar
ω: vector
$C(1/n)\Sigma L(x_i, d_i)$ n i=1 is the experimental error.

v. Naïve Bayes
It is called the Naïve Bayes because the probability calculation of each hypothesis is simplified to make their calculation slower. Naïve Bayes is a simple but powerful algorithm with the ability to create a predictive model. Naïve Bayes is a binary (two class) division algorithm with multiple classification problems. This method is easy to understand when defined using binary or input values.

vi. Boosted Random Forest
The random forest algorithm is strong toward noise and with the help of random training it reaches higher elevations. Conversely, it demands a greater number of decision-making trees and usage of a smaller number of decision trees leads to decrease the performance and efficiency [7]. Hence, it is highly difficult to maintain the normality in computer hardware usages. Due to this kind of reason, the concept of breeding is presented into the wild forest. The ultimate purpose of using this solution is mainly to expand the normal retention even if there is a smaller number of decision trees by using the basic truth that usually successive training creates balancing decision trees for the provided training samples [9].

Based on the various machine learning algorithms, Figure 12.9 shows the performance of these algorithms in various parameters like F1 score, accuracy, precision, and recall. It is observed that F1 score and accuracy similar in all the algorithms but the precision and recall are inversely proportional. For practicing, the future health predictions of COVID-19–recovered patient's precision and recall will give good results.

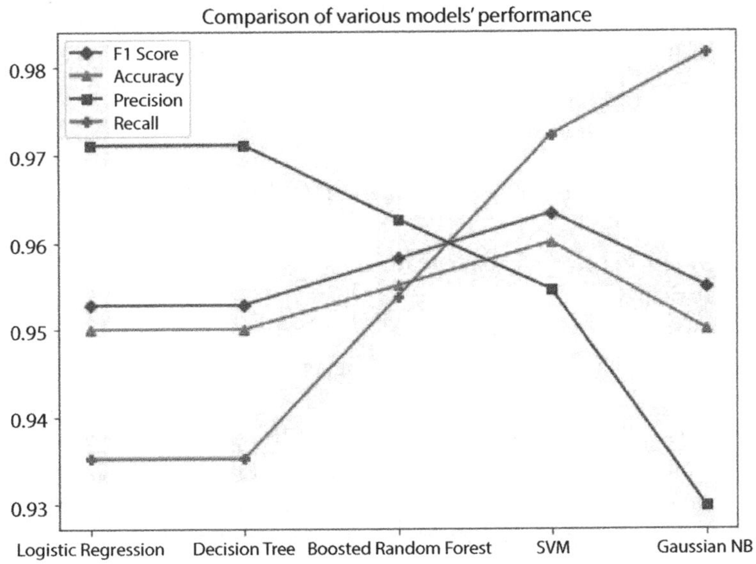

Figure 12.9 Comparison of various machine learning algorithms.

12.6 Conclusions and Future Work

COVID-19 is already confirmed as a deadly infection distressing people worldwide, except for the common indications that happen in COVID-19. Also, in recent studies and surveys, COVID-19 develops into a complicated condition that affects the other parts in a human body. The infection seems to be in the hub of numerous mutations. It is creating various side effects and affecting various organs of an individual infected by the virus. When the condition of the COVID-19 patient is severe, it develops into ARDS, infected tremor, clotting malfunction, and demise. Slight and sensible cases of COVID-19 appear to have a decent rescue rate. Millions of infected patients worldwide diagnosed by COVID-19 have been hospitalized. They are given very good treatment and discharged from hospital after recovery. But, the main issue is that the long-term effects of COVID-19 are not known to anyone. In order to obtain real-time results for this COVID-19 condition, to collect the follow-up studies, it is important to maintain the patient's conditions that it will help to diagnose infection in the early stages and allow timely medical intervention. In addition, this study may provide sufficient comparison data between the infected patients worldwide to regulate the effects of corona on an extensive kind of persons. Aside from physical manifestations, it is worth looking at the negative effects of this epidemic. People all over the world have been living together and are afraid of getting the virus. Recovered corona patients may be stigmatized, contributing to the mental and emotional problems of depression. For millions of people all over the world who have been recovered from corona, it is significant to take all the necessary precautions and collect information for its future epidemic. But, the main problem is that COVID-19 has been included in the fast transmission capabilities.

It is essential to have a follow-up study in infected and later recovered patients from COVID-19 and offer suitable administration of this widespread pandemic in terms of psychological and social factors. As COVID-19 causes more fear in international level, it is highly important to obtain a detailed analysis toward the patients' conditions after the disease recovery. This can also help patients to come together to find harmony, even after if any of the patients' friend or the family members are not infected or at risk. This research focused on the corona infected persons after they are recovered and discharged from hospital in a good health condition. Research focused on different age groups and number. Promising findings were obtained.

This research concludes that children below the age group of 18 are less affected by COVID-19. The casualty rate of COVID-19 was less in children comparatively incidence rate [22]. People in the age group of 50 and above are more affected by COVID-19, and also after recovery, they face serious health issues. When compared, food habits also play a major role in the health of people. It is also observed from the data collected, less number of women are affected from COVID-19 when compared to men. Also, after recovery from COVID-19 also, less number of women is having health issues when compared to men. These findings are confirmed when it is possible to collect data from regions of the worldwide and also if the data is huge.

In future work, it is planned to include blood type and some other factors that may cause COVID-19. There is no related literature for the health issues of COVID-19–recovered patients; this research referred various news articles and websites to conduct this research. Also, the discussion with various health workers and doctors concludes that patients after recovering from COVID-19 visit hospitals for various health issues.

This research in future will focus on huge volumes of data from various countries.

References

1. https://www.ncbi.nlm.nih.gov/pmc/articles/PMC7184970/
2. https://www.aljazeera.com/news/2020/04/countries-reported-coronavirus-cases-200412093314762.html
3. https://poseidon01.ssrn.com/
4. https://labblog.uofmhealth.org/rounds/rare-covid-related-inflammatory-disease-affecting-children
5. https://www.youtube.com/watch?v=NZcidVmk7Bk
6. https://medium.com/analytics-vidhya/data-preprocessing-techniques-in-data-mining-4c90666f6f7a
7. https://medium.com/@harishdev516/concepts-of-data-preprocessing-ce9773c9e462
8. https://medium.com/data-science-everywhere/data-preprocessing-a-practical-guide-1b1ce3e884d8
9. https://medium.com/@vinitasilaparasetty/guide-to-data-preprocessing-for-data-science-d2447a09ba4c
10. https://machinelearningmastery.com/smote-oversampling-for-imbalanced-classification
11. Ludvigsson, J. F., Case report and systematic review suggest that children may experience similar long-term effects to adults after clinical COVID-19. *Acta Paediatrica*, 110, 3, 914–921, 2021.
12. Rai, D. K., Sharma, P., Kumar, R. Post covid 19 pulmonary fibrosis. Is it real threat?. *Indian J. Tuberc.*, 68, 3, 330, 2021.
13. Das, K.M., Lee, E.Y., Singh, R. *et al.*, Follow-up chest radiographicfindings in patients with MERS-CoV after recovery. *Indian J. Radiol. Imag.*, 27, 3, 342–349, 2017.
14. Da, Ben L., Gene Y. Im, Schiano, T.D., Coronavirus disease 2019 hangover: a rising tide of alcohol use disorder and alcohol-associated liver disease. *Hepatology*, 72, no. 3, 1102–1108, 2020.
15. Nusair, S., Abnormal carbon monoxide diffusion capacity inCOVID-19 patients at time of hospital discharge. *Eur. Respir. J.*, 56, 1, 2001832. PMID: 32703822; PMCID: PMC7376285, Jul 23 2020.
16. Mo, X., Jian, W., Su, Z., Chen, M., Peng, H., Peng, P., ... & Li, S., Abnormal pulmonary function in COVID-19 patients at time of hospital discharge. *Eur. Respir. J*, 55, 6, 2020.
17. Chen, R., Gao, Y., Chen, M., Jian, W., Lei, C., Zheng, J., Li, S., Impaired pulmonary function in discharged patients with COVID-19: more work ahead. *Eur. Respir. J.*, 56, 1, 2002194. PMID: 32586883; PMCID: PMC7315814, 2020 Jul 23, https://doi.org/10.1183/13993003.02194-2020.
18. Liu, J., Zheng, X., Tong, Q., Li, W., Wang, B., Sutter, K., Yang, D., Overlapping and discrete aspects of the pathology and pathogenesis of the emerging human pathogenic coronaviruses SARS-CoV, MERS-CoV, and 2019-nCoV. *J. Med. Virol.*, 92, 5, 491–494, 2020.

19. Wu, C., Chen, X., Cai, Y., Zhou, X., Xu, S., Huang, H., Zhang, L. *et al.*, Hanping Huang, Li Zhang et al. "Risk factors associated with acute respiratory distress syndrome and death in patients with coronavirus disease 2019 pneumonia in Wuhan, China. *JAMA Intern. Med.*, 180, 7, 934–943, 2020.
20. Carfì, A., Bernabei, R., Landi, F., Persistent symptoms in patients after acute COVID-19, in: *Jama*, vol. 324, pp. 603–605, 2020.
21. Zhao, Y.M., Shang, Y.M., Song, W.B. *et al.*, Follow-up study of the pulmonary function and related physiological characteristics of COVID-19 survivors three months after recovery. *EClinicalMedicine.*, 25, 100463, 2020, https://doi.org/10.1016/j.eclinm.2020.100463.
22. Mallineni, S.K., Bhumireddy, J.C., Nuvvula, S., Dentistry for children during and post COVID-19 pandemic outbreak, in: *Children and Youth Services Review*, p. 105734, 2020.
23. Lin, J.E., Asfour, A., Sewell, T.B., Hooe, B., Pryce, P., Earley, C., Shen, M.Y., *et al.*, Neurological issues in children with COVID-19. *Neurosci. Lett.*, 743, 135567, 2021.
24. Alhiyari, M.A., Ata, F., Alghizzawi, M.I., Bilal, A.B.I., Abdulhadi, A.S., Yousaf, Z., Post COVID-19 fibrosis, an emerging complicationof SARS-CoV-2 infection. *IDCases*, 23, e01041, 2021.
25. Lake, M.A., What we know so far: COVID-19 current clinical knowledge and research, in: *Clinical Medicine*, vol. 20, p. 124, 2020.
26. Liu, K., Chen, Y., Lin, R., Han, K., Clinical features of COVID-19 in elderly patients: A comparison with young and middle-aged patients. *J. Infect.*, 80, 6, e14-e18, 2020.
27. Ata, F., Almasri, H., Sajid, J., Yousaf, Z., COVID-19 presenting with diarrhoea and hyponatraemia, in: *BMJ Case Reports CP*, vol. 13, p. e235456, 2020.
28. Shams, A., Ata, F., Mushtaq, K., Munir, W., Yousaf, Z., Coronary thrombosis in a young male with COVID-19. *IDCases*, 21, e00923, 2020, http://dx.doi.org/10.1016/j.idcr.2020.e00923.
29. Davenport, T. and Kalakota, R., The potential for artificial intelligence in healthcare. *Future Healthc. J.*, 6, 2, 94–98, 2019, https://doi.org/10.7861/futurehosp.6-2-94.
30. Pawar, U., O'Shea, D., Rea, S., O'Reilly, R., Explainable AI in Healthcare. *2020 International Conference on Cyber Situational Awareness, Data Analytics and Assessment (CyberSA)*, Dublin, Ireland, pp. 1–2, 2020.
31. Jiang, F., Jiang, Y., Zhi, H., Dong, Y., Li, H., Ma, S., Wang, Y., Dong, Q., Shen, H., Wang, Y., Artificial intelligence in healthcare: past, present and future. *Stroke Vasc. Neurol.*, 2, 4, 230–243, 2017.
32. Velásquez, R.M.A. and Lara, J.V.M., Knowledge management in two universities before and during the COVID-19 effect in Peru. *Technol. Soc*, 101479, 2020.
33. Jiang, B., You, X., Li, K., Li, T., Zhou, X., Tan, L., Interactive Analysis of Epidemic Situations Based on a Spatiotemporal Information Knowledge Graph of COVID-19, in: *IEEE Access*, 2020.

34. Georgiou, K., Mittas, N., Angelis, L., Chatzigeorgiou, A., A preliminary Study of Knowledge Sharing related to Covid-19 Pandemic in Stack Overflow. *2020 46th Euromicro Conference on Software Engineering and Advanced Applications (SEAA)*, Portoroz, Slovenia, pp. 517–520, 2020.
35. Vinod, D.N. and Prabaharan, S.R.S., Data science and the role of Artificial Intelligence in achieving the fast diagnosis of Covid-19. *Chaos, Solit. Fractals*, 140, 110182, 2020.
36. Pasha, D.F., Lundeen, A., Yeasmin, D., Fayzul M., Pasha, K., An analysis to identify the important variables for the spread of COVID-19 using numerical techniques and data science. *Case Studies in Chemical and Environmental Engineering*, 3, 100067, 2021.
37. Swapnarekha, H., Behera, H.S., Nayak, J., Naik, B., Role of intelligent computing in COVID-19 prognosis: A state-of-the-art review. *Chaos Solit. Fractals*, 138, 109947, 2020.
38. SShi, F., Wang, J., Shi, J., Wu, Z., Wang, Q., Tang, Z., He, K., Shi, S., Shen, D., Review of artificial intelligence techniques in imaging data acquisition, segmentation, and diagnosis for COVID-19. *IEEE Rev. Biomed. Eng.*, 14, 4–15, 2020.
39. Alotaibi, S.R., Applications of artificial intelligence and big data analytics in m-health: a healthcare system perspective. *J. Healthc. Eng.*, 2020 (2020).
40. Fox, J., South, M., Khan, O., Kennedy, C., Ashby, P., Bechtel, J., OpenClinical.net: Artificial intelligence and knowledge engineering at the point of care *BMJ Health Care Inform.*, 27, 2, 2020.
41. Leung, C.K., Chen, Y., Shang, S., Deng, D., Big Data Science on COVID-19 Data, in: *2020 IEEE 14th International Conference on Big Data Science and Engineering (BigDataSE)*, pp. 14–21, 2020, December.
42. Feigenbaum, Edward A. The art of artificial intelligence. 1. *Themes and case studies of knowledge engineering*. Stanford Univ CA Dept of Computer Science, 1977.
43. Sengupta, P.P., Huang, Y.-M., Bansal, M., Ashrafi, A., Fisher, M., Shameer, K., Gall, W., Dudley, J.T., Cognitive machine-learning algorithm for cardiac imaging: a pilot study for differentiating constrictive pericarditis from restrictive cardiomyopathy. *Circulation: Cardiovascular Imaging*, 9, 6, e004330, 2016.
44. Fiszman, M., Chapman, W.W., Aronsky, D., Evans, R.S., Haug, P.J., Automatic detection of acute bacterial pneumonia from chest X-ray reports. *J. Am. Med. Inform. Assoc.*, 7, 6, 593–604, 2000.
45. Narula, S., Shameer, K., Salem Omar, A.M., Dudley, P.P., Sengupta, P.P., Machine-learning algorithms to automate morphological and functional assessments in 2D echocardiography. *J. Am. Coll. Cardiol.*, 68, 21, 2287–2295, 2016.
46. Madanian, S., Airehrour, D., Kumar, N. *et al.*, Smart cap for visually impaired in disaster situations, in: *Proceedings of the ISCRAM Asia Pacific*, Wellington, New Zealand, 2018.

47. Pramanik, M., II, Lau, R.Y.K., Demirkan, H., Azad, M.A.K., Smart health: big data enabled health paradigm within smart citie. *Expert Syst. Appl.*, 87, 370–383, 2017.
48. Madanian, S., Parry, D.T., Airehrour, D., Cherrington, M., mHealth and big-data integration: promises for healthcare system in India. *BMJ Health Care Inform.*, 26, Article ID e100071, 2019.
49. Gulshan, V., Peng, L., Coram, M., Stumpe, M.C., Wu, D., Narayanaswamy, A., Venugopalan, S., Widner, K., Madams, T., Cuadros, J., Kim, R., Development and validation of a deep learning algorithm for detection of diabetic retinopathy in retinal fundus photographs. *Jama*, 316, 22, 2402–2410, 2016.
50. Esteva, A., Kuprel, B., Novoa, R.A., Ko, J., Swetter, S.M., Blau, H.M., Thrun, S., Dermatologist-level classification of skin cancer with deep neural networks. *nature*, 542, 7639, 115–118, 2017.

13

Knowledge Engineering Challenges in Smart Healthcare Data Analysis System

Agasba Saroj S. J.[1], B. Saleena[2*] and B. Prakash[2]

[1]*Srija Kid's World, Fahaheel, Kuwait, (Alumni, VIT, Chennai Campus, India)*
[2]*School of Computer Science and Engineering, VIT, Chennai Campus, India*

Abstract
Diabetes mellitus is one of the significant risk-prone diseases in the world. More than diabetes alone, the consequences caused because of diabetes are entirely unknown to many, even to some who have diabetes. The symptoms caused for each disease are unaware and left unnoticed, leading to several chronic diseases, mainly due to being severely diabetic. The proposed intelligent system (IS) is intended for diabetic and pre-diabetic patients to perform a self-test based on their symptoms. This IS is designed so that the patients can analyze their diabetic consequence disease by responding to the set of questions asked based on the symptoms on what the patients are undergoing and validates with the systems knowledge base and provides an appropriate suggestion. An inference engine processes the analysis, and the results (disease) are displayed along with the treatments, causes, and any further investigation suggestions. This system greatly helps diabetic patients analyze their risk level based on their health condition and helps them determine the appropriate doctors for further consultation on a need basis. This particular system is based on the patient's responses as a survey. The results are further analyzed using the developed base system, such that the patient's data leads to suitable consultation.

Keywords: Diabetic diseases, intelligent system, knowledge base, symptoms, prolog

13.1 Introduction

Intelligent systems (ISs) for diabetes are proposed to deal with complex issues by considering types of learning, implied mostly as though then principles than through unsurprising procedural code. An IS will be

**Corresponding author*: saleena.b@vit.ac.in

A. Jaya, K. Kalaiselvi, Dinesh Goyal and Dhiya Al-Jumeily (eds.) Handbook of Intelligent Healthcare Analytics: Knowledge Engineering with Big Data Analytics, (285–308) © 2022 Scrivener Publishing LLC

isolated as duplet subsystems: the knowledge base and inference engine. The knowledge base determines the facts and rules. The inference engine spreads the rules to the distinguished facts to find out the new facts. IS for diabetes is an artificial intelligence–based system that diabetic patients can use for directing and looking at their current indispensable health data from their homes [1]. Diabetes, a drawn-out illness, requires consistent clinical consideration and self-organization, guidance for patients to avoid serious disarray, and diminishing the peril of challenges related to their health condition in the long run. The most well-known type of diabetes begins from a dynamic insulin discharge deformity. The body does not create sufficient insulin where the insulin delivered by blood is not adequate for the body. IS are computerized algorithms that deal with issues in a non-procedural way by using human beings' health data to duplicate human reasoning and thinking. This approach is called a knowledge-based system or inference-based program. This proposed method is a rule-based IS utilizing artificial intelligence to manage the health condition of diabetic patients.

Clinical features such as diabetes mellitus were depicted 3,000 years back by the old-fashioned Egyptians. The articulation "diabetes" was initially established by Araetus of Cappodocia (81–133AD). Later, the word "mellitus" (nectar sweet) was incorporated by Thomas Willis (Britain) in 1675 in the rouse of regaining the pleasantness of urine and blood of sufferers (initially observed by the old Indians). It was particularly in 1776 that Dobson (Britain) directly off the bat insisted on the proximity of excess sugar in urine and blood as an explanation behind their pleasantness. In the current time, the authentic setting of diabetes blended with the advancement of preliminary medicine. A huge accomplishment all through diabetes is the establishment at work of the liver in glycogenesis, and the possibility that diabetes is a direct result of excess glucose creation Claude Bernard (France) in 1857. The activity of the pancreas in the pathogenesis of diabetes was found by Mering and Minkowski (Austria) 1889. Later, this disclosure built up the reason for insulin partition and clinical use by Banting and Best (Canada) in 1921. This description makes the dietary organization's verifiable background and extraordinary and unending traps of diabetes.

India is considered one of the most affected by diabetes mellitus. Quick money-related improvement and food portion changes, which was a nearby extended weakness for Indian individuals, have incited the precarious augmentation in the power of diabetes mellitus in India throughout the ongoing decennium. Type 2 diabetes mellitus, particularly with Asian Indian people, is depicted by an energetic starting and occasion at low BMI levels. The available health data recommend a lack of protection of Asian Indian people to the complexities of diabetes mellitus compared to that of white masses. Novel interference using quickly available resources and development assurance changes the thought of patients with diabetes mellitus in India. The similar nature of these problems is ordinary to most making countries of the world. The activities picked up from India's contribution to diabetes mellitus are presumably going to be of huge overall essentialness.

Diabetes is a very acute disease requiring extra attention and regular checkup to avoid its consequences. A combination of diabetes and its diabetic-related consequences creates a severe threat that requires periodic monitoring and continuous health checkup. The study also points out that the chances of neuronal dysfunction, impaired pulmonary function, or increased risk of cancer are complications due to chronic diabetes patients [2]. Another recent study conducted among researchers revealed that the cause of major health complications such as retinopathy, amputations, and nephropathy is mainly due to Type 1 and Type 2 diabetes mellitus [3]. The purpose of this study was to comprehend the current information, insights, and worries of people living with diabetes, explicitly zeroing in on diabetes, its unexpected drawn-out problems, and its relationship with complicated diseases. The study conducted by Albert J Farias and his team revealed that young individuals with diabetes are at an expanded danger of affecting pancreatic malignancy [4]. The research also shows that it is uncertain about associating diabetes-related complications with the risk of pancreatic cancer.

In recent times, in solving these kinds of medical problems, an emerging artificial intelligence helps to a greater extent for understanding the hidden knowledge discovered by defining the appropriate rules and facts [1, 5, 6]. An intelligent healthcare system was developed using Prolog to make it more effective and efficient. An IS is a connected system-based choice instrument that utilizes the two realities and heuristics to take care of troublesome choice issues in light of information gained from an intelligent source [7].

13.1.1 Motivation

Over 80% of diabetes demises happen in low- and middle-salary nations. In developed countries, many people with diabetes are over the time of retirement, though in developing nations, the most frequently affected people are aged between 35 and 64 years. World Health Organization (WHO) projects that diabetes would become the seventh, driving reason for an overall demise rate during the 2030s. The survey shows that an absolute demise rate due to diabetes anticipates ascending by over half in the upcoming 10 years. The researchers are worried that India has the grievous benefit of being the "Diabetes Capital" of the world. Additional concerning is the way that diabetes commonness in recent decades has expanded fourfold. Another intriguing marvel is that Indians who move to other nations grow extremely high pervasiveness pass of 10% to 20%. The development of living standards is one of the major causes of the increase in diabetes and similar weakness. According to another recent survey [8] as depicted in Figures 13.1 and 13.2, the national prevalence of diabetes is 10.9%, pre-diabetes is 14%, gestational diabetes is 5.1%,

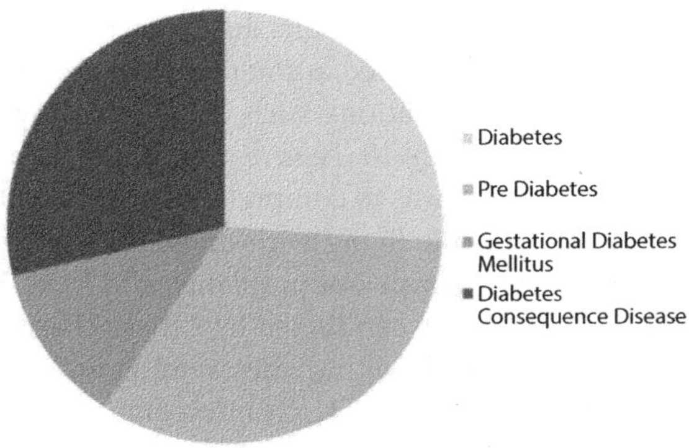

Figure 13.1 Diabetes survey as per the category.

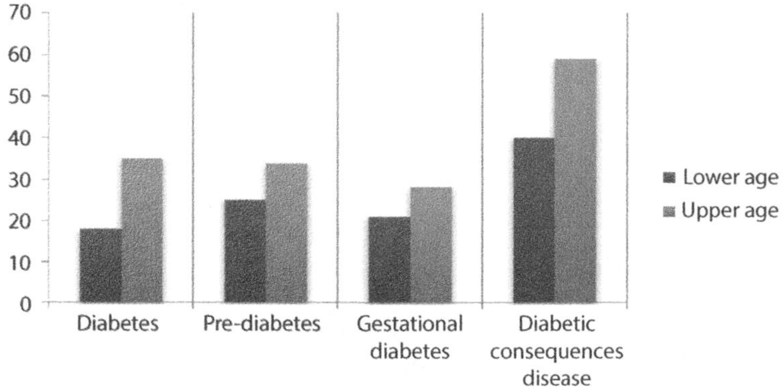

Figure 13.2 Diabetes survey as per the age range.

and diabetic consequences diseases are 12% of India's population (1.37 billion). The most affected age group for diabetes is 18–35 years, pre-diabetes is 25–34 years, gestational diabetes is 21–25 years, and diabetic consequences diseases are 40–59 years. In recent times, to handle these issues, computerized support helps the situation to reduce or control the current risky condition in the future. The ISs being developed help the patients by checking their health conditions associated with diabetes from home.

13.2 Ongoing Research on Intelligent Decision Support System

To compose the rules for health maintenance knowledge-based decision system and deliver through revelations and advice through the forward binding inferential methods, which is thinking from realties to the ends coming about because of those realities, was clarified by Ping-Tsai Chung and Bing-Xing Chen in 2011 [9]. Further work was described in three stages: First, the system will incorporate the warning of influenza illness and the sickness' consulting into the structure. Second, the framework actualizes and builds the far-reaching examination of the danger of a heart-associated malady. Third, the system will plan and build up a protected web service for healthy life.

An intelligent clinical decision support system was introduced by Gabriel Spiridon *et al.* [10] for evaluating and foreseeing cardiovascular confusions in patients with diabetes mellitus, organizing a lot of rules in the middle of different hazard elements, hereditary tests, clinical history, dietary and ways of life factors, and the degree and seriousness of cardiovascular entanglement in patients with diabetes mellitus. They also created a prescient representation and development of cardiovascular difficulties for patients with diabetes mellitus. Based on the broadening of clinical rules, the essential principles they characterize with the new arrangements of rules are evaluated with the support of a neural system.

Jyothi *et al.* [11] illustrated the effect of significant hazards in diabetes in gestational diabetes, and this system will help in four significant advances: First, this system will help in the early discovery of hazards. Second, this system will give guardians and specialists time to adapt to inconveniences. Third, this system's outcome will be of extraordinary importance for the clinical industry. Fourth, this system will defeat human blunders that are made during manual diagnostics. It was disclosed to foresee when the children will experience the ill effects of diabetes who are conceived from gestational diabetes pregnancy.

An Advanced Bolus Calculator for Diabetes (ABC4D) was proposed by Peter Pesl *et al.* [12], which gives customized and versatile insulin suggestions. ABC4D executes an in silico–approved calculation dependent on CBR, which is separated in the middle of different diets and ways of life, to guarantee that solitary safe adaption of the bolus mini-computer frameworks is executing to part the CBR cycle into a patient stage and a clinical stage for management.

Sanyukta *et al.* [13] proposed a method that monitors the patient's very own propensity, by considering the calorie admission levels. By utilizing Diabot, the client can find solutions to their inquiries and distinguish the sorts of diabetes they may have. The client can check for proposals given by the master system, along these lines keeping up their wellbeing. The patient can check their pulse and internal heat level. Diabot is a solid human service bot that will be extremely helpful for diabetic patients. Further work should be possible by incorporating more sensors in this proposed structure, similar to a blood pressure sensor, an inbuilt glucose meter sensor, and so forth. By including more sensors, this structure will have the option to take more limits from the patient, and thus, it can deliver increasingly exact outcomes.

A web-based expert system provides a self-administered diagnosis for candidiasis conditions [14]. The system provides a diagnosis in conjunction with laboratory test verification before treatment advice and remedial measures are recommended. Requirement gathering analysis and specification were carried out through informal questionnaires and one-on-one sessions with healthcare providers. Simple production rules, as well as coding questions and answers, were employed for inferences. In the present situation of pandemics, the rapid collection and testing of patients' specimens are recommended. Hence, if a rule-based expert system with productions and datasets can be produced globally, then artificial intelligence solutions would prove to be combative in ameliorating the dangers of the escalating pandemic.

The whole KBS was trained with three algorithms they are J48, PART, and JRip [15]. The trained KBS showed the final result as JRip rule induction is the second most performing model next to the J48 model whereas PART was the last performing classifier. Therefore, rules generated by the J48 model were used for prototype KBS for diagnosis and treatment of diabetes. The knowledge about diabetes was represented using Clisp.net. However, they find it difficult to update the rules of the KBS as the rules were generated automatically and manually. Further work could be integrating automatic generating rules from the database to the knowledge base to generate the rule so that data mining techniques can become easy.

The main aim of this medical expert system is to detect early-stage detection of kidney disease using the IS to save lives, the comprehensive review of medical expert systems, and the best methods used to recognize chronic kidney disease [16]. This paper compares the methods based on the performance measures like accuracy, precision, sensitivity, and F-measure and found that the J48 algorithm gave the highest performance measures.

Samit *et al.* [17] analyzed the impact of coronavirus infection on worsening glycosylated hemoglobin and increased diabetes-related complications. The study finding suggests that the lockdown impact is directly proportional to the deterioration of glycemic control and diabetes-related complexities, which will put extra load on the overburdened medical care framework and increment COVID-19 diseases. Wu *et al.* [18] conducted a study based on the Hong Kong Diabetes Surveillance Database (HKDSD) to determine the occasions of coronary heart disease (CHD), stroke, cardiovascular breakdown, and hyperglycaemic emergency by analyzing the patients' discharge summary records with the help of diagnosis codes. The Joinpoint algorithm has been adopted to depict frequency patterns based on age and sex. The study findings depict that the event rates of diabetes-related complications were reduced significantly and no proper evidence was reported about the increase in Hong Kong until 2016. The study also revealed that improvement is applicable for all age subgroups except young aged people with diabetes, which shows the necessity of health measures to improve quality of life care to reduce diabetes-related health complications among young people. Reza *et al.* [19] performed a comparative analysis of Euclidean distance and nearest neighbor algorithm in an expert system for diagnosing diabetes mellitus. The study reveals that the Euclidean distance method performs better in terms of accuracy of predicting diabetes mellitus than the nearest neighbor method.

13.3 Methodology and Architecture of the Intelligent Rule-Based System

The proposed venture's principle target is to plan a specialist framework for the diagnosis of diabetes and furnish the patient with a proportional consultancy as such given by a specialist. Hence, the patient has simple access to medicinal services. From this proposed IS, there might be a need to include a brilliant and easy-to-understand system for diabetic patients, and a professional platform for diagnosing diabetes using artificial intelligence.

The proposed system comprises the following functional and non-functional requirements.

Functional requirements display the IS from a user perspective (patient) and describe the system's capabilities in terms of behavior or operations. A regular IS may be as per the following:

- Mandatory prerequisite: The patient should be aware of his symptoms what he/she is feeling or undergoing.
- Desirable prerequisite: The patient can feel one or more symptoms, and he might be confused so the IS asks questions of all the symptoms that a person can feel being a diabetic.
- Optional prerequisite: The patient might know the disease but might not know what treatments or investigations to take or to which department the patient should visit.
- Possible future improvement: The patient gets an appointment with the department of a doctor so that the emergency is avoided.

Non-functional requirements portray framework traits, to explain this with an example, security, reliability, maintainability, scalability, and usability. These attributes are referred to as abilities. These requirements are just as critical as the functional requirements as they ensure the usability and efficiency of the system. Once a systems start to fail to meet any one of these attributes, it can result in system failures.

13.3.1 Proposed System Design

The architecture diagram of the proposed IS for diabetes is illustrated in Figure 13.3. The system considers symptoms, knowledge, and rules for processing and provides output in terms of medications or consultations. The primary user of this system is the patient.

Symptoms: For all the 15 cases, certain verified symptoms are specified for each case. For type 1 diabetes, the symptoms are excess thirst, frequent urination, frequent urge to urinate, excessive hunger, weight loss

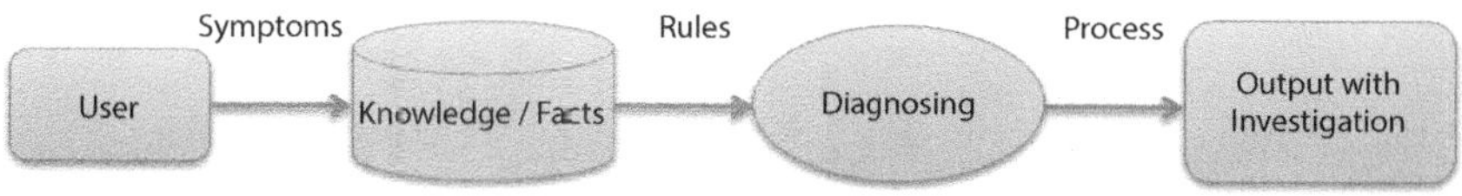

Figure 13.3 Architecture diagram of the intelligent system for diabetes.

developed, weakness, fatigue and shortness of breath. These symptoms are asked as a question to the patient and answer patient.

Knowledge/Facts: Those are the pre-determined data for verification. Each case is assigned, and their symptoms are analyzed, and then, the particular disease, treatment, and consultation will be displayed.

Rules: Major component of this system as it controls the design and analyzes the exact disease concerning the patient's symptoms and displays its treatment and consultation.

Process: The two main algorithms are forward chaining and backward chaining. The forward-chaining works from known facts (symptoms) and moves forward by applying inference rules to extract the disease. The backward chaining works from the goal (disease) move backward by using inference rules to determine the facts (symptoms) that satisfy the goal.

Output: This IS provides the patient with the disease specifically analyzed by their symptoms and suggests what treatment to take, what investigations to take further, and to which department they have to consult further.

The overall process flow of the proposed system is illustrated in Figure 13.4. Initially, the system asks for the user name and then the user gives inputs like his list of symptoms what he is undergoing according to the questions being asked. When this process completes, the data is analyzed. The inference engine then compares the received data to its set of rules present in the system's knowledge base. The symptoms are classified and identified to find disease and followed by its treatments, investigations, and which department of doctor that the patients should refer further.

13.3.2 Algorithms Used

13.3.2.1 Forward Chaining

It is an information-driven strategy for getting a specific objective from a given information base and set of inference rules. Inference rules are

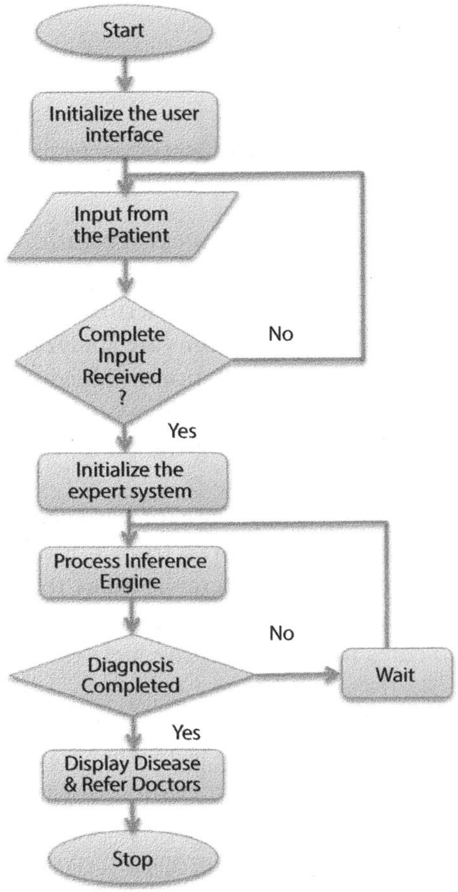

Figure 13.4 Process flow of proposed intelligent system for diabetes.

applied by coordinating facts to the precursors of result relations in the knowledge base. The utilization of interface rules brings about new information (from the consequents of the relations coordinated), which is then added to the inference base.

13.3.2.2 Backward Chaining

It is an objective-driven strategy for getting a specific objective from a given information base and set of inference rules. Inference rules are applied by coordinating the objective of the pursuit to the consequents of the relations store in the knowledge base.

The knowledge base of the IS for diabetes is stored with a distinct hypothesis for each type of consequences of diseases caused by diabetes; the symptoms are referred; according to the disease caused due to diabetes, when the patient gives input to the intelligent engine, it compares the input (symptoms) with the data in the knowledge base and infers output (disease) according to the hypothesis. The IS for diabetes has set rules for 15 specific diseases caused due to diabetes. They are as follows:

i. Type 1 diabetes mellitus
ii. Type 2 diabetes mellitus
iii. Gestational diabetes mellitus
iv. Diabetic retinopathy
v. Diabetic nephropathy
vi. Diabetic neuropathy
vii. Osteoporosis
viii. Tia/stroke
ix. Diabetic foot
x. Heart disease
xi. Urinary tract infection
xii. Cellulitis
xiii. Onychomycosis
xiv. Glaucoma
xv. Progressive hearing loss

13.4 Creating a Rule-Based System using Prolog

Prolog consists of several clauses. Each clause is either a fact or a rule. After a Prolog program is loaded (or consulted) in a Prolog interpreter, users can submit (response) to the goals or queries, and the Prolog interpreter will give results (answers) according to the facts and rules [20, 21]. When a prolog program loads, the user submits/responds to the goals or queries, and here, it is structured in a way as below (these are the facts for type_one_diabetes).

symptom(Patient,excess_thrist) :- verify(Patient," have a excess_thrist (y/n) ?").

symptom(Patient,frequent_urination) :- verify(Patient," have a frequent_urination (y/n) ?").
symptom(Patient,frequent_urge_to_urinate) :- verify(Patient," have a frequent_urge_to_urinate (y/n) ?").
symptom(Patient,excessive_hunger) :- verify(Patient," have a excessive_hunger (y/n) ?"). symptom(Patient,weight_loss_developed) :- verify (Patient," have a weight_loss_developed (y/n) ?").
symptom(Patient,weakness) :- verify(Patient," have a weakness (y/n) ?").
symptom(Patient,fatigue) :- verify(Patient," have a fatigue (y/n) ?").
symptom(Patient,shortness_of_breathe) :- verify(Patient," have a shortness_of_breathe (y/n) ?").

Each goal or query is linked with the fact, so that the query is asked to the user by the system, to make sure that the user responds to the questions, which they require to answer, and to analyze the result.

A fact or a hypothesis is literal. A functor whose syntax is the same for predicates and arithmetic functions has arity e.g., the number of arguments. arity=1 is unary, arity=2 is binary or for a higher degree, just arity. Fact or rule are structured as **"hypothesis(Patient,type_one_diabetes)"**.

"hypothesis(Patient,type_one_diabetes)" is the clause and each clause is linked/associated with a fact and that is **"symptoms(Patient,excess_thrist)", symptoms(Patient,frequent_urination)","symptoms(Patient, frequent_urge_to_urinate)"** and so on for one disease, all the possible symptoms to diagnose a single diabetic consequences. This is how a hypothesis is written for all the diabetic consequences listed in this proposed work.

The arity of a functor is the number of arguments it takes. For example, the arity of hypothesis/2, as in hypothesis(Patient,type_one_diabetes), is 2, as it takes two arguments—patient and type_one_diabetes.

By the **"hypothesis(Patient, type_one_diabetes)"** the queries related to type_one_diabetes are asked by the system, and accordingly, the result is displayed.

Facts and Rules for Type one Diabetes

```
symptom(Patient,excess_thrist) :- verify(Patient," have a excess_thrist
(y/n) ?").
symptom(Patient,frequent_urination) :- verify(Patient," have a fre-
quent_urination (y/n) ?").
symptom(Patient,frequent_urge_to_urinate) :- verify(Patient," have
a frequent_urge_to_urinate (y/n) ?").
symptom(Patient,excessive_hunger) :- verify(Patient," have a exces-
sive_hunger (y/n) ?"). symptom(Patient,weight_loss_developed) :-
verify(Patient," have a weight_loss_developed (y/n) ?").
symptom(Patient,weakness) :- verify(Patient," have a weakness (y/n) ?").
symptom(Patient,fatigue) :- verify(Patient," have a fatigue (y/n) ?").
symptom(Patient,shortness_of_breathe) :- verify(Patient," have a
shortness_of_breathe (y/n) ?").

hypothesis(Patient,type_one_diabetes) :-
    symptom(Patient,excess_thrist),
    symptom(Patient,frequent_urination),
    symptom(Patient,frequent_urge_to_urinate),
    symptom(Patient,excessive_hunger),
    symptom(Patient,weight_loss_developed),
    symptom(Patient,weakness),
    symptom(Patient,fatigue),
    symptom(Patient,shortness_of_breathe),
    write('Treatment :'), nl,
    write('1: Diet Control'), nl,
    write('2: Frequent blood sugar monitoring'), nl,
    write('3: Healthy Weight'), nl,
    write('4: Insulin Therapy'), nl,nl,nl,
    write('Further Investigations: '), nl,
    write('1: HbA1c'), nl,
    write('2: Fasting Blood Sugar'), nl,
    write('3: Post-Prandial'), nl,nl,nl,
    write('You can consult to Diabetologist '), nl.
```

Results - "**write('Treatment :')**", is a simple output format in Prolog to display the results according to the facts and rules.

The user responds to corresponding goals or queries; according to the response given by the user, the fact or hypothesis works with the arity and proceeds to the next goal or query. Until the system reaches the desired result associated with the hypothesis, the goal or query is asked/thrown. The prolog interpreter eliminates and decides the next goal/query which relies on the response received from the user. Accordingly, the interpreter works and decides the results.

Here, for type 1 diabetes, the user responds to the query, the system throws a query of **"Mira, Do you have excess thirst ?"** and the user responds **"y/n"** to the system. The system receives the input from the user in the below format for type_one_diabetes and the prolog interpreter proceeds to the next questions (fact).

CASE 1: Type_one_diabetes. (when a user responds to all the facts associated with type one diabetes)

"Mira, Do you have a excess_thrist (y/n) ?" – y

"Mira, Do you have a frequent_urination (y/n) ?" – y

"Mira, Do you have a frequent_urge_to_urinate (y/n) ?" – y

"Mira, Do you have a excessive_hunger (y/n) ?" – y

"Mira, Do you have a weight_loss_developed (y/n) ?" – y

"Mira, Do you have a weakness (y/n) ?" – y

"Mira, Do you have a fatigue (y/n) ?" – y

"Mira, Do you have a shortness_of_breathe (y/n)?" – y

Once the user responds to all the facts that are associated with type one diabetes, the interpreter will give the result as "Mira, You probably have type one diabetes" along with simple/basic treatment to follow, further investigations, and which department of doctor to consult.

Output –

Treatment:
1: Diet Control
2: Frequent blood sugar monitoring
3: Healthy weight
4: Insulin Therapy

Further Investigations:
1: HbA1c
2: Fasting blood sugar
3: Post-prandial

You can consult a Diabetologist

CASE 2: Gestational_diabetes (when a user responds to "n" to certain symptoms and "y" to certain symptoms, it leads to the diabetic consequences based on the response)

When the user gives "n" to the query, it asks for the next disease query; on that way, here, it for the next disease, gestational diabetes. The user responds to the system for gestational diabetes symptoms and answers "y". Below are the facts, rules, and hypotheses for gestational diabetes (code and output).

```
symptom(Patient,vomiting) :- verify(Patient," have a vomiting (y/n) ?").
symptom(Patient,weight_change) :- verify(Patient," have a weight_change (y/n) ?").
symptom(Patient,nausea) :- verify(Patient," have a nausea (y/n) ?").
symptom(Patient,unusal_thrist) :- verify(Patient," have a unusal_thrist (y/n) ?").
symptom(Patient,fatigue) :- verify(Patient," have a fatigue (y/n) ?").
symptom(Patient,itching_in_vulva) :- verify(Patient," have a itching_in_vulva (y/n) ?").
symptom(Patient,abnormal_vaginal_infections) :- verify(Patient," have a abnormal_vaginal_infections (y/n) ?").
    hypothesis(Patient,gestational_diabetes_mellitus) :-
    symptom(Patient,vomiting),
    symptom(Patient,weight_change),
    symptom(Patient,nausea),
    symptom(Patient,unusal_thrist),
    symptom(Patient,fatigue),
    symptom(Patient,itching_in_vulva),
    symptom(Patient,abnormal_vaginal_infections),
    write('Treatment : '),
    nl,
    write('1: Monitoring Blood Glucose Level'),
    nl,
```

```
write('2: Healthy Diet'),
nl,
write('3: Mild Exercise'),
nl,nl,nl,
write('Further Investigations: '),
nl,
write('1: HbA1c'),
nl,
write('2: Fasting Blood Sugar'),
    nl,
    write('3: Post-Prandial'),
    nl,
    write('4: Abdomen Scanning'),
    nl,nl,nl,
    write('You can consult to Diabetologist and Gynaecologist'),
    nl.
```

System Display –

"Mira, Do you have an excess_thrist (y/n) ?" – y

"Mira, Do you have frequent_urination (y/n) ?" – n

"Mira, Do you have a slow_healing_wound (y/n) ?" – n

"Mira, Do you have vomiting (y/n) ?" – y

"Mira, Do you have a weight_change (y/n) ?" – y

"Mira, Do you have nausea (y/n) ?" – y

"Mira, Do you have a unusal_thrist (y/n) ?" – y

"Mira, Do you have a fatigue (y/n) ?" – y

"Mira, Do you have an itching_in_vulva (y/n) ?" – y

"Mira, Do you have an abnormal_vaginal_infections (y/n) ?" – y

Once the user responds to all the facts associated with gestational diabetes, the interpreter will give the result as "Mira, You probably have gestational diabetes." along with simple/basic treatment to follow, further investigations, and which department of doctor to consult.

Output –
Treatment:
1: Monitoring Blood Glucose Level
2: Healthy Diet
3: Mild Exercise
Further Investigations:
1: HbA1c
2: Fasting blood sugar
3: Post-prandial
4: Abdomen Scanning

You can consult a Diabetologist and Gynaecologist

CASE 3: Unable to diagnose (when a user responds to the combination of "y" and "n" and when the user undergoes non-diabetic consequences, this case arises)

If in case the person undergoes a combination of symptoms that are not related to diabetes or its consequences, then the system asks a query from all the diseases in the order of the responses received from the user, and an example is shown below:

```
symptom(Patient,excess_thrist) :- verify(Patient," have a excess_thrist (y/n) ?").
symptom(Patient,frequent_urination) :- verify(Patient," have a frequent_urination (y/n) ?").
symptom(Patient,slow_healing_wounds) :- verify(Patient," have a slow_healing_wounds (y/n) ?").
symptom(Patient,vomiting) :- verify(Patient," have a vomiting (y/n) ?").
symptom(Patient,blurry_vision) :- verify(Patient," have a blurry_vision (y/n) ?").
symptom(Patient,gradual_loss_of_vision) :- verify(Patient," have a gradual_loss_of_vision (y/n) ?").
symptom(Patient,loss_of_appetite) :- verify(Patient," have a loss_of_appetite (y/n) ?").
symptom(Patient,tingling_sensation) :- verify(Patient," have a tingling_sensation (y/n) ?").
```

```
symptom(Patient,bone_pain) :- verify(Patient," have a bone_pain (y/n) ?").
symptom(Patient,trouble_speaking) :- verify(Patient," have a trouble_speaking (y/n) ?").
symptom(Patient,foot_ulcer) :- verify(Patient," have a foot_ulcer (y/n) ?").
symptom(Patient,chest_pain_discomfort) :- verify(Patient," have a chest_pain_discomfort (y/n) ?").
symptom(Patient,pressure_in_chest_or_arms) :- verify(Patient," have a pressure_in_chest_or_arms (y/n) ?").
symptom(Patient,burning_while_urination) :- verify(Patient," have a burning_while_urination (y/n) ?").
symptom(Patient,red_area_of_skin) :- verify(Patient," have a red_area_of_skin (y/n) ?").
symptom(Patient,discolouration_of_nails_white_black_yellow_green) :- verify(Patient," have a discolouration_of_nails_white_black_yellow_green (y/n) ?").
symptom(Patient,cloudy_vision) :- verify(Patient," have a cloudy_vision (y/n) ?").
symptom(Patient,redness) :- verify(Patient," have a redness (y/n) ?").
symptom(Patient,frequently_asking_others_to_repeat_themselves) :- verify(Patient," have a frequently_asking_others_to_repeat_themselves (y/n) ?").
symptom(Patient,trouble_following_conversations) :- verify(Patient," have a trouble_following_conversations (y/n) ?").
hypothesis(_,"I'm Sorry, I am unable to diagnose the disease").
```

System Display –

"Mira, Do you have an excess_thrist (y/n) ?" – y

"Mira, Do you have frequent_urination (y/n) ?" – n

"Mira, Do you have a slow_healing_wound (y/n) ?" – n

"Mira, Do you have vomiting (y/n) ?" – n

"Mira, Do you have a blurry_vision (y/n) ?" – y

"Mira, Do you have a gradual_loss_of_vision (y/n) ?" – n

"Mira, Do you have a loss_of_appetite (y/n) ?" – n

"Mira, Do you have a tingling_sensation (y/n) ?" – n

"Mira, Do you have a bone_pain (y/n) ?" – n

"Mira, Do you have a trouble_speaking (y/n) ?" – n

"Mira, Do you have a foot_ulcer (y/n) ?" – n

"Mira, Do you have a chest_pain_discomfort (y/n) ?" – y

"Mira, Do you have a pressure_in_chest_or_arms (y/n) ?" – n

"Mira, Do you have a burning_while_urination (y/n) ?" – n

"Mira, Do you have a red_area_of_skin (y/n) ?" – n

"Mira, Do you have a discolouration_of_nails_white_black_yellow_green (y/n) ?" – n

"Mira, Do you have a cloudy_vision (y/n) ?" – y

"Mira, Do you have redness (y/n) ?" – n

"Mira, Do you have a frequently_asking_others_to_repeat_themselves (y/n) ?" – y

"Mira, Do you have a trouble_following_conversations (y/n) ?" – n
Once the user responds to all the facts and they are not related to a diabetic-related issue then the interpreter will give the result as "Mira, I'm Sorry, I am unable to diagnose the disease.".

Output –
Mira, I'm Sorry, I am unable to diagnose the disease

The system asks the queries in the format of "do you" and the question is posted as given below:

"ask(Patient,Question):-
    Write(Patient),write(',do you'),write(Question),".

This query is displayed to the user with the help of **"interface(',do you',Patient,Question),"**.

pt(Patient) :- Hypothesis(Patient,Disease),
    Interface(Patient', you probably have', disease, '.'),
    Write(Patient),write(', you probably have '),write(Disease), write('.'),
    undo,end.

The above piece of code is linked with the patient's name and their responding queries to display the disease to that patient.

13.5 Results and Discussions

The main objective of this proposed system is the detection of the symptoms. Based on the detection of symptoms, the system will suggest the patients to the appropriate doctors for consultation and also the possible treatment for the disease mainly diabetes. This IS identifies the types of consequences of diabetes like type 1 diabetes, type 2 diabetes, gestational diabetes mellitus, diabetic retinopathy, diabetic nephropathy, diabetes neuropathy, osteoporosis, transient ischaemic attack/stroke, diabetic foot, heart disease, urinary tract infection, cellulitis, onychomycosis, glaucoma, and progressive hearing loss. This system can be very feasible and easily implemented through low cost, commodity, machinery, and software.

IS for diabetes has been developed using Prolog programming. This medicinal purpose system is mainly developed using Prolog because it is effortless to peruse and note programs that assemble the structure. It has built-in file handling, functional for representing succession, trees, and so on. The development of IS using Python or any other programming language is difficult because of incorporating rules and facts with the associated algorithms.

The following are the system requirements considered as part of the implementation process.

The proposed work is being implemented using a pre-determined database of symptoms of the diseases that are likely to be strains of diabetes. The answers provided by the patients can be cross verified with the symptoms to provide the optimal solution of consultation, diagnosis, and treatment. The proposed work's accuracy rate is comparatively higher than the other similar work of different platforms. The current system is a rule-based framework that orders diabetes on the introduced indications utilizing fuzzy logic and the system is structured for only one disease. The proposed IS for diabetes can diagnose the type and consequences caused by diabetes based on the symptoms and diagnose fifteen cases.

An intelligent rule-based system's starting screen is shown in Figure 13.5. The application asks patients name, and the system follows a set of questions and patient should answer the questions by giving input as either "y" or "yes" if they are facing that symptom or "n" or "no" if the patient is not facing that particular symptom as shown in Figure 13.6. The system asks certain questions of symptoms until they process the hypothesis in the knowledge base, and then, according to the rules, the

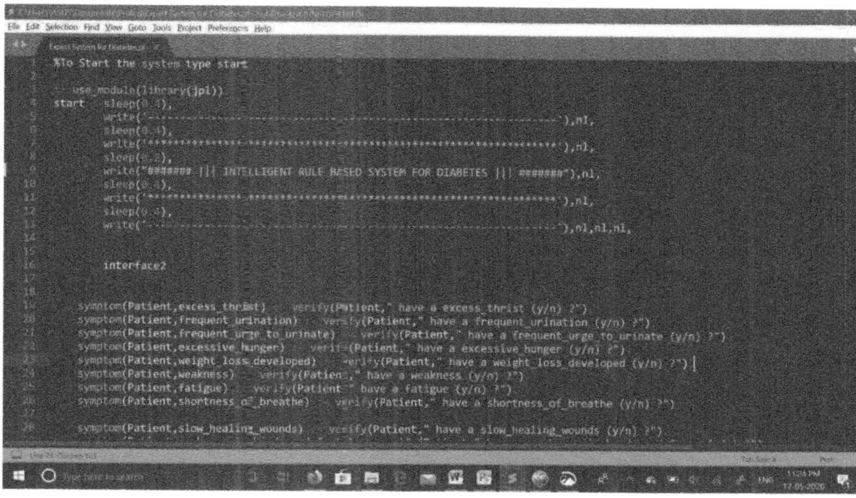

Figure 13.5 Facts for type_one_diabetes.

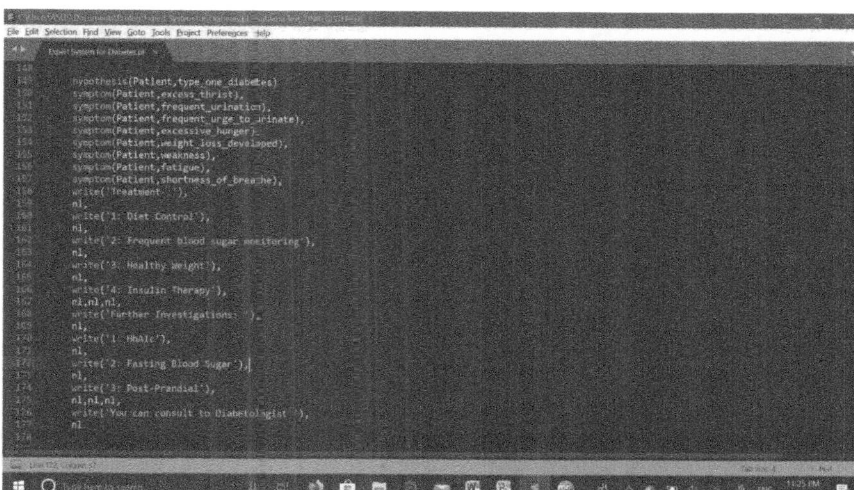

Figure 13.6 Rules for type_one_diabetes.

disease is predicted and displayed as shown in Figure 13.7 along with the treatments, further investigations, and department of the doctor to consult as shown in Figure 13.8. Figures 13.5 and 13.6 are the facts and rules of an intelligent rule-based system shown for the case type 1 diabetes.

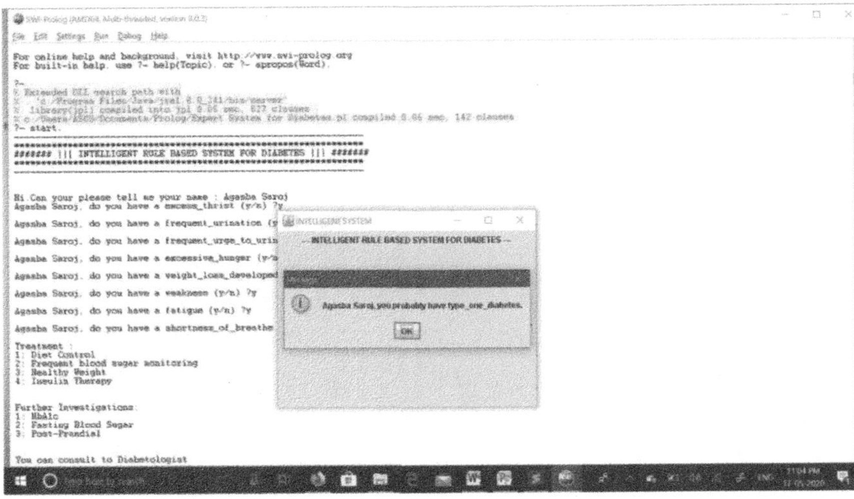

Figure 13.7 Predicted output for type_one_diabetes.

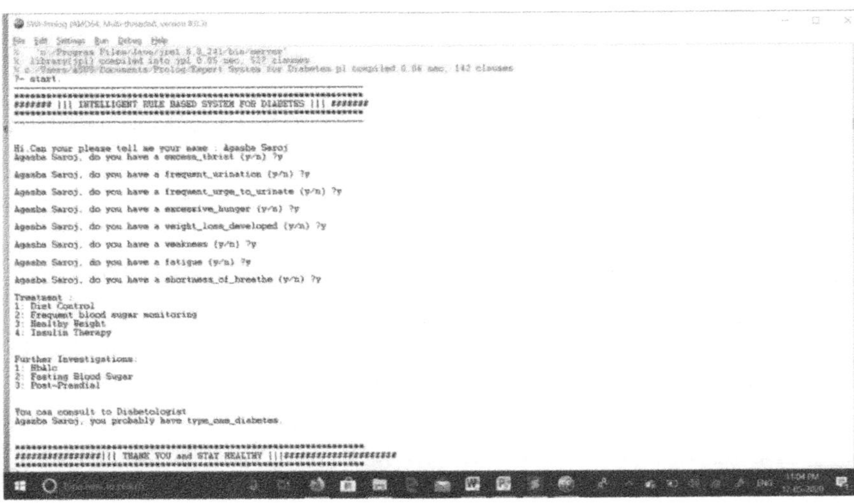

Figure 13.8 Intelligent system's complete output for type_one_diabetes.

13.6 Conclusion

The implemented system works on gathering data from the patients through questionnaires based on the identified disease symptoms. The preliminary consultations of the patient with the doctors can be suggested, and also, the symptoms will be cross analyzed/verified by our knowledge base to identify the alleged disease that we are claiming as per the suggestion of the doctor and the concerned department. Also, suitable treatment will be suggested by this IS for diabetes being the merit and a game-changer among such applications. Intelligent system for diabetes helps the patients to know their diseases by analyzing their symptoms using this system. Diabetes mellitus is a highly risky disease because it has many consequences if untreated or unmonitored. The consequences of diabetes deal with all parts of the body and damage all the organs. With this IS, the patients can discover their diabetes-related effects and what further investigations to be preventive. This system is mainly for people who have diabetes and pre-diabetic. This IS for diabetes was developed using Prolog based on the facts and rules. It gives the result based on the symptoms of the patients

13.7 Acknowledgments

The authors would like to thank Dr. Roshini Rabindran, MBBS, DSMCH, for giving us inputs for framing the database and clarifying all our doubts about this domain.

References

1. Behera, A., Use of artificial intelligence for management and identifications of complications in diabetes. *Clin. Diabetol.*, 10, 2, 221–225, 2021.
2. Mauricio, D., Alonso, N., Gratacos, M., Chronic Diabetes Complications: The Need to Move beyond Classical Concepts. *Trends Endocrinol. Metab.*, 31, 4, 287–295, 2020.
3. Sanz-Nogues, C., Mustafa M., Burke, H., O'Brien, T., Coleman, C.M., Knowledge, Perceptions and Concerns of Diabetes-Associated Complications among Individuals Living with Type 1 and Type 2 Diabetes Mellitus. *HealthCare*, 8:1, 1–12, 2020
4. Farias, A.J., Wu, A.H., Porcel, J., Marchand, L.L., Wilkens, L.R., Monroe, K.R., Maskarinec, G., Pandol, S.J., Setiawan, V.W., Diabetes-Related Complciations and Pancreatic Cancer Incidence in the Multiethnic Cohort. *JNCI Cancer Spectr.*, 4:5, 1–8, 2020.
5. Sadeghi, M.R. and Leung, H.K.K., Predicting adverse outcomes due to diabetes complications with machine learning using administrative health data. *NPJ Digit. Med.*, 4, 24, 1–24, 2021.
6. Gilbert, M.J. and Sun, J.K., Artificial Intelligence in the assessment of diabetic retinopathy from fundus photographs. *Semin. Ophthalmol.*, 35, 7–8, 2020.
7. Dheir, I.M., Soliman A. Mettleq, A., Elsharif, A.A., Abu Al-qumboz, M.N., Abu-Naser, S.S., Knowledge Based System for Diabetes Diagnosis Using SL5 Object. *Int. J. Acad. Pedagogical Res. (IJAPR)*, 3:4, 1–10, 2019.
8. https://main.icmr.nic.in/content/reports

9. Chung, P. and Chen, B., A knowledge-based decision system for healthcare diagnosis and advisory. *2011 IEEE International Conference on Systems, Man, and Cybernetics*, Anchorage, AK, pp. 2357–2362, 2011.
10. Spiridon, G., Sârbu, A., Carstoiu, D., Ion, F., Computerised decision system for diabetes mellitus and associated complications — CODES. *2018 IEEE International Conference on Automation, Quality and Testing, Robotics (AQTR)*, pp. 1–4, 2018.
11. Gudu, J., Gichoya, D., Nyongesa, P., Muumbo, A., Development of a Medical Expert System as an Expert Knowledge Sharing Tool on Diagnosis and Treatment of Hypertension in Pregnancy. *Int. J. Biosci. Biochem. Bioinforma.*, 2, 5, 297–300, 2012.
12. Pesl, P., Herrero, P., Reddy, M., Xenou, M., Oliver, N., Johnston, D., Toumazou, C., Georgiou, P., An Advanced Bolus Calculator for Type 1 Diabetes: System Architecture and Usability Results. *IEEE J. Biomed. Health Inform.*, 20, 11–17, 2016.
13. Suman, S., Badala, S.P., Thakuri, D.M., Manandhar, A.R., Martis, R.L., DIABOT-An Expert System in management of Diabetics using Artificial Intelligence. *Int. J. Comput. Appl.*, 182, 4, 33–36, July 2018.
14. Eteng, I.E., Chinedu, U.L., Chinasa, A., Utibeima, O., The Development of a rule-based systems for the diagnosis of a candidiasis. compusoft. *Int. J. Adv. Comput. Technol.*, 9, 9, 3848–3855, 2020.
15. Eyasu, K., Jimma, W., Tadesse, T., Developing a Prototype Knowledge-Based System for Diagnosis and Treatment of Diabetes Using Data Mining Techniques. *Ethiop. J. Health Serv.*, 30, 1, 115–121, 20202020.
16. Kaur, B., Nikita, H., Sadawarti and Singla J., A Comprehensive Review of Medical Expert Systems for Diagnosis of Chronic Kidney Diseases. *2019 International Conference on Smart Systems and Inventive Technology (ICSSIT)*, Tirunelveli, India, pp. 1008–1013, 2019.
17. Ghosal, S., Sinha, B., Majumder, M., Misra, A., Estimation of effects of nationwide lockdown for containing coronavirus infection on worsening of glycosylated haemoglobin and increase in diabetes-related complications: A simulation model using multivariate regression analysis. *Diabetes Metab. Syndr.*, 14, 4, 319–323, 2020.
18. Wu, H., Lau, E.S.H., Yang, A., Ma, R.C.W., Kong, A.P.S., Chow, E., So, WY., Chan, J.C.N., Luk, A.O.Y., Trends in diabetesrelated complications in Hong Kong, 2001–2016: a retrospective cohort study. *Cardiovasc. Diabetol.*, 19, 60, 1–11, 2020.
19. Zubaedah, R., Xaverius, F., Jayawardana, H., Hidayat, S.H., Comparing Euclidean distance and nearest neighbour algorithm in an expert system for diagnosis of diabetes mellitus. *Enferm. Clin.*, 30:2, 2020.
20. http://www.cse.unsw.edu.au/~billw/prologdict.html
21. https://wiki.visual-prolog.com/index.php?title=Fundamental_Visual_Prolog_-_GUI

14

Big Data in Healthcare: Management, Analysis, and Future Prospects

A. Akila[1]*, R. Parameswari[1] and C. Jayakumari[2]

[1]Department of Computer Science, School of Computing Sciences, Vels Institute of Science, Technology and Advanced Studies, Chennai, India
[2]Middle East College, al Rusayl, Sultanate of Oman

Abstract
Machine learning (ML) is a subsection of artificial intelligence that combines or merges statistics and probabilities and finds best solution to facilitate computers to study from prior examples and to detect hard-to-detect patterns in large, noisy, or complex datasets. These characteristics are mainly beneficial in medical applications that need intricate proteomic and genomic measurements. ML techniques like support vector machine, neural networks, K-nearest neighbor, AdaBoost, logistic regression, and random forest are normally applied in cancer identification and discovery. ML has been applied in cancer prognosis and prediction. In this chapter, the top performing algorithms for the analysis of dataset features were discussed. The data used to train the ML models could be huge, and the extraction from these big data is a challenging job.

Keywords: Breast cancer, mitosis, prediction, F1-score, accuracy, recall

14.1 Introduction

Machine learning (ML) and big data are the two major areas in information technology. Nowadays, most of the scientists and researchers used ML techniques to mix out of a number of smart solutions that can be eventually help in diagnosing and treating an illness. Patients are place to gain the most, as technology will help them improve their outcomes by evaluating

**Corresponding author*: akila.scs@velsuniv.ac.in

A. Jaya, K. Kalaiselvi, Dinesh Goyal and Dhiya Al-Jumeily (eds.) Handbook of Intelligent Healthcare Analytics: Knowledge Engineering with Big Data Analytics, (309–326) © 2022 Scrivener Publishing LLC

the best treatment options for them. Machine learning techniques are capable of detecting diseases more accurately at an earlier stage, reducing the number of readmissions in hospitals and clinics.

One of the major reasons of death in women is breast cancer. As a result, early detection is critical. Breast cancer detections are less invasive method technique a digitized result, making prediction easier. There are various algorithms that can be used to train the dataset and obtain accurate results. In this chapter, five classification algorithms are used to find their accuracy. The classification models trained on the dataset had various levels of accuracy. The most accurate model was K-nearest neighbor (KNN). The classification model such as random forest and boosted trees showed the similar accuracy. Therefore, the most accurate classifier can be used to detect the tumor so that the cure can be found in early stage.

14.2 Breast Cancer: Overview

The human body is made up of a large number of cellular entities that wear out and are substituted on a regular basis by new cells. This process is known as mitosis. When an enduring change in the DNA structure, known as a mutation, occurs, cancer develops. Breast cancer is identified in broadly among women in the world, and it is also the principal reason of cancer death in women. The subsequent leading source of death in women is breast cancer. Breast cancer is more generic in women with genealogical record of the disease. The risk increases with age as well. Mammograms can help to identify in the preliminary stage of breast cancer, improving the chances of a patient's survival. Radiologists are being tasked with manually analyzing an increasing number of images. After a certain number of images have been analyzed, the process of diagnosing them becomes tedious and, as a result, more sensitive to error.

Changes in Gene Structure

(i) Primary Stage:
During the chemotherapy process, cells interact with one another through reception organs. These changes in DNA structure of gene configuration occurred during these primary steps.
- Proto-Oncogenes:
Oncogenes, which send messages to cells to multiply repeatedly, are the primary cause of cancerous cell growth.

- Cell Cycle Clock Malfunctions:
  The cell cycle is regulated by a collection of interacting proteins found in the nucleus of the cell. If the cell's DNA is destroyed, then the tumor suppressor destroys it.

(ii) Secondary Stage:

- Tumor forms:
  The tumor is a compilation of cells that exists without the help of extracellular matrix and uses angiogenesis to establish a blood vessel network.
- Benign Tumor:
  - Only present in one part of the body.
  - It is possible to remove the tumor surgically.
- Malignant Tumor:
  - Acts as a predecessor to cancer.
  - It is not possible to remove the tumor surgically.

14.3 State-of-the-Art Technology in Treatment of Cancer

Chemotherapy treatment is provided as anticancer (cytotoxic) drugs that penetrates to entire body and prevents cancer cells from growing, such as taxol. It has efficient in diminishing breast cancer cells that will reach other organs.

In the current situation, cancer can be treated in the following ways.

- Immunology
- Radiotherapy

14.3.1 Chemotherapy

- Destroys all cancerous and non-cancerous cells.
- Causes anemia, which worsens hypoxia in the tumor.
- Nanosensors are used in Nano devices, which is chemotherapy treatment.

14.3.2 Radiotherapy

Multi-fractionated irradiation, radiotherapy along with chemotherapy, a radiosentitizer, or hyperthermia has also been applied to increase the therapeutic outcome of radiotherapy. Cancer cells in an aerobic system have radio sensitivity three times greater than hypoxic cells. It could improve radio sensitivity if nanotechnology-based artificial blood can turn a hypoxic cell into an oxygenated cell.

14.4 Early Diagnosis of Breast Cancer: Overview

Identifying the breast cancer in prior in women is critical for the disease's successful management. In addition, there is currently no treatment for breast cancer; early detection may result in a satisfactory result.

There are some early warning strategies for breast cancer such as screening of mammography, analyzing clinical breast cancer data and breast sensitivity. There no evidence and no longer advised for breast self-examination, which was not proven for good result.

14.4.1 Advantages and Risks Associated with the Early Detection of Breast Cancer

There are more treatment options and a higher chance of survival if breast cancer is detected early. In the past years, women who are diagnosed with breast cancer early have a survival rate of 93% or higher.

Early detection is significant since it could be easier to take care of irregular tissue or cancer if discovered early. When the symptoms are identified, medicating for the cancer will be difficult as the aggressiveness of cancer will be more. One of the major early identification of cancer is through screening tests and decreases the possibility of death due to cancer.

These may have unintended and undesirable consequences. Other possible adverse effects of breast cancer screening include pressures during the operation and radiation exposure during the mammography exam. Having multiple X-rays can represent a risk, even if the amount of radiation in a mammogram is low.

Breast cancer can be detected early with the use of breast screening. The sooner a disease is identified, the better the odds of survival. Early detection of breast cancer reduces the need for a mastectomy (breast removal) or chemotherapy.

Early breast cancer is frequently treated with surgery, radiation treatment, chemotherapy, hormone treatment, and/or HER2-targeted therapy.

Breast cancer early detection is crucial since it leads to a wider range of treatment options, a better probability of survival, and a higher quality of life. Although there is no 100% guarantee that you will not develop breast cancer, early identification provides you the highest chance of a successful treatment. Although early detection of breast cancer can help improve results, every screening program comes with its own set of risks. These risks include the possibility of a false positive or false negative diagnosis. The risks and benefits of any screening test should be weighed before continuing.

14.4.2 Diagnosis the Breast Cancer

- Breast ultrasound

  Sonograms are accurate pictures of areas within the breast created by a computer that uses sound waves.

  The most common reason for a breast ultrasound is to determine if an issue discovered by a mammogram or physical examination of the breast is a cyst filled with fluid or a solid tumor. Breast ultrasound is seldom used to detect breast cancer. This is due to the fact that it can miss some early cancer symptoms.
- Diagnostic mammogram

  A mammogram is an x-ray that is used to look at the breast. Diagnostic mammograms are performed when a screening mammography gives abnormal findings or when any indicators of breast cancer urge the clinician to examine the tissue.
- Magnetic resonance imaging (MRI)

  MRI can be a valuable screening tool when combined with mammography and breast ultrasound. According to new research, MRI can detect certain small breast abnormalities that mammography is unable to detect. It could also help people with breast implants or younger people with thick breast tissue detect breast cancer.
- Breast biopsy

  The breast biopsy could be applied to analysis the breast lump, which is viewable in a mammogram or ultrasound or that your doctor experiences (palpates) all through a clinical breast exam. The small hollow needle is used by the radiologist or surgeon to get tissue samples from the breast mass with the help of ultrasound.

14.5 Literature Review

Several studies using various statistical methods have been conducted on the prediction problem. We only found a few studies that were linked to medical diagnosis after conducting a literature review. However, all of the contributions made in this field are progressing day by day. The algorithms used are mostly data mining algorithms that have been tested on a dataset.

Big data is a term used to describe a large collection of data that is still growing exponentially over time. It includes data that is structured, unstructured, or semi-structured. Today, big data is increasingly employed in healthcare for disease prediction. One of the most frequent malignancies in women is breast cancer. It is the second leading cause of death for women in the United States and Asia. We have a better chance of healing this condition if we catch it early on. For this experiment, we utilized the R tool to develop the KNN algorithm for determining classification accuracy. We consider the original Wisconsin breast cancer dataset, which was downloaded from the UCI ML repository [1].

The author using a radial basis function (RBF) with support vector machine (SVM) classification to present quantitative diagnosis of breast tumors by morph metric classification of micro environmental myoepithelial cells in p63 immunohistochemistry images in 2017 [2].

In 2016, the author presented a wrapper-based feature selection approach for wavelet co-occurrence feature (WCF) in mammogram classification using the genetic fuzzy system (GFS) [3]. A computer-aided theragnosis (CAT) system was demonstrated that used spectroscopy techniques in conjunction with an advanced state-of-the-art texture method, the local binary pattern (LBP) method, to categorize healing reaction before time after the start of neo adjuvant chemotherapy to classify treatment response. The author presented a model to detect the coexistence of two common breast cancer subtypes (HER2, basal-like) in the same patient using four different types of technologies: reverse phase protein arrays (RPPA), complete RNA expression (RNA), DNA methylation, and micro RNA sequencing (miRNA) [4, 5]. The author demonstrated feasibility of developing a new image-based clinical marker to evaluate breast tumor response to neoadjuvant chemotherapy [6]. The author presented an integrated unsupervised feature learning (UFL) framework for the detection of basal cell carcinoma (BCC) [7]. A multi-scale super pixel classification method for automatic detection of areas of interest in whole slide histopathological images was introduced [8].

The author suggested using four-sized assessment windows to compute the mean, standard deviation, skewness, and kurtosis characteristics using a bank of Gabor filters. The author completed two major experiments: a categorization of abnormal/normal breast tissue and the skill to identify diverse breast cancer types [9].

The author identified a ML technique that uses adaptive boosting and selective pruning to optimize cost and performance when using ultrasound imaging to differentiate solid breast masses [10]. The proposed a CAD system for detecting breast cancer tumor grades in ultrasound imagery. To make feature selection and model parameter optimization easier, a SVM classifier was created using a genetic algorithm (GA) [11]. Using a nonlinear Hebbian learning algorithm, a combined breast cancer risk assessment model was proposed for breast cancer prediction and features of the screening mammogram involving cancer subtype [12]. Breast cancer growth has predict using SVM, 10-fold validation method, 10-fold cross-validation method, receiver operating characteristic (ROC) curves, and multi-scaled feature analysis techniques from digital mammogram images [13]. Many classification algorithms were used to automatically detect benign and malignant images from histopathology [14]. The author used a random forest classifier along with feature selection to detect breast cancer [15]. Breast cancer dataset is to compare artificial neural network (ANN), decision tree, and SVM methods for predicting breast cancer and found SVM to be the best classifier for predicting the disease [16].

The major issue in healthcare is predicting the status of the readmission of the patient due to severe chronic diseases. Many researches carried out in predicting the disease itself. So, Samah Alajmani and Hanan Elazhary proposed a comparative study of many ML algorithms for predicting the status of readmission due to the level of disease [17].

14.6 Machine Learning Algorithms

ML is a variety of artificial intelligence (AI) that makes software applications to enhance their prediction accuracy without being specifically programmed to do so. Supervised learning, unsupervised learning, and reinforcement learning are the various categories of ML. Some of the existing machines learning algorithms are discussed in the following section.

The existing ML algorithms are compared with the Breast Cancer Wisconsin from UCI Repository dataset with 683 instances which is shown in Figure 14.1. The variables considered are type, clump thickness,

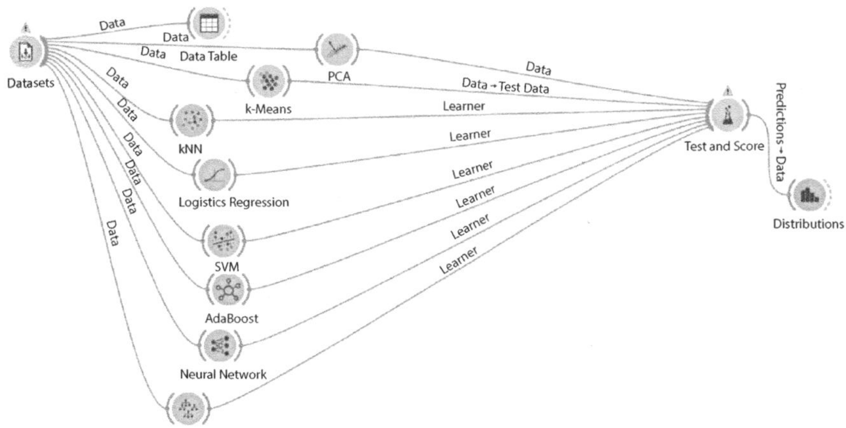

Figure 14.1 Prediction of breast cancer using machine learning algorithms using Orange tool.

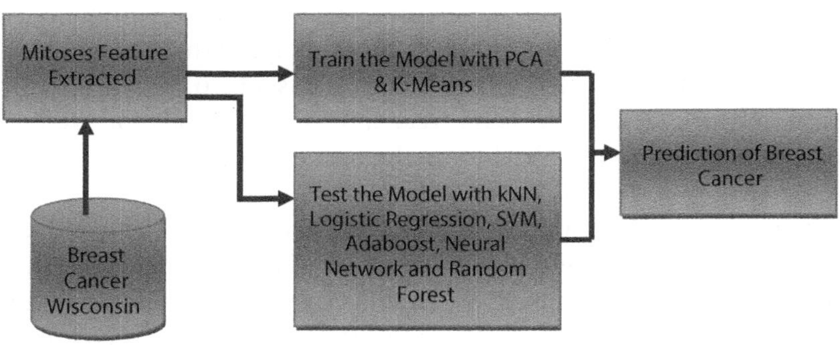

Figure 14.2 Prediction of breast cancer using machine learning algorithms.

unif_cell_size, unif_cell_shape, Marinial_adhesion, Single_cell_size, Bare_nuclei, Bland_chromatine, Normal_Nucleoli, and mitoses. The type could be benign and malign. In the identification of the breast cancer level, the mitotic cell count is a vital biomarker for finding the aggressiveness, prognosis, and grade of breast cancer. So, the mitoses are considered for evaluating the mean and standard deviation to predict the presence of breast cancer. The process of the Breast Cancer Prediction process using various Machine Learning algorithms is shown in Figure 14.2.

14.6.1 Principal Component Analysis Algorithms

The principle objective of principal component analysis (PCA) is to control the boundary of an informational index comprised of a few factors that are related with one another, either emphatically or pitifully, while saving, however, much difference that possible. The same can be done by modifying the features into alternate arrangement of factors known as head segments (or basically PCs), which are symmetrical and organized so that the measure of fluctuation held in the primary factors diminishes as we pass down the request. Accordingly, the first head part saves the greatest fluctuation found in the first components. The head segments are symmetrical since they are the eigenvectors of a covariance framework.

The dataset to which the PCA technique will be applied must be scaled. The relative scaling has an effect on the outcomes. It is a way of summarizing data in layman's terms. Consider a dining table with a few wine bottles. Color, strength, age, and other characteristics are used to characterize each wine. However, since all of them will quantify similar properties, redundancy will occur. In this case, PCA will summarize each wine in the stock that has fewer characteristics. When viewed from its most insightful perspective, PCA may give the user with a lower-dimensional image, a ridge or "shade" of this object.

14.6.2 K-Means Algorithm

In data mining, the K-means algorithm initiates with a set of arbitrarily chosen centroids that provide as the starting points for each cluster and then carry out iterative (repetitive) manipulations to optimize the centroid positions.

It stops generating and optimizing clusters if one of the following conditions occurs:

- The centroids have stabilized—their values have not changed as a result of the clustering.
- The defined number of iterations has been reached.

14.6.3 K-Nearest Neighbor Algorithm

The classification and regression problems could be resolved using KNN, which is in the category of supervised ML algorithm. The KNN algorithm assumes that objects that are identical are close to one another. To put it another way, things that are connected are close together.

The KNN Algorithm

Step 1: Put the data on the hard drive.

Step 2: Assign the number of neighbors to K.

Step 3: For every data point in the dataset,

- Compute the discrepancy between the query example and the current example.
- To an organized set, add the example's distance and index.

Step 4: Using the distances, arrange the structured set of distances and indices from least to highest (in chronological order).

Step 5: From the sorted list, select the first K entries.

Step 6: Find the labels for the selected K entries.

Step 7: Return mean of the K labels if regression found.

Step 8: If the K labels are classified, return the mode of the K labels.

14.6.4 Logistic Regression Algorithm

- Logistic regression is one of the most common ML algorithms used in the supervised learning approach. It is used to predict a categorical dependent variable based on a set of independent variables.
- Logistic regression is used to predict the behavior of a categorical dependent variable. As a result, the output must be discrete or categorical in nature. It can be Yes or No, 0 or 1, true or false, and so on, but it also provides probabilistic values that are somewhere in the middle of exact values like 0 and 1.
- Linear regression and logistic regression are somewhat similar in terms of how they are used. For regression problems, linear regression is used, whereas for classification problems, logistic regression is used.

14.6.5 Support Vector Machine Algorithm

A common supervised learning approach for handling classification and regression problems is the SVM. In ML, supervised learning algorithms are commonly used for classification algorithms. The purpose of the SVM algorithm is to find the optimal line or decision boundary for categorizing n-dimensional space into classes in the future, so that new data points may be easily placed in the right category. Hyperplanes are the best boundary option. SVM is used to choose the acute points/vectors that contribute to the construction of the hyperplane. The algorithm is known as a SVM, and help vectors are the extreme cases.

14.6.6 AdaBoost Algorithm

The AdaBoost algorithm, which stands for Adaptive Boosting, is a boosting technique used as an ensemble method in ML. Weights are reassigned to each instance, with wrongly graded instances receiving higher weights. Adaptive boosting is the term for this. Boosting is a technique used in supervised learning to minimize bias and variance. It is based on the principle of learners' sequential development. With the exception of the first, each subsequent learner is created from previously created learners. To put it another way, weak students become strong students. Although the AdaBoost algorithm is based on the same principles as boosting, it functions in a slightly different way.

14.6.7 Neural Networks Algorithm

ANNs are thought to function in a similar way to the human brain. ANNs are conceptually influenced by brain neural networks, but their practical implementation in ML is far from reality. Multiple inputs are fed into an ANN, which produces a single output. It's worth noting that ANNs are inspired by animal brains, but they're not the same as biological neural networks.

The RBF is used in neural network algorithms, which can be used for strategic purposes. There are some other neural network models, like the one we just discussed. Continue reading this article for an introduction to neural networks and their working model. You'll see how they work and how they're applied to real-world issues.

14.6.8 Random Forest Algorithm

Random forest, a well-known ML method that can handle both classification and regression issues in ML, employs the supervised learning technique. It is based on ensemble learning, which is a methodology for addressing a complicated problem by combining numerous classifiers and enhancing the model's accuracy by mixing numerous classifiers.

Random forest is a classifier that combines and averages a number of decision trees on different subsets of a dataset to increase the dataset's predictive accuracy, according to the phrase. The random forest, instead of depending on a single decision tree, uses the forecasts from each tree to calculate the final performance based on the majority of votes.

14.7 Result and Discussion

The Orange tool is used to implement current ML methods. Orange is an open-source data visualization, ML, and data mining toolset. It comes with a visual programming front-end for examining qualitative data analysis and alluring data visualization. Orange is a visual programming software package for data visualization, ML, data mining, and data analysis that leverages component-based visual programming. Widgets are orange components that can be used for everything from basic data display to subset selection and preprocessing, as well as empirical evaluation and predictive modeling of learning algorithms.

14.7.1 Performance Metrics

14.7.1.1 ROC Curve

An ROC curve is a graph that shows how well a classification model performs across all categorization levels [18]. On this curve, two parameters are plotted:

- True positive rate (TPR)
- False positive rate (FPR)

TPR is a synonym for recall and is therefore defined as true positive divided by the sum of true positive and false negative. False positive rate (FPR) is defined as false positive divided by the sum of true negative and false positive.

$$\textit{True Positive Rate} = \frac{\textit{TruePositive}}{\textit{TruePositive} + \textit{FalseNegative}} \tag{14.1}$$

$$\text{False Positive Rate} = \frac{\textit{False Positive}}{\textit{True Negative} + \textit{False Positive}} \tag{14.2}$$

On a ROC curve, TPR vs. FPR at various categorization levels are displayed. More items are identified as positive as the rating threshold is dropped, resulting in a rise in both False Positives and True Positives.

14.7.1.2 Accuracy

One metric for assessing classification models is accuracy. Informally, accuracy refers to the percentage of right predictions carried out by our model. The following is the formal concept of accuracy: Number of accurate predictions divided by total number of predictions equals to accuracy.

To measure accuracy in terms of positives and negatives for binary classification, the true positive plus true negative divided by the sum of true positive plus true negative plus false positive plus false negative.

$$Accuracy = \frac{TruePositive + TrueNegative}{TruePositive + TrueNegative + FalsePositive + FalseNegative} \tag{14.3}$$

14.7.1.3 Precision and Recall

Precision is a good statistic to employ when the cost of false positives is high. Consider the task of recognizing spam mail. When an email that is not spam (true negative) is mistakenly labeled as spam, it is called a false positive (predicted spam). If the precision of the spam detection model is low, then the email user will miss crucial emails.

$$Precision = \frac{TruePositive}{TruePositive + FalsePositive} \tag{14.4}$$

Recall measures how many real positives our model captures by identifying it as positive (true positive). With this knowledge, we can see that, when false negative has a large cost, recall will be the model metric we use to select the optimal model.

$$Recall = \frac{TruePositive}{TruePositive + FalseNegative} \tag{14.5}$$

For illustration, in the identification of ill or healthy patients, the diagnosis center will face severe consequences if a fraudulent transaction (actual positive) is discovered to be non fraudulent (predicted negative).

Detecting ill patients is similar. If a sick patient (actual positive) is tested and predicted to be well (predicted negative) and if the illness is infectious, then the cost of false negative would be extremely high.

14.7.1.4 F1-Score

When looking at the performance measures precision and recall in research, the other measure, F1, which is a function of precision and recall, cannot be overlooked.

$$F1 = 2\,X \frac{Precision * Recall}{Precision + Recall} \tag{14.6}$$

You will need the F1-score if you need to beat a balance of precision and recall. The difference between F1-score and accuracy is that, in most commercial scenarios, a large number of true negatives will contribute greatly to accuracy, which we do not focus on very often. While false negative and false positive generally have business cost (physical and intangible), F1-score may be a good statistic to utilize if the balance between precision and recall is found, and there is an unequal class distribution (large number of actual negatives).

14.8 Experimental Result and Discussion

To train the classifier, the PCA and K-means algorithm are used with bin width as 5 and considering the normal distribution as fitted distribution. The mean and the standard deviation values for the benign and malign are computed with the Orange tool, which is shown in Figure 14.3.

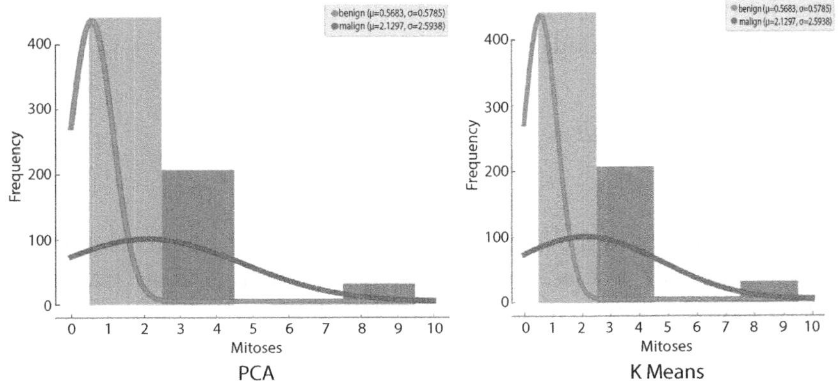

Figure 14.3 Mitoses distribution in PCA and K-means algorithm.

Figure 14.4 shows the mitoses distribution of the other ML algorithms where the dataset is testing for prediction of grade of cancer. The neural networks show the benign mean as 0.5659, and standard deviation 0.5782 and malign mean as 2.1275 and standard deviation as 2.5880. The result of the overall performance comparison using

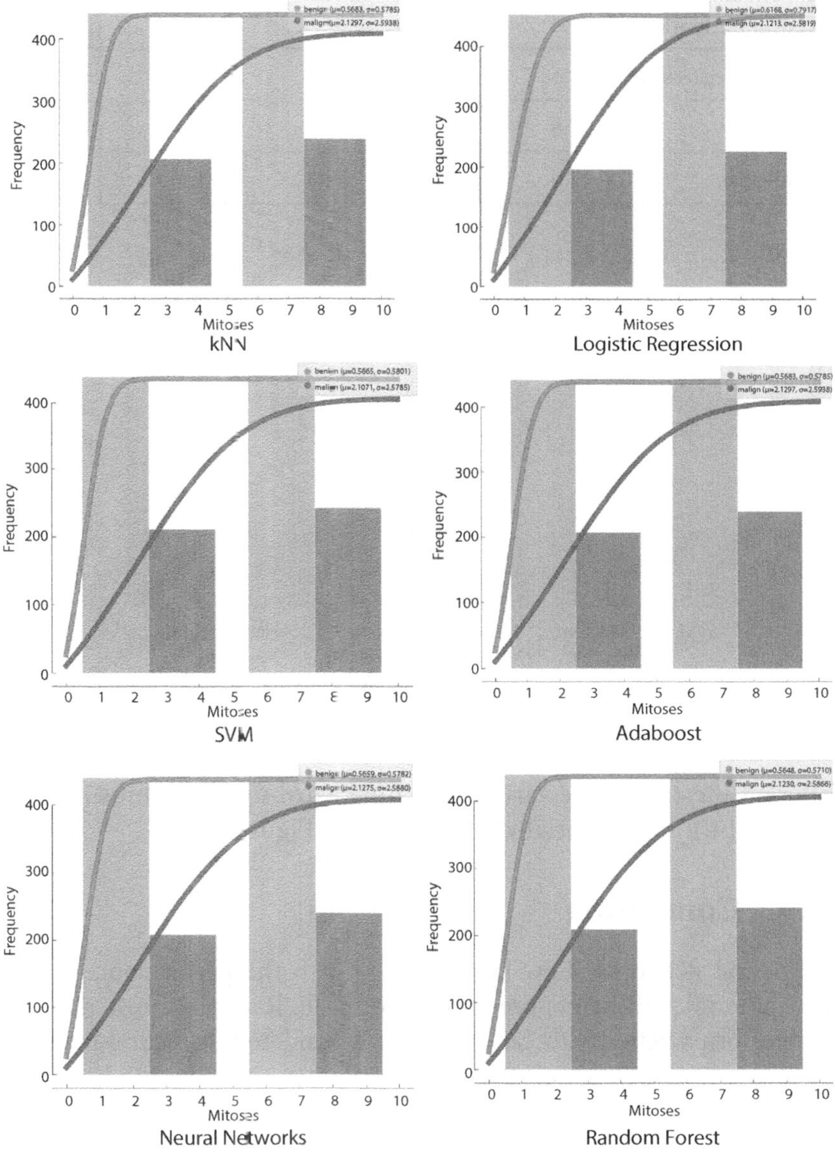

Figure 14.4 Mitoses distribution in machine learning algorithms.

Table 14.1 Performance of the various machine learning algorithms.

Machine learning algorithms	AUC (%)	CA (%)	F1 (%)	Precision (%)	Recall (%)
SVM	99.3	96	97	96.9	97.1
Random forest	98.7	96.9	97.6	98	97.3
Neural networks	99.4	97.2	97.9	97.5	98.2
Logistic regression	99.5	95.8	96.8	95.2	98.4
kNN	97.8	94.6	95.9	93.6	98.4
AdaBoost	92.4	93.6	95.1	94.1	96.2

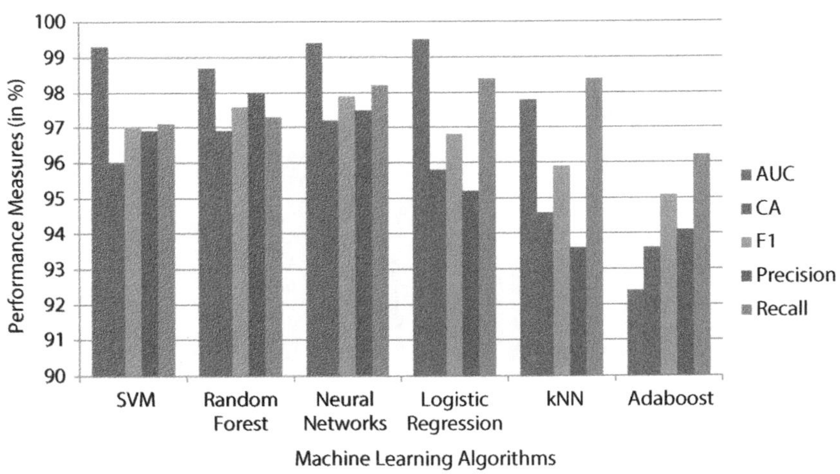

Figure 14.5 Performance comparison of various machine learning algorithms.

different metrics of the various machine learning algorithms is shown in Table 14.1 and pictorially represented in Figure 14.5.

14.9 Conclusion

The neural networks show better results with the performance metrics AUC, accuracy, F1-score, precision, and recall. Random forest algorithm also performs better with cancer dataset considered. In neural networks, the area under the curve is 99.4%, the accuracy is 97.2%, the F1-score is 97.9%, the precision rate is 97.5%, and recall rate is 98.2.%. The prediction of breast cancer with the variable mitosis is carried out. The other variables could be considered for the prediction, and the result could be analyzed to identify the suitable variable for the breast cancer prediction. The same set of the algorithms could be compared with the different datasets, and the optimal ML algorithm could be identified.

References

1. Shailaja, K. and Jabbar, M., Prediction of Breast Cancer Using Big Data Analytics. *Int. J. Eng. Technol.*, 7, 223, 1–3, 2018.
2. Yamamoto, Y. *et al.*, "Quantitative Diagnosis Of Breast Tumours By Morphometric Classification Of Micro Environmental Myoepithelial Cells Using A Machine Learning Approach". *Sci. Rep.*, 7, 1, 1–12, 2017.
3. Pawar, M.M. *et al.*, "Genetic Fuzzy System (GFS) Based Waveletco-Occurrence Feature Selection Inmammogram Classification For Breast Cancer Diagnosis'. *Perspect. Sci.*, 8, 247–250, 2016.
4. Gangeh, M.J. *et al.*, "Computer Aided Theragnosis Using Quantitative Ultrasound Spectroscopy And Maximum Mean Discrepancy In Locally Advanced Breast Cancer". *IEEE Trans. Med. Imaging*, 35, 3, 778–790, march 2016.
5. Verma, R. *et al.*, "Detecting Multiple Subtypes Of Breast Cancer In A Single Patient". *IEEE International Conference On Image Processing (ICIP)*, 2016.
6. Aghaei, F. *et al.*, "Applying A New Quantitative Global Breast MRI Feature Analysis Scheme To Assess Tumor Response To Chemotherapy". *J. Magn. Reson. Imaging*, 44, 5, 1099–1106, 2016.
7. Arevalo, J. *et al.*, "An Unsupervised Feature Learning Framework For Basal Cell Carcinoma Image Analysis". *Artif. Intell. Med.*, 64, 2, 131–145, 2015.
8. Bejnordi, B.E. *et al.*, *"A Multi-Scale Superpixel Classification Approach To The Detection Ofregions Of Interest In Whole Slide Histopathology Images"*, Research Gate, 2015.
9. Torrents-Barrena, J. *et al.*, "Computer-Aided Diagnosis Of Breast Cancer Via Gabor Wavelet Bank And Binary-Class SVM In Mammographic Images". *J. Exp. Theor. Artif. Intell.*, 28, 1-2, 295–311, 2015.
10. Venkatesh, S.S. *et al.*, "Going Beyond A First Reader: A Machine Learning Methodology For Optimizing Cost And Performance In Breast Ultrasound". *Ultrasound Med. Biol.*, 41, 12, 3148–3162, 2015.
11. Chen, D.-R. *et al.*, "Computer-Aided Assessment of Tumor Grade for Breast Cancer In Ultrasound Images". *Comput. Math. Methods Med.*, 2015, 6, 2015.
12. Subramanian, J.S. *et al.*, "An Integrated Breast Cancer Risk Assessment And Management Model Based On Fuzzy Cognitive Maps". *Comput. Methods Programs Biomed.*, 1180, 3, 280–297, 2015.
13. Sun, W. *et al.*, "Using Multiscale Texture And Density Features For Near-Term Breast Cancer Risk Analysis". *Int. J. Of Med. Phys. Res. And Pract.*, 42, 6 Part 1, 2853–2862, 2015.

14. Spanhol, F.A. *et al.*, "A Dataset For Breast Cancer Histopathological Image Classification". *IEEE Trans. Biomed. Eng.*, 63, 7, 1455–1462, 2015.
15. Nguyen, C. *et al.*, "Random Forest Classifier Combined With Feature Selection For Breast Cancer Diagnosis And Prognostic". *J. Biomed. Sci. Eng.*, 6, 551–560, 2013.
16. Ahmad, L.G. *et al.*, "Using Three Machine Learning Techniques for Predicting Breast Cancer Recurrence". *J. Health Med. Inform.*, 4, 124, 3, 2013.
17. Alajmani, S. and Elazhary, H., Hospital Readmission Prediction using Machine Learning Techniques. *Int. J. Adv. Comput. Sci. Appl.*, 10, 4, 212–220, 2019.
18. Classification: ROC Curve and AUC. *Mach. Learn. Crash Course*, n.d.). Retrieved June 5, 2021. from https://developers.google.com/machine-learning/crash-course/classification/roc-and-auc.

15

Machine Learning for Information Extraction, Data Analysis and Predictions in the Healthcare System

G. Jaculine Priya* **and S. Saradha**†

Department of Computer Science, VELS Institute of Science Technology & Advanced Studies (VISTAS), Pallavaram, Chennai, Tamilnadu, India

Abstract

Information extraction is key irrespective of any domain. All these years, there have been numerous techniques leveraged to extract information. One of those techniques is "machine learning" (ML). By leveraging ML techniques, information can be extracted and analyzed in a better way. In this work, to gather information, such algorithms based on ML have been used. There are many areas in healthcare where ML is used. The applications of ML include disease prediction, medicine manufacture, improving security, eliminating fraud practices research, and e-record. Also, electronic health record is used widely in all healthcare systems. A healthcare system is said effective only when the outcome of the services for particular health issue is good and the timely service for any health issue brings hope. So, in this work, all the algorithms in ML, which can be applied in healthcare, have been discussed.

Keywords: Machine learning, artificial intelligence, healthcare, security, supervised, un-supervised, medicine, electronic health record

15.1 Introduction

Medical care could be a multi-dimensional structure with the exclusive emphasis on the eagerness, willpower, and to treat people from a fitness or weakness perspective. Health experts from health offices and financial

**Corresponding author*: priyajaculine@gmail.com
†*Corresponding author*: saradha.research@gmail.com

A. Jaya, K. Kalaiselvi, Dinesh Goyal and Dhiya Al-Jumeily (eds.) Handbook of Intelligent Healthcare Analytics: Knowledge Engineering with Big Data Analytics, (327–344) © 2022 Scrivener Publishing LLC

organizations are two important sectors which support the medical services framework. The scope where health experts focus on is extremely large and includes dentistry, medicine, maternity care, nursing, brain research, physiotherapy, and various others. Criticality of the situation services for medical is required at different levels [4]. This information is leveraged as the first purpose of counsel and can be a replacement for strong attention requiring gifted experts, advanced clinical checkup and treatment, and extraordinarily amazing analytic or surgeries. At all these stages, the well-being experts are accountable for different sorts of data; for example, patient's clinical history, clinical information, and other private or individual clinical information [5].

Thus, for storing all the details of patients, the research and institutes of health introduced a system called electronic health record (EHR). This EHR holds all the information about the patients and anytime anywhere, and the data can be transmitted or retrieved. With the quick reception of EHRs, it is fascinating to gather data and information from EHRs to help computerized frameworks at the purpose of care and to empower optional utilization of EHRs for clinical and translational exploration. Discriminated [10]

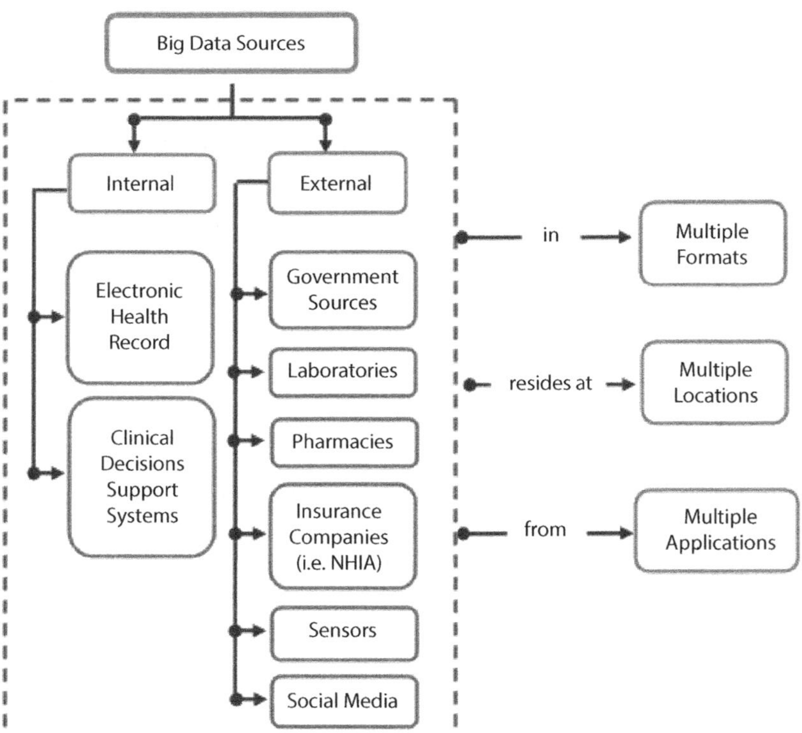

Figure 15.1 Healthcare data sources.

with organized information, free content is a more common and expressive technique to record clinical occasions and encourage correspondence among the consideration group in the medical services climate. One basic segment to encourage the utilization of EHR [1] information for clinical choice help, quality improvement, or clinical and interpretation research is the data extraction task, which naturally extricates and encodes clinical data from text. Figure 15.1 depicts the sources of healthcare data.

15.2 Machine Learning in Healthcare

Machine learning (ML) [2, 19] offers itself to certain cycles better compared to other people. Calculations can furnish moment help to disciplines with measures that are reproducible or predictable. Additionally, those with enormous picture data sets, like radiology, cardiology, and pathology, are solid applicants. ML can be prepared to see pictures, recognize irregularities, and highlight regions that need consideration, in this way, improving the exactness of every one of these cycles. Developed ML will help the family expert.

ML is developed with the objective to improve efficiency, reliability, and accuracy. ML has all objectives and purposes unlimited applications in the medical services industry. Today, ML is assisting with smoothing out authoritative cycles in clinics, guide and treat irresistible infections, and customize clinical medicines [21]. The process of data handling [11] is shown in Figure 15.2.

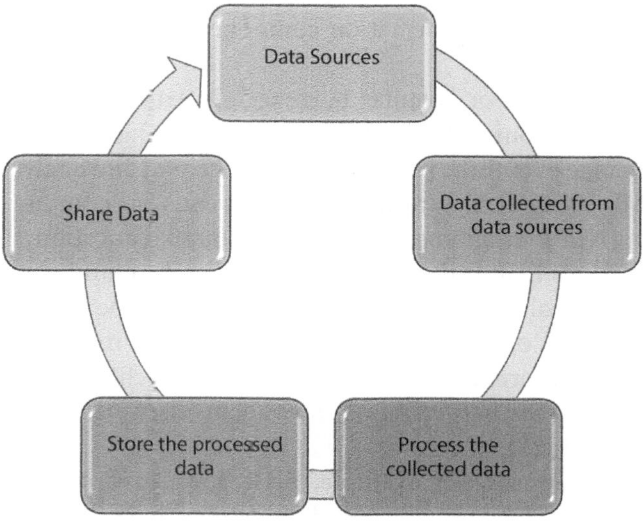

Figure 15.2 Process of data handling.

In radiology, intense learning in medicinal services identifies difficult samples certainly and supports radiologists with settle down on smart adoptions graphing images; for instance, regular radiographs, radiology, MRI, and CT reports [8]. The presentation of ML-based programmed innovation and exploration structures has proved to be indistinguishable to that of a skilled radiologist. Google's ML applications in medical services were prepared to identify disease and accomplished 89% precision, on similarity or better than radiologists. These are only a pair of illustrations of the several services of ML in medical care. ML is a kind of man-made reasoning when PCs are customized to learn data without human mediation [9]. In ML, the advancement of the basic calculations depends on computational insights. PCs are given information and afterward the PCs "learn" from that information. The information really "educates" the PC by uncovering its perplexing examples and hidden calculations. The better-quality illustration of information was given by the machine, and, to be more exact, the machine yield develops [27].

ML plays an extremely vital part in helping not only the clinicians but also the patients to look at the data from multiple points of view. Some of the most popular use cases in the medical industry are autocharging clinical expenses, clinical choice assistance, and improving medical procedures. There are also numerous examples where ML and medical services approach has been found useful to improve medication. For instance, the exact NLP innovation that is utilized to decide reliability for buyer or else assumption examination of somebody's web-based media post would now be able to be utilized to peruse a patient's outline to separate significant information components like the patient's drugs and therapy plans.

Human language, or "regular language", is extremely intricate, lacks consistency, and consolidates a huge measure of language and doubtfulness. To change over these reports into more helpful and analyzable information, ML in medical care frequently depends on common language preparing (NLP) programs. Supreme profound education in medical care applications that utilization NLP needs nearly a kind of clinical ML. Unstructured health services information for ML speaks to practically 82% of data held or "bolted" in electric well-being record frameworks. These are not information components rather archives or manuscript records which in the past could not investigated lacking of human perusing the material [27].

Required practices, for example, EHR, have just prepared medical care frameworks for applying big data devices for cutting edge information examination. Computer-based intelligence/ML instruments are

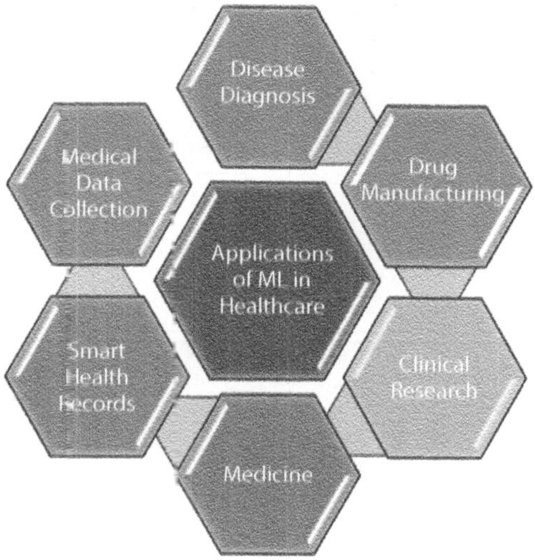

Figure 15.3 Applications of ML.

bound to enhance this stream [12]. They are required to improve the nature of robotization and smart dynamic in essential/tertiary patient consideration and public medical services frameworks. This could be the greatest effect of AI gadgets as it can conceivably change the personal satisfaction for billions of individuals around the planet [28]. AI methods take tremendous waves through medical services, in any occurrence, fuelling operational discussion of whether AI specialists will eventually succeed human doctors well along. With the consent that humanoid physicians would not be supplanted by machines with the predictable prospect, still, this technology can definitely help doctors and physicians to help make better decisions or may be in certain cases where we can practically leverage AI, like radiology field [14]. Since it is getting more and more easy to access medical information and to go with it the multi-stride advancements, it has allowed AI to make realistic advancements in the medical field [13]. The applications of ML are shown in Figure 15.3.

15.3 Types of Learnings in Machine Learning

We can classify ML into four broad categories. They are shown in Figure 15.4. Each and every type has its own properties as the function of these types varies depending on the data set.

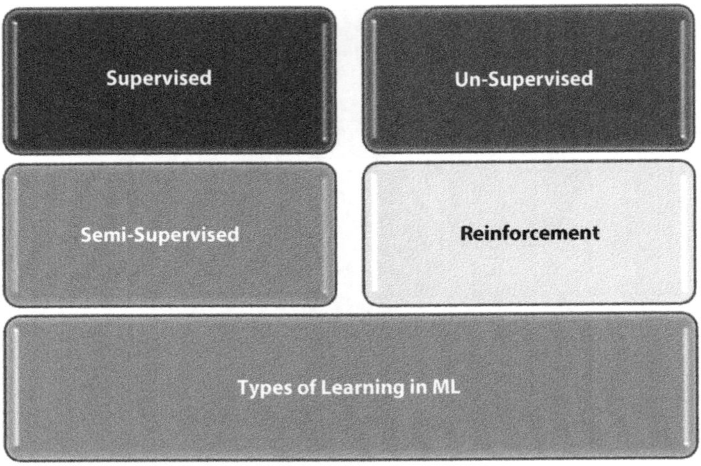

Figure 15.4 Types of learning in ML.

15.3.1 Supervised Learning

Model is trained based on a labeled data set, and then, it is called supervised learning [29]. Labeled data set will have both input and output parameters. In Table 15.1, the classification has been made with customer data set. With this data set, one can predict whether the buying chance of particular customer is high or low.

Table 15.1 Classification and prediction of customer data set.

User ID	Age	Gender	Salary	Purchase
1	34	M	35000	1
2	21	M	19000	0
3	45	F	50000	0
4	19	M	15000	1
5	29	F	32000	0
6	21	F	19000	1
7	40	M	48000	1
8	23	M	21000	1
9	37	M	36000	0
10	20	F	18000	0

From Table 15.1, it is clear that, by seeing this prediction of purchase, one can identify the probability of purchase, that is, 1 represents the possibility of purchase and 0 represents that there is no possibility of purchase. Classification and regression are the two types of supervised learning. In the former (classification type), you can expect output as discrete values. However, in the latter (regression model), result would be in a sequence [2, 6, 12, 16, 20, 29].

15.3.2 Unsupervised Algorithms

In the unsupervised learning, the target to the model will not be given during preparation which would mean only input parameter values are provided to the training prototype. The model figures a way by itself to learn on its own [2, 6, 12, 16, 20, 29]. In Table 15.2, the clustering has been made with customer data set. From the same data set, one can group the data accordingly.

In the training data, the user feed may be of two types: either unstructured data or unlabeled data. Data, if it is in unstructured format, might miss a lot of information or it could be noisy or, sometimes, unknown and irrelevant. On the other hand "unlabeled data" does not have an output

Table 15.2 Clustering and association of customer data set.

User ID	Age	Gender	Annual income	Total spendings %
1	34	M	35000	50
2	21	M	19000	43
3	45	F	50000	21
4	19	M	15000	13
5	29	F	32000	56
6	21	F	19000	78
7	40	M	48000	50
8	23	M	21000	23
9	37	M	36000	24
10	20	F	18000	49

value and will only have value for input parameters. The clustering and association are the two types of unsupervised learning. Clustering is a technique where the ML model applies logic to group data in a data set by looking at various patterns. Association technique is a rule-based ML technique that finds out some very useful relations between parameters of a large data set. This is possible by implementing techniques to associate relations between data in huge data sets.

15.3.3 Semi-Supervised Learning

Semi-supervised learning is a combination of both supervised and unsupervised techniques. One can utilize these strategies when managing an information which is somewhat marked and rest enormous bit of it is unlabeled. Solo procedure can be utilized to anticipate marks and, afterward, feed these names to managed methods. This strategy is generally relevant if there should be an occurrence of picture informational collections where, normally, all pictures are not named [2, 6, 12, 16, 20].

15.3.4 Reinforcement Learning

In this technique [6, 12, 16, 29], model continuously expands its exhibition utilizing a reward feedback to get familiar with the conduct or example. These calculations are explicit to a specific issue, for example, Google self-driving vehicle and AlphaGo, where a bot rivals human and even itself to improving and better entertainer of Go Game. Each time, when the client feed in information, they learn and add the information to its existing source of data, thereby learning incrementally. In this way, the more it learns, the better it gets prepared and, henceforth, becomes more experienced. Delegates notice input. Agent plays out an activity by settling on certain choices. After its exhibition, agent gets compensate and, in like manner, fortify and the model stores in state-activity pair of data.

15.4 Types of Machine Learning Algorithms

ML calculations are portrayed as learning an objective capacity (f) that best guides input factors (X) to a yield variable (Y): Y = f(X) [22].

15.4.1 Classification

Classification is a supervised learning type because the class label is already known. It builds models of data classes. This is built to make prediction about the class labels. Learning and classification steps are two steps in classification. In the classification step, the class label for particular data is predicted. The performance evaluation is done with the parameter accuracy. Next is the learning step. Here, the data is analyzed with the class label that was given already, and classification is created on pre-defined procedure. The classification technique can be applied in healthcare in diagnosing diseases based on historical data.

15.4.2 Bayes Classification

Bayesian Classification comes under Classification Analysis. The prob-Bayesian classification comes under classification analysis. The probability of a given tuple will be predicted that to which particular class it belong to. For decision-making process, posterior and prior probability is meant for Bayes classification. If the hypothesis is made from the given information using posterior probability, then, that is, the attribute values that are then given for prior probability, nevertheless of the attribute values, the hypotheses were given. Bayes theorem is based on the decision theory and probability. The method is derived from the classification analysis. Based on the hypothesis of a particular method, this will work.

15.4.3 Association Analysis

Association analysis will search for remarkable relations between the objects in database, and it determines correlation relationship between them and it identifies the pattern. For example, shopping basket analysis can be taken in to account. At the same time, if any product is the most frequent choice of customer, then find that product from the shop. The product may be either paper and pen or fruits and vegetables. This can be found by placing the product on shelves of the store. The two parameters that are used for finding out the associated items are support and confidence. Support is nothing but two products are purchased at a time. If two products got purchased one by one, then it is known as confidence.

The threshold value is decided by the domain experts. If a pattern has minimum support and minimum confidence, then the pattern mined is considered as interesting. This is known as market basket analysis. Find the frequent item sets that are nothing but all the items. Several approaches are in usage for discovering repeated item set like apriori algorithm and vertical data format.

15.4.4 Correlation Analysis

It is the augmentation of affiliation rules. The help and certainty boundaries yield tiresome examples to the clients. The help and certainty are improved with another measure called connection examination, which helps to mine interesting examples.

15.4.5 Cluster Analysis

Cluster analysis is used to partition a data record into clusters or collection of objects. There are various algorithms for doing clustering. It comes under unsupervised learning because the tag data is not recognized. Clustering methods find out the data that are related or dissimilar from each other and, with that data, the analysis of characteristics is done. Cluster analysis is the preliminary processing step for relating different algorithms like classification and aspect subcategory assortment. For outlier detection, cluster analysis can be applied similar to extraordinary acquisitions cutting-edge by getting approval for pass industries. This can be applied in picture identification, internet browsing, and security.

15.4.6 Outlier Analysis

Cluster and outlier analysis are connected to for one another. During cluster analysis, data objects which is different form the other objects is called outlier detection. Outlier analysis methods are classified into numerical, proximity, clustering, and classification based. The altered kinds of outliers are collective, universal, and contextual noise. Collective outlier is nothing but if the documents possess dissimilar qualities from the whole information. If the outlier represents the information that gets deviated from the data set, then it is called global outlier. If the data objects that diverges with respect to context, then it is said to be contextual outlier. It is based on the parameters like time, day, and location. The application of outlier analysis is revealing debit card fraud risks, novelty detection, etc.

15.4.7 Regression Analysis

In this method, the analysis is supervised, and it identifies the item sets with some relationships that are correlated or liberated of each other. The aforementioned usually predicts the sales, temperature forecast, profit, etc. Also, this consumes information which is previously identified. Once an response is delivered, the regression procedure resolve associate the idea and estimated cost, and the mistake remains considered to acquire to the exact outcome [23]. This method can be applied in marketing.

15.4.8 K-Means

K-means is a widespread and extensively used cluster analysis method in which cluster of related items is bundled.

15.4.9 Apriori Algorithm

In this calculation, the continuous item set mining method and affiliation procedures are applied on value-based information bases. The circumstance will find incessant item sets and feature general turns of events.

15.4.10 K Nearest Neighbor

K-nearest neighbor (KNN) is primarily leveraged in regression and data classification analysis. The KNN leverages data to classify them based on similarity. The approach KNN is based on majority vote to its neighbor. Figure 15.5 shows the example for KNN.

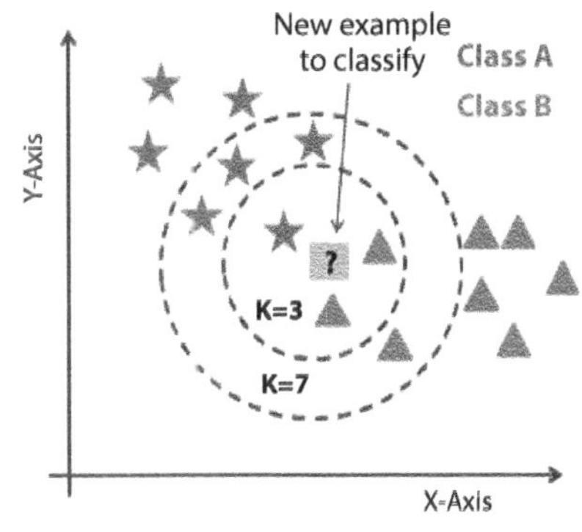

Figure 15.5 Example for KNN.

15.4.11 Naive Bayes

Naive Bayes works on an assumption that the features in a data set are completely independent of each other. The reason to call it naive is because, most of the time, the assumption is not true.

15.4.12 AdaBoost

AdaBoost is an AI meta-calculation, which is utilized to expand execution. AdaBoost is significant to funny information and exceptions.

15.4.13 Support Vector Machine

"Support vector machine" is one and only of the incomparable far and wide supervised learning calculations, which is reused for classification and regression issues. This is a very popular algorithm to satisfy classification requirements. The way a "support vector machine" works is that it looks at the data set, which is mostly multidimensional and then draws a line to classify them into new classes. Any new data that comes in is then predicted based on the side of the line it falls into and the class is then determined. SVM basically uses the vectors to classify its data and hence its name support vector machine [20]. In Figure 15.6, the categories of hyperplane are depicted.

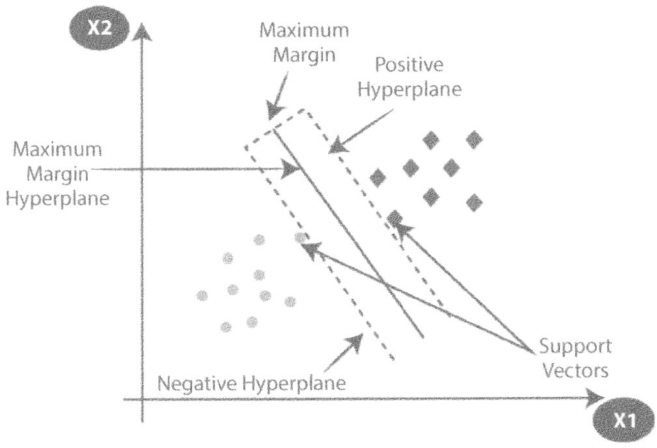

Figure 15.6 Categories of hyperplane.

15.4.14 Classification and Regression Trees

A decision tree applies to both classification and regression problems, and basically, the model is arrived in a tree structure. The root node in any decision tree is nothing but the best predictor within a given data set or range. In other words, given a variable (a), the decision tree makes a series of decisions by matching the value of "a" across the tree and branching based on a factor "y" until it finds a favorable point [22].

15.4.15 Linear Discriminant Analysis

"Logistic regression" is a grouping calculation generally restricted to just two-class characterization issues. In the event that there are multiple classes, the "linear discriminant analysis" calculation is the favored straight characterization strategy [22].

15.4.16 Logistic Regression

Logistic regression [28] resembles direct relapse in that the objective is to discover the qualities for the coefficients that weight each information variable. In contrast to straight relapse, the expectation for the yield is changed utilizing a non-direct capacity called the strategic capacity. Strategic relapse is another method acquired by AI from the field of measurements. It is the go-to strategy for paired characterization (issues with two class esteems).

15.4.17 Linear Regression

Linear regression is definitely a popular ML algorithm which operates such as—given a variable as input it predicts the output. The output variable is also known as dependent variable. A good example of a linear regression implementation is based on variable like area/square feet and location of a property, and we can predict the price of it (outcome) [22].

15.4.18 Principal Component Analysis

"Principal component analysis (PCA)" is a technique that is utilized to simplify information to analyze and foresee by weakening the number of factors. In other words, a multidimensional huge data set is reduced into a smaller data set but, at the same time, maintaining the most of the critical information from the parent data set. Every segment is a direct mix of the first factors and is symmetrical to each other. Symmetry between parts shows that the relationship between these segments is zero [25].

15.5 Machine Learning for Information Extraction

The innovative advances have presented to us the likelihood to get to huge sums of text based data, either in the Internet or in specific assortments. Be that as it may, individuals cannot peruse and process this data any quicker than previously. To make it helpful, it is frequently needed to place this data in some kind of organized design, for instance, in a social information base [18]. Like the direct separator approach, it is upheld on the possibility that taking a gander at the words blends around the fascinating data (for example, the unique circumstance) is enough to get familiar with the necessary extraction designs. Linear separator is leveraged to pull information from documents that are semi-structured [15].

Information extraction (IE) measure extricates valuable organized data from the unstructured information as elements, relations, items, occasions, and other numerous types. The pulled out data from information that is unstructured is leveraged to get ready information to be examined. What it means is that the more efficient the change to the unstructured information in data extraction, the better the results in terms of examining the information. There are different methods available for IE for different type of information, for example, video, audio, text, and pictures [7].

15.5.1 Natural Language Processing

Natural language processing (NLP) is a field in AI with the capacity of a PC to comprehend, dissect, control, and possibly create human language [26]. Normal language handling (NLH) is a subfield that helps in converting language used by humans to a machine readable one. In other words, it is the approach to analyze characteristic language information and program it such a way that machines can also analyze and investigate. NLP combined with other ML techniques makes it very powerful allowing PC to learn on their own by extracting and analyzing data and making predictions that were previously not possible [30].

15.6 Predictive Analysis in Healthcare

The term "analyzing medical data" and "predictive analytics" developments showed a great development of premium from years, in light of the fact that the cycle of examination in the well-being territory does not stop exactly at the level of the capacity to oversee huge data sets; however, it surpasses this to the capacity to recover future information, which is empowered by numerous researchers and experts [16]. So, for predicting the disease from a large database, many algorithms were in use. Predictive analytics and ML in healthcare are quickly turning into probably the most-talked about, maybe most-advertised subjects in medical care investigation. ML is a very much contemplated discipline with a long history of accomplishment in numerous enterprises [17]. Medical care can take in important exercises from this past progress to kick off the utility of prescient examination for improving patient consideration, constant sickness the board, emergency clinic organization, and store network efficiencies. The chance that right now exists for medical care frameworks is to characterize what "prescient examination" signifies to them and how might it be utilized most viably to make upgrades [24]. This predictive analysis process is shown in Figure 15.7.

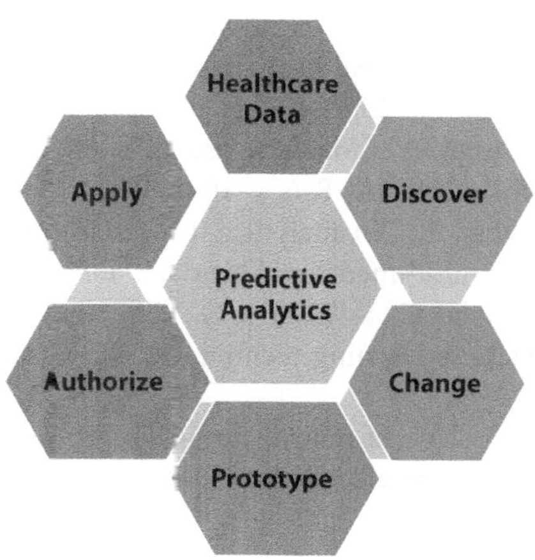

Figure 15.7 Process of predictive analytics.

15.7 Conclusion

In this chapter, various concepts like ML, Predictive Analytics, NLP has been discussed. The learning types in ML are deliberated with appropriate examples. The types of ML algorithms are discussed. The various data sources and data handling methods are shown clearly with appropriate examples and images. As ML plays a vital role in healthcare systems using EHR is depicted and the applications of ML are discussed which clearly shows the importance of ML in healthcare systems. Technologists should attempt for a bright future where the power of ML algorithms benefits large number of common people to boost their basic health.

References

1. Ghassemi, M. *et al.*, "A review of challenges and opportunities in machine learning for health." *AMIA Summits on Translational Science Proceedings 2020*, p. 191, 2020.
2. Qayyum, A., Qadir, J., Bilal, M., Al-Fuqaha, A., "Secure and Robust Machine Learning for Healthcare: A Survey." *IEEE Rev. Biomed. Eng.*, 14, 156–180, 2021..
3. Ali, F. *et al.*, "A smart healthcare monitoring system for heart disease prediction based on ensemble deep learning and feature fusion.". *Inf. Fusion*, 63, 208–2225, 2020.
4. Saranya, G. and Pravin, A., "A comprehensive study on disease risk predictions in machine learning.". *Int. J. Electr. Comput. Eng.*, 10, 4, 4217, 2020.
5. Dash, S. et al., "Big data in healthcare: management, analysis and future prospects." *J. Big Data*, 6, 1, 1–25, 2019.
6. Dhillon, A. and Singh, A., "Machine learning in healthcare data analysis: a survey." *J. Biol. Today's World*, 8, 6, 1–10, 2019.
7. Adnan, K. and Akbar, R., "An analytical study of information extraction from unstructured and multidimensional big data." *J. Big Data*, 6, 1, 1–38, 2019.
8. Mohan, S., Thirumalai, C., Srivastava, G., "Effective heart disease prediction using hybrid machine learning techniques.". *IEEE Access*, 7, 81542–81554, 2019.
9. Savova, G.K. *et al.*, "Use of natural language processing to extract clinical cancer phenotypes from electronic medical records.". *Cancer Res.*, 79, 21, 5463–5470, 2019.
10. Wang, Y. *et al.*, "Clinical information extraction applications: a literature review." *J. Biomed. Inform.*, 77, 34–49, 2018.

11. Kaur, J. and Mann, Dr. K. S., "AI based HealthCare Platform for Real Time, Predictive and Prescriptive Analytics using Reactive Programming". *10th International Conference on Computer and Electrical Engineering.*
12. Panch, T., Szolovits, P., Atun, R., "Artificial intelligence, machine learning and health systems." *J. Glob. Health*, 8, 2, 2018.
13. Jiang, F. *et al.*, "Artificial intelligence in healthcare: past, present and future." *Stroke Vasc. Neurol.*, 2, 4, 230–243, 2017.
14. Nilashi, M. *et al.*, "An analytical method for diseases prediction using machine learning techniques.". *Comput. Chem. Eng.*, 106, 212–2235, 2017.
15. Shickel, B., Tighe, P.J., Bihcrac, A., Rashidi, P., Deep EHR: a survey of recent advances in deep learning techniques for electronic health record (EHR) analysis. *IEEE J. Biomed. Health Inform.*, 22, 5, 1589–1604, 2017.
16. Nithya, B., "An analysis on applications of machine learning tools, techniques and practices in healthcare system." *Int. J. Adv. Res. Comput. Sci. Softw. Eng.*, 6, 6, 1–8, 2016.
17. Boukenze, B., Mousannif, H., Haqiq, A., "Predictive analytics in healthcare system using data mining techniques." *Comput. Sci. Inf. Technol.*, 1, 1–9, 2016.
18. Tellez –Valero, A. *et al.*, "A Machine Learning Approach to Information Extraction". *Conference Paper in Lecture Notes in Computer Science*, 2014.
19. https://www.healthcatalyst.com/
20. https://www.javatpoint.com/machine-learning-support-vector-machine-algorithm
21. https://builtin.com/artificial-intelligence/machine-learning-healthcare
22. https://builtin.com/data-science
23. https://www.upgrad.com/blog/what-is-text-mining-techniques-and-applications/
24. https://www.healthcatalyst.com/predictive-analytics
25. https://www.dataquest.io/blog/top-10-machine-learning-algorithms-for-beginners/
26. https://towardsdatascience.com/natural-language-processing
27. https://www.foreseemed.com/blog/machine-learning-in-healthcare
28. https://towardsdatascience.com/
29. https://www.geeksforgeeks.org/
30. https://en.wikipedia.org/wiki/Natural_language_processing

16

Knowledge Fusion Patterns in Healthcare

N. Deepa[1]* and N. Kanimozhi[2]

[1]Department of Computer Science & Engineering, Saveetha School of Engineering, Saveetha Institute of Medical and Technical Sciences, Saveetha University, Chennai, India

[2]Department of Computer Science & Engineering, GKM College of Engineering and Technology, Chennai, India

Abstract

Computer science as vast technological development such as Fourth Industrial Revolution completely improves the growth in education, marketing, healthcare, employment, etc., which gives essential interaction worldwide. The ultimate use of Industry 4.0 is to increase flexibility in healthcare and fill the gap in business interaction. In condition such as poverty, tetra billion amount of data accessing are global issues that can be handled by Technology 4.0. Artificial Intelligence (AI) and Internet of Things (IoT) along with data science are relying on technology for optimizing and innovation for improving throughout the universe. Automation with available framework uses IoT devices such as Sensors, Arduino, and Raspberry Pi that commercially involve for reducing the redundant information and complexity from healthcare analysis. Not only does the healthcare system collect the users' database, the healthcare environment is the place where massive databases are accessed with extreme resources as well. This work focuses on knowledge fusion patterns in healthcare to make clear that massive databases that are extracted from healthcare environment have challenges of data fusion, are retrieved from the IoT devices, router, and other networking LAN communication devices followed with cloud services. Patients' information such as electronic health records (EHRs) are mostly accessed, this brings complexity in time, dependency issues, and processing time for data connectivity in real time. We propose a distributed cask data application framework with a hierarchical ecosystem, which performs a data fusion in a virtualized manner. In different resources of data that are mixed with sensors and IoT devices,

**Corresponding author*: deepa23narayanan@gmail.com

A. Jaya, K. Kalaiselvi, Dinesh Goyal and Dhiya Al-Jumeily (eds.) Handbook of Intelligent Healthcare Analytics: Knowledge Engineering with Big Data Analytics, (345–364) © 2022 Scrivener Publishing LLC

the accurate and efficient multi-access can be used. Real-time applications using hierarchical order based on storage processes and to avoid storage limitation from various applications are implemented through order of fusion levels. To improve the performance of accurate healthcare data and avoid poverty due to inefficient healthcare, Industry 4.0 fine-grained decisions through multiple access such as initial level, middle level, and higher level of event processing through improved distributed framework should be by initialized.

Keywords: Artificial Intelligence (AI), data fusion, improved distributed framework, Internet of Things (IoT), Industry 4.0, healthcare, sensors

16.1 Introduction

One of the significant research domains is healthcare, which is being improved by technologies such as Internet of Things (IoT) that can support the healthcare core functions. This is the method where the hospitals following traditional methods are being converted as smart environments working digitally. They move toward next generation by incorporating sensors that are being interconnected and processing techniques of big data [23].

NIST, which stands for National Institute of Standards and Technology, has defined cloud computing as a model that can help in provisioning of resources. NIST defines the roles as consumer, provider, auditor, broker, and carrier [16, 23]. The deployment models in cloud are public, private, community, and hybrid cloud, and the delivery models are IaaS (Infrastructure as a Service), PaaS (Platform as a Service), and SaaS (Software as a Service). Cloud computing services are providing more benefit to the users because of its features such as decentralization and sharing of resources [3]. As decentralized mode of work was introduced, there was a transition from works carried out on site toward the cloud. The computing power offered to the users is unlimited, and the access method enables the user to work from anywhere. Based on the choice of users, the resources can either be included to their working platform or they can also be made free. The resources such as CPU and memory along with storage can be requested to the specific cloud provider. Services from Amazon Cognito or Elastic Beanstalk are also provided to the user for making applications to work effectively [3, 13].

Hospital medical management has started wireless access through various devices. These router connected devices instantly monitor the activities of every person, and also the detection [6] of patients' history such as blood pressure (BP) level, body temperature, and oxygen measurement is under surveillance of IoT-based devices. Embedded MRI scans, CTC scans, and X-ray are medical factors as image classification [10]. There are various features monitored for high critical patients for prescribing them. Hospital equipment that updates the biometric data of patients' authorization, hospital staff information, and also their duty charts can be viewed from any router connected devices such as android phones, tablets, laptops, and personal computers. When emergency 911 situations are fetching the entire database from all over GIS (Geographical Information System) locations for necessary actions based on the requirements for the human life, the database can be accessed for connectivity and dependency of each table that are stored in the server as a knowledge experience [4].

Ubiquitous decisions might be taken from all multiple resources where the patient information, doctor information, mode of payment, and highly secured information are stored. Single data that are from one source are not able to fetch the clear information. By streaming the data fusion, the samples connected from patients not only help in diagnosis but also detect the indication of human body temperature that can help to automate the lifespan of other patients also [2]. There are some tedious processes when the spatial data accesses are validated from individual servers that are available in various GPS (Global Positioning System) such that the distributed network is established according to the problem identified.

Doctors who are specialized and interconnected based on the input features; one single data will not be sufficient to give diagnosis. When this single access, diagnosis becomes a trouble for the healthcare ecosystem; the organization performs the access priority order to bring the convenience and performance level in a better manner. There are three levels of hierarchical access such as initial, middle, and higher level for making decisions based on the diagnosis system [22].

However, the upcoming technologies and opportunities for processing the huge amount of information are analyzed from trillions of smart devices that are interconnected with each other. For simple integration and available performance based on the multiple accesses that can represent the integrity, performance and accuracy are maintained to distribute the data fusion in a consistent way. The IoT devices that are labeled with the healthcare system that are accessed by various cloud service providers are collecting unsupervised data from many GIS position. In such a way, load balancing is also not stable since the mathematical analysis on resources is from soft computing [7, 19]. Ubiquitous devices can be analyzed for data science. There exists the current approach that has simple processing from cloud providers as public devices stored with the multiple issues such that network lagging and restricted bandwidth are analyzed issues. Not only the flexibility of data processing but also the integrity of the data was not encrypted due to the unbalanced load from the huge database as data science [9].

Thereafter, big data that incorporates its access through cloud becomes more expensive. This proposed system that is automation of data fusion in a distributed level of framework can determine the best multi-access process. This management in the proposed ecosystem relied on the performance of data in multiple databases; the initial level becomes the primary to relate the relevant data and avoid redundant data. Also, the fusion of updated information from last initial data that transferred and aggregated the mutual data can be connected to the middle class of data in a hierarchical order [24]. The complex data that are captured from sensors and IoT-based sensor information that are taken from the smart ecosystem of healthcare can be computed based on the decision that describes the enabling event processing.

The computation is performed for the i) data fusion, ii) hierarchical framework for complex analyzed events, and iii) increasing the performance and accuracy through the centralized distributed approach.

The proposed work is organized as follows:

a) Related work from the framework of healthcare ecosystem that uses the IoT devices updated databases and used multi-access data fusion methodology as efficient solution;
b) Materials and methods with three level hierarchical frameworks for healthcare ecosystem;
c) Results and discussion show the implementation results and comparison for healthcare datasets using the IoT devices and Mobile devices connected with the all other hierarchical order;
d) Conclusion and future work describe the limitations along with the scope in future for further study purpose.

The chapter is organized as follows: Section 16.2 discusses on the existing work. Section 16.3 deals with the materials and methods. Section 16.4 explains the proposed system. Section 16.5 discusses the results obtained. Section 16.6 presents the conclusion and future work.

16.2 Related Work

Latest technologies such as data science [14], cyber security, and applications for native and iOS along with network-based communication are working toward intervention of Industry4.0. Many embedded techniques such as mobile computing, soft computing, and sensor related data are automated with data centers that are using cloud services [23]. Environment is dependent on service-oriented framework for healthcare, business aspects, and fetching of information from various devices using sensors, storing accesses, and other wearable devices that are connected to internet for smart access. Patients who handled sensible disease-based data are monitored through smart wearable devices that can be integrated with highly secured cyber based systems. Not only the sensitive information such as patient ID and type of causes is more important to be reliable on both server side and hospital-based access from doctors, which should be handled carefully [5].

16.3 Materials and Methods

Table 16.1 describes the classification model that has data fusion from various knowledge engineering aspects to increase the quality of information and reduce the loss of data, less time for extraction of parameters, and computation calculation strategies as well. Many patterns have been insisted along with the processed parameters that help to identify the single/multiple databases that deal with the distributed approach. These combinations of data sciences, IoT devices, and classification are discussed in Table 16.1.

16.3.1 Classification of Data Fusion

Cloud service providers have an N number of choices for customer active expectations such that the services as infrastructure might be a better option for healthcare management that is connected over the network all over the place. The mode of connectivity in the cloud for various addresses of virtual servers is updating the IoT-based devices to be automated according to the user's access from their network-enabled devices [21]. The properties with five parameters are grouped into different types that are shown in Table 16.1.

In this type, the resources that are categorized from the virtual sensors that are enabled in a particular region are updated. Also, the related sensors and network access are based on router-based connectivity, which are forwarded from the pages as per the user's functional usage [20]. When the parameters are allotted according to the memory or processing speed, then the resources is assigned accordingly. Initial resources that are from infrastructure providers are virtualized from the third party who have enabled their network as per the customer's requirements.

Table 16.1 Classification of healthcare data for accurate multiple data integrity.

		Classification model		
S. no.	**Type**	**Initial level**	**Middle level**	**Higher level**
1.	Data preprocessing	Middle	Initial	Middle
2.	Application platform	Initial	Middle	Higher
3.	Data analysis according to decision	Unsupervised data	Parameter based	Decision process
4.	Data management for handling dynamic collective information	Time series data	GIS data	Time series and GIS
5.	Data cardinality and resources based on user's choice	Choice of access on single location	Area around all over country	Area over internet world

Equipment is instantly making its entries from the devices where the interfaces or gateway are not direct access for micro controllers. Based on the smart board connectivity, each pin is responsible for data sharing, data processing, data fusion, data reduction, etc. Based on the sensor, data that are interlinked with the MBS (Mobile Base Station), BSs (Base Stations), and cloudlets are also monitoring [15] the gateway according to the data-centers' support for data-efficient access [18]. So that the distributed access using the proposed system as a hierarchical framework for hospital accessed dataset is computed according to the requirements.

Data centers and cloud services are accessed remotely where metropolitan cities are locating nearby hospitals and collecting the relevant information through the sensor and IoT-based inputs. Transferred results based on balance load are according to the network range and provide high connectivity such as 4G for service providers. In latest trend, technology such as 5G and Industry 4.0 can collaborate and provides a great topology that can brings a great bandwidth and reduces the network latency along with frequency drop as well. Much exploration over network supports the efficient healthcare management that revolves around the infrastructure-based service in an effective manner.

Classification model represents the framework for the top approach on healthcare systems, which are accessed in hierarchical manner such as initial, middle, and high levels.

16.3.2 Levels and Its Working in Healthcare Ecosystems

Three levels of hierarchical order based on top level cloud center access can be brief as follows.

16.3.2.1 Initial Level Data Access (ILA)

All internet-enabled devices are collecting data from smart IoT embedding devices that can relatively mix the data from multi-access data fusions. The ultimate role of processing the patient's ID is to know the severity of health conditions. By joining the live data from any IoT devices of individual patients along with the central database, it cross-checks the records of their health progress and gives an alarm to the top level to set a checkup from the GIS located doctors such that the event can be processed and informed doctors to take necessary steps for the particular patient. In such progress, fusion brings the integrated sensor (Figure 16.1) and main database to get updated and summarize the hospital in charges to provide the exact treatment [17].

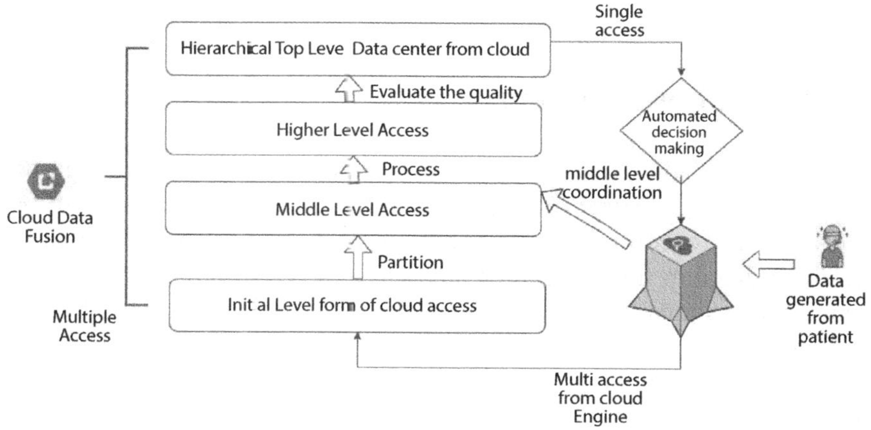

Figure 16.1 Data fusion hierarchical framework for big data and IoT devices.

16.3.2.2 Middle Level Access (MLA)

Once the alerts have been initiated from the initial level of process from the search engine, the distributed cask data initiate the necessary steps to be taken as action to the MLA. When the hospital restored the data from previous partitioned data fusion, the important step is to match the exact data from the current location of the database matched patients. In such a way, the patient ID, the building number along with the floor details, and name of the hospital with room number are also located from the sensor that are updating sequentially. Based on the diseases, they are grouped and organized to treat the comfort of doctors and caretakers. According to the knowledge shared by the sensor fusion data, multiple information will allot them the tablets, medications, and treatment based on demands. If the doctors demand, then the next available person will give diagnosis for the problem [27].

16.3.2.3 High Level Access (HLA)

As the data from HLA are accessed globally, the piconets from Bluetooth and access points from all data centers are viewable on the internet. Based on the limit of state and country, the data can be accessed according to the need. In particular, the global position system helps the emergency (911) situations to coordinate comfortably and avoid the network and traffic control to avoid the critical issues for the patients. Heart transplantation and other necessary first priority cases will be handled in the HLA mode [26].

16.4 Proposed System

The proposed approach which takes the process, partition, and coordination among the supervised data from the raw data is accessed in bi-directional procedure. The cardinality of each data fusion label will be easily updated in the cloud after the last updated, modified, or removed from the primary database. When the requirements are fulfilled or patients' data are not available, this instance may slow the data fusion during decision-making according to the levels of distribution. Figure 16.2 represents the TLCA healthcare ecosystems.

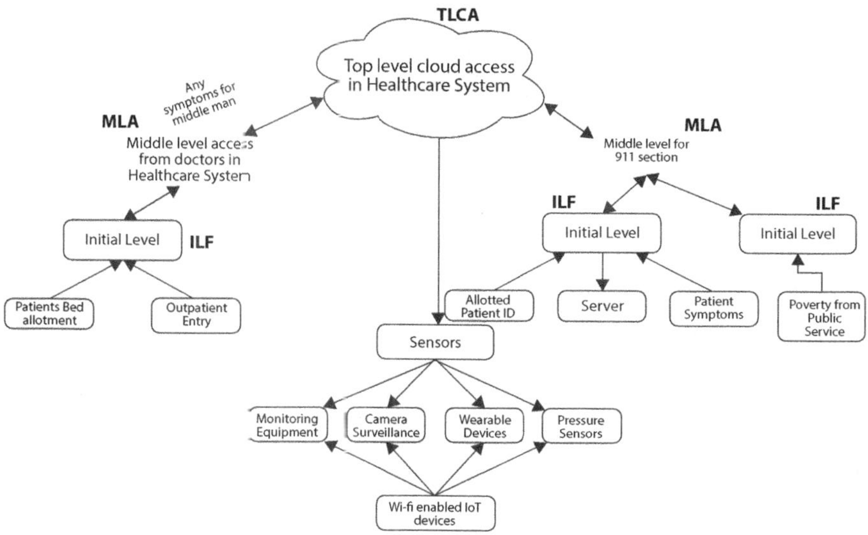

Figure 16.2 Proposed architecture TLCA in healthcare ecosystem.

16.4.1 Objective

The main objectives of the proposed research work are as follows:

(i) Healthcare dataset observed from various IoT devices such as Sensors and Aurdino where data processing has been processed and redundant data can be reduced to avoid data complexity;
(ii) To classify the data fusion that is monitored from healthcare datasets that are stored and utilized as multi-accessed frameworks from cloud services;
(iii) To propose a distributed cask data application which works in a hierarchical way that implies data fusion from a virtual machine;
(iv) To increase the level of performance when distributed all over the cloud and avoid poverty by providing availability of experts in all places over countries.

Using Top Level Cloud Access (TLCA) in the healthcare system, the three-level classification model observes from Wi-Fi–enabled IoT devices that are obtained from monitoring equipment, camera surveillance, wearable devices through critical patients, and pressure sensors to observe the oxygen level. To reduce the repeated data while processing that makes complexity when accessed by IoT devices, aggregated data about patients diagnosis can be maintained balanced.

For example, symptoms of a patient with COVID-19 are stored in initial level by collecting the parameters as follows:

a) Allotted patients ID
b) Patient name

These data are stored in a primary database that can be retrieved as single access and no data fusion is there. Once updated about the patients along with the diagnosis from the available specialist, they will be monitored to make their presence better. In the middle-man symptoms, sometimes, it is not enough for the doctor who has not updated with the remaining initial level data:

- Bed allotment entry data.
- Inpatient/outpatient references.

During the MLA, the data access point loss if the single fusion were used. To avoid the emergency situation during travel from long distance, multi-access hierarchical framework can be introduced. In these hierarchical top level approaches using the distributed cask system, the stored information can be accessed firstly, and integrity can be guaranteed since the data are fetched from various sources [7].

The device that is ready to operate, and parameters based on the updates are analyzed only through online mode, since networks are router based and piconet with master/slave access point which are storing the data as per the utilizations. Batch processing to reduce the memory allocated and energy saving from the electronic utilities such as healthcare data are able to manage [25]. When there are many biometric entries using various sensors such as hospital related employees, others are using their entries and exits from multiple gates such that many sensors are updating the interface which is collecting all the updated data that are operated from different locations. As per the access point, data from the connectivity have sensor data that can be stored in the main database and the owned corporate database as well. Some enabled IoT devices [1] that have restrictions for accessing chips, micro-controlled board, and raspberry pi are handled by the owner of the network for security purposes [8].

When the personal data are set with the limit of threshold value such as BP level, oxygen level, and heart rate level for member matching, the event prepares the streaming through the sensor fusion background.

16.4.2 Sample Dataset

Healthcare systems using data fusion in a hierarchical way used the dataset created by analyzing the multiple resources to handle the different regions. The dataset (Table 16.2) is collected from enabled IoT devices such as how long the diseases have caused and encountered, and identified and assigned a unique id. Based on the encountered data, individual patient ID has been generated [12]. Basic parameters such as age, gender, race, admission number, source of admission, discharge ID, time spent in hospital, specialization of doctors, and payment mode are gathered for aggregation of data. Technologies such as RDBMS (Relational Database Management), NoSQL (Not Only SQL), and HDFS (Hadoop Distributed File System) are used for collecting the information from clinical aspects and IoT devices. During the collection procedure, no data are grouped based on type and, sometimes, image processing, and also kind of sources for gathering patients data. Images such as CT (computed tomography), MRI (magnetic resonance imaging), and X-ray for keen knowledge and accuracy for prediction of interpretation for diagnosis can be maintained [12].

Based on the pattern from customer hierarchical data, the level of availability suggested from the sensor devices. The streaming is dynamic and changes eventually according to various entries from Raspberry Pi and insisted as initial level (ILA). After the processed data from cloud are arranged based on the cloud centers and then hospital servers, side event processes for fixing the database preprocessing These multi-data fusion established a network connection such as confidential data of patients as biometric ID and doctor prescription to diagnose the patient. Figure 16.3 shows the comparison of each data from sensor fusion according to the pattern matching from predicted and actual analysis.

16.5 Results and Discussion

Cloud using virtual machines has storage and allocation of memory space and virtually supports CPU process on its own, which matches the multiple organized models for classifying the results using various technologies such as mobile devices, fog computing, and media available dataset for understanding the performance level. Also, the data that are updated through signal processing, image data, and raw data are utilized for saving life spans and human lives by reaching the doctors and donors on right

Table 16.2 Healthcare sample dataset for preprocessing and data fusion.

encounter_id	patient_nbr	race	gender	age	admission_type_id	discharge_disposition_id	admission_source_id	time_in_hospital	payer_code	medical_specialty
2278392	8222157	Caucasian	Female	[0-10)	6	25	1	1	?	Pediatrics-Endocrinology
149190	55629189	Caucasian	Female	[10-20)	1	1	7	3	?	?
64410	86047875	African American	Female	[20-30)	1	1	7	2	?	?
500364	82442376	Caucasian	Male	[30-40)	1	1	7	2	?	?
16680	42519267	Caucasian	Male	[40-50)	1	1	7	1	?	?
35754	82637451	Caucasian	Male	[50-60)	2	1	2	3	?	?
55842	84259809	Caucasian	Male	[60-70)	3	1	2	4	?	?
63768	114882984	Caucasian	Male	[70-80)	1	1	7	5	?	?
12522	48330783	Caucasian	Female	[80-90)	2	1	4	13	?	?
15738	63555939	Caucasian	Female	[90-100)	3	3	4	12	?	Internal Medicine
28236	89869032	African American	Female	[40-50)	1	1	7	9	?	?
36900	77391171	African American	Male	[60-70)	2	1	4	7	?	?

(Continued)

Table 16.2 Healthcare sample dataset for preprocessing and data fusion. (*Continued*)

encounter_id	patient_nbr	race	gender	age	admission_type_id	discharge_disposition_id	admission_source_id	time_in_hospital	payer_code	medical_specialty
40926	85504905	Caucasian	Female	[40-50)	1	3	7	7	?	Family/General Practice
42570	77586282	Caucasian	Male	[80-90)	1	6	7	10	?	Family/General Practice
62256	49726791	African American	Female	[60-70)	3	1	2	1	?	?
73578	86328819	African American	Male	[60-70)	1	3	7	12	?	?
77076	92519352	African American	Male	[50-60)	1	1	7	4	?	?
84222	108662661	Caucasian	Female	[50-60)	1	1	7	3	?	Cardiology

<table>
<tr><th></th><th>num_medications</th><th>time_in_hospital</th><th>num_medications|time_in_hospital</th></tr>
<tr><td>1</td><td>18</td><td>3</td><td>54</td></tr>
<tr><td>2</td><td>13</td><td>2</td><td>26</td></tr>
<tr><td>3</td><td>16</td><td>2</td><td>32</td></tr>
<tr><td>4</td><td>8</td><td>1</td><td>8</td></tr>
<tr><td>5</td><td>16</td><td>3</td><td>48</td></tr>
</table>

Predict Actual	0	1	All
0	1465	1	1466
1	167	0	167
All	1632	1	1633

Accuracy is 0.89
Precision is 0.91
Recall is 0.86

Figure 16. 3 Comparison of features to calculate the prediction of data fusion accuracy, precision and recall.

time. When multi-level approaches take a remotely distributed cloud access, many people in poverty can also get availability of experts from GPS who have no financial hurdles to support them.

Figure 16.4 represents the data fusion from IoT devices monitoring, as data have been formed based on procedure to perform at high speed and accurately. As diagnoses are designed based on the middle level and established the time representation whether the GIS shows the nearby location, any biometric-based matches can be fined for the emergency consultation.

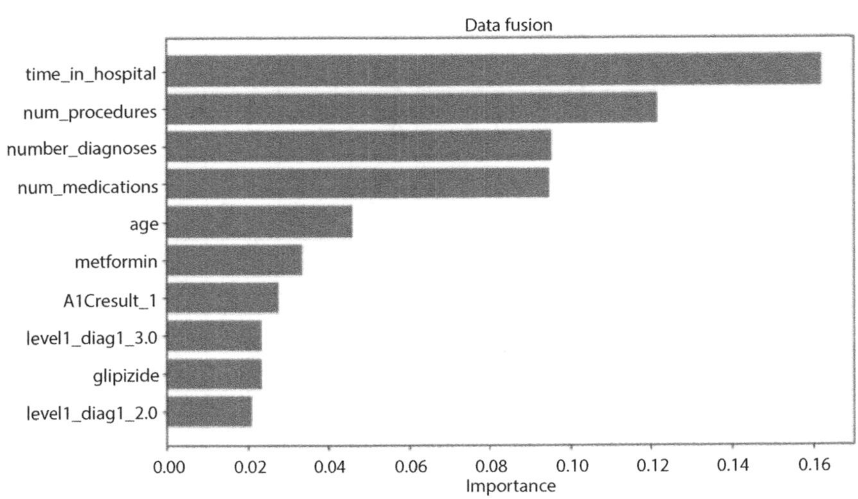

Figure 16.4 Data fusion along with sensor fusion using TLCA healthcare system.

Let us consider the simple Sensor fusion entities as follows

$$SSF = \{Temp, Timst, devid\}$$

where Temp - defines the static temperature value
Timst - defines the timestamp dynamic changes from GIS
devid - unique identity from the IoT devices.

Such that ILA approach the object matched with pattern are according to the patient's body temperature

$$ILA = [SSF \text{ as } aa; (SSF \text{ as } bb)+; SSF \text{ as } cc] \quad (16.1)$$

clean
(aa.Temp < = 42^ cc.Temp > 44 ^ bb.Temp > aa.temp ^(cc.Timst=<69))

The thermometer captures the rise of temperature where the affected patient has no symptoms and instantly changes and causes the temperature variation that has been registered. Equation (16.1) shows the initial level of temperature identified and updates in healthcare data through hierarchical framework (Figure 16.5).

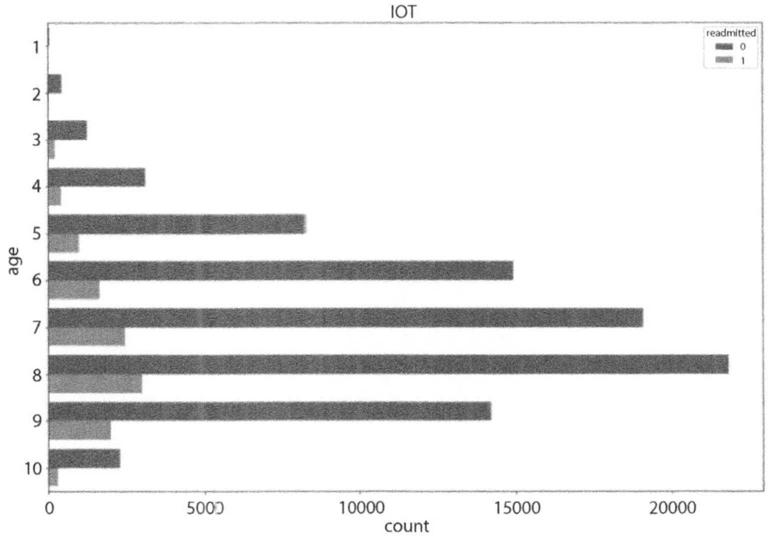

Figure 16.5 Comparison of IoT devices count based on data aggregation.

Similarly, MLA also defines the event processing for finding the patients availability in ward or room that fetches from network nodes. Matching of pattern shows that, if the registered ID and health data fusion ID paired in equal, then the health condition also matched for the same person. For HLA, the event is processed based on the symptoms that are correlated with the pattern identified from sensor fusion accordingly to confirm the severity of body changes according to the timeline [11].

When traditional systems are used, the cloud-based updates use the interlinked data for understanding only lower level access such as patients' basic information alone. When there is a time stamp variation, then the internet fails to update sometimes, so do avoid the drawback of traditional systems: the multi-sensors that work on each unsupervised data in person by updating the same from IoT devices to cloud data centers that avoid the lower performance and increase the access in faster manner. Also, when the combination of organized GIS also performs based on the centralized database, this sets the latency of the network to be balanced without time delay. Figure 16.6 shows the instant updates based on the frequency of occurrence in initial level.

The results are performed by comparing the classification model from the proposed system as shown in Figure 16.7 to obtain the facilities through the centralized access. Also, based on the requirements, the fusion among each level such as initial, middle, and higher level decision-making can be determined. When the repetition of dynamic updates has been removed and patients' symptom-based inputs alone are considered as parameters, then the analysis in the middle level can increase the performance and also provide high integrity. For the diagnosis of the patients, the sensor fusion can avoid traffic management, and critical situations are handled. Also, the

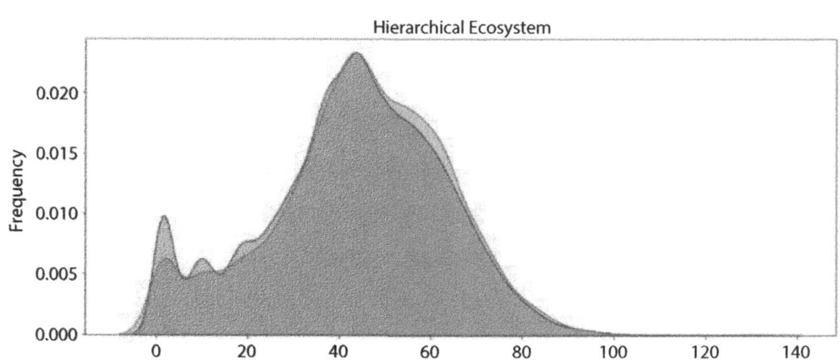

Figure 16.6 Number of procedure based on hierarchical ecosystem vs frequency.

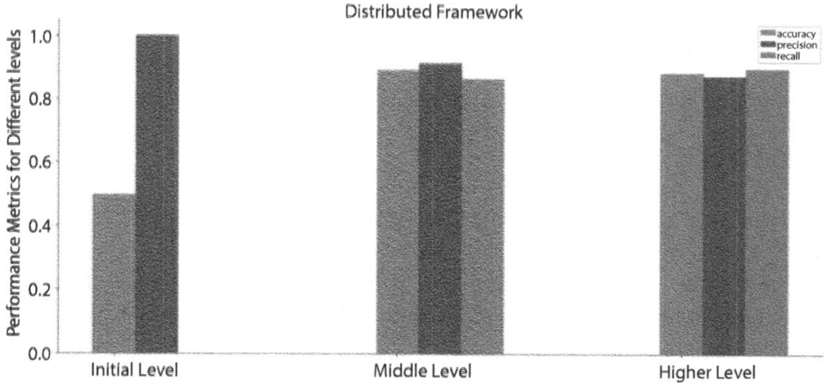

Figure 16.7 Accuracy, precision and recall (%) based on distributed framework.

operation performed as per the identity of stages makes the proposed system as highly accurate compared to the traditional approach.

16.6 Conclusion and Future Work

Industry 4.0 has its involvement based on mobility enabled sensor data that are available in healthcare and enterprise data. When the patients, doctors, or customers' ubiquitous devices are enabled with the time series data, storage and pending decisions are incorporated. By the limitation of existing frameworks from the smart healthcare system, the data fusion becomes the problem. For splitting dataset, cloud-enabled data are sometimes not reached according to the processing speed and performance. Even though internet from LAN cable, various routers that connect the access points from the remote sensors, data centers, providers, or edge computation are prioritized and brought for main access for the importance of patients' needs. Whereas the cloud service framework is not able to update the speed according to the level of categories for deciding the criticality from patients' needs. So, the framework can be distributed and cask bi-direction on piconet can also be enabled. When multiple channels are accessed in a top-down approach, backtracking can be avoided and configured based on multi-access (fusion) for taking the right choice of processing the IoT-based devices. In the future, the smart healthcare ecosystem can be updated with central process data to centralize the data using block chain technology.

Also, the ecosystem can be organized in a hierarchical framework for increasing the accuracy for client and server side enabled channels. IoT devices in poverty-based people can be used vastly and avoid over throughput along with cost.

References

1. Rao, A. *et al.*, "A Novel Cardiac Arrest Alerting System using IoT". *Int. J. Sci. Technol. Eng.*, 3, 10, 78–83, 20172017.
2. Agrawala, A.K., *"Revolutionary Advances in Ubiquitious, Real-Time Multicomputers and Runtime Environments,"* 2001, https://doi.org/10.21236/ada388037.
3. Ahson, S.A. and Ilyas, M., *Cloud Computing and Software Services: Theory and Techniques*, CRC Press, Florida, 2010.
4. Vito, A., Digregorio, G., Sechi, G., "A Decision Based Support System Based on GIS Technology," in: *Encyclopedia of Healthcare Information Systems*, 2008, https://doi.org/10.4018/978-1-59904-889-5.ch050.
5. Aljuaid, H. and Parah, S.A., "Secure Patient Data Transfer Using Information Embedding and Hyperchaos." *Sensors*, 21, 1, 1–20, 2021. https://doi.org/10.3390/s21010282.
6. Anand, J. *et al.*, "Smart Indication System for Spinal Cord Stress Detection". *Int. J. Recent Technol. Eng.*, 8, 3, 6164–6168, 20192019.
7. Bertino, E., Verma, D., Calo, S., "A Policy System for Control of Data Fusion Processes and Derived Data." *2018 21st International Conference on Information Fusion (FUSION)*, 2018, https://doi.org/10.23919/icif.2018.8455563.
8. Lou, R.C. and Chen, O., "Multisensory Data Fusion for Ubiquitous Robotics Services," in: *Sensor and Data Fusion*, 2009, https://doi.org/10.5772/6577.
9. Kotouza, M.T., Psomopoulos, F.E., Mitkas, P.A., "A Dockerized Framework for Hierarchical Frequency-Based Document Clustering on Cloud Computing Infrastructures." *J. Cloud Comput.*, 9, 1–17, 2020. https://doi.org/10.1186/s13677-019-0150-y.
10. "Labetalol Reduces BP and Heart Rate in Patients With Hypertension." *InPharma*, 11, 1–1, 1986. https://doi.org/10.1007/bf03299537.
11. Li, H. and Tan, J., "Body Sensor Networks Based Sensor Fusion for Cardiovascular Biosignal Predictions." *Proceedings of the 2nd International Workshop on Systems and Networking Support for Healthcare and Assisted Living Environments - HealthNet '08*, 2008, https://doi.org/10.1145/1515747.1515752.
12. Messina, A.R., "Data Fusion and Data Mining Analysis and Visualization," in: *Data Fusion and Data Mining for Power System Monitoring*, 2020a, https://doi.org/10.1201/9780429319440-11.
13. National Research Council, Division on Engineering and Physical Sciences, Computer Science and Telecommunications Board, and Committee on

Sustaining Growth in Computing Performance, *The Future of Computing Performance: Game Over or Next Level?*, National Academies Press, Washington, 2011.
14. Niharika, M., Deepa, N. Devi., T., "Image segmentation and detection for healthcare data in deep learning". *Test Eng. Manage.*, 81, 11–12, 5402–5407, 2019.
15. Niranjana, S. *et al.*, "Smart Monitoring System for Asthma Patients". *Int. J. Electron. Commun. Eng.*, 7, 5, 5–9, 20202020.
16. Nist, *NIST-SP 500-291, NIST Cloud Computing Standards Roadmap*, Springer, Switzerland, 2013.
17. Purwar, S., Tripathi, R., Ranjan, R., Saxena, R., "Classification of Thalassemia Patients Using a Fusion of Deep Image and Clinical Features." *2021 11th International Conference on Cloud Computing, Data Science & Engineering (Confluence)*, 2021, https://doi.org/10.1109/confluence51648.2021.9377054.
18. Qi, J., Yang, P., Hanneghan, M., Waraich, A., Tang, S., "A Hybrid Hierarchical Framework for Free Weight Exercise Recognition and Intensity Measurement with Accelerometer and ECG Data Fusion." *Conference Proceedings: Annual International Conference of the IEEE Engineering in Medicine and Biology Society*, Conference 2018 July, pp. 3800–3804, 2018.
19. Schultz, G.G., Black, C.W., Saito, M., "GIS Framework for Hierarchical Bayesian Based Crash Data Analysis." *T&DI Congress 2014*, 2014, https://doi.org/10.1061/9780784413586.046.
20. Shrestha, S. and Drozdenko, B., "Smart Rural Framework Using IoT Devices and Cloud Computing." *2019 IEEE Green Technologies Conference (GreenTech)*, 2019, https://doi.org/10.1109/greentech.2019.8767142.
21. Raajesh, S., "Tele-Health in Population Health Management Using IoT Devices in Cloud Based Environment." *J. Adv. Res. Dyn. Control Syst.*, 12, 212–221, 2020. https://doi org/10.5373/jardcs/v12sp7/20202100.
22. Sun, H., Li, X., Liu, Y., Chai, Y., "A Hierarchical Federated Integration Framework for Trust E-Commerce Cloud," in: *Information Science and Management Engineering*, 2013, https://doi.org/10.2495/isme130801.
23. Thuemmler, C. and Bai, C., *Health 4.0: How Virtualization and Big Data Are Revolutionizing Healthcare*, Springer, New York, 2017.
24. Tritt, A.J., Rubel, O., Dichter, B., Ly, R., Kang, D., Chang, E.F., Frank, L.M., Bouchard, K., "HDMF: Hierarchical Data Modeling Framework for Modern Science Data Standards." *2019 IEEE International Conference on Big Data (Big Data)*, 2019, https://doi.org/10.1109/bigdata47090.2019.9005648.
25. Wang, J.-B., Lian, Y.-F., Chen, K., "Virtualization Detection Based on Data Fusion." *2012 International Conference on Computer Science and Information Processing (CSIP)*, 2012, https://doi.org/10.1109/csip.2012.6308876.
26. Xu, H., "Machine Learning Based Data Analytics for IoT Devices." pp. 80, 2017, https://doi.org/10.32657/10356/72342.
27. Yao, W., Dumitru, O., Lorenzo, J., Datcu, M., *"Data Fusion on the CANDELA Cloud Platform,"* https://doi.org/10.5194/egusphere-egu2020-19933.

17

Commercial Platforms for Healthcare Analytics: Health Issues for Patients with Sickle Cells

J.K. Adedeji*, T.O. Owolabi and R.S. Fayose

Department of Physics and Electronics, Adekunle Ajasin University, Akungba, Akoko, Ondo State, Nigeria

Abstract

This research has applied the principles of convolutional deep learning algorithm and image classification to solving problems relating to sickle cell disease, causing cardiovascular disorder in patients living with it. The demise due to ignorance of many who lives with this disorder in developing countries in recent is on the high side. Hence, there is the need to have an advanced computerized biological image processing device that can provide effective information and assist in monitoring the health status of SCD patients in our communities.

In order to achieve this, the CNN classifier algorithm with fully connected layers coded in Python on a PyTorch platform was used on 300 images. The pictures of the red blood cells were taken with electron microscope from 60 normal people and re-sized up to 224 × 224 dpi in resolution. The training epoch was 10 to ensure non-overfitting and pre-trained using the ResNet algorithm. The values of the training and validation sets after adjustment are 0.020 accuracy error and the dropout at 0.022036 best value accuracy.

The chapter concluded that the image classifier is suitable for monitoring the health status of sickle cell anemia patients and processing of any complex biological images to obtain information relevant to any abnormality especially patients in sickle cell crisis.

***Keywords*:** Convolutional neural network, algorithm, fully connected, sickle cells, cardiovascular disorders, image classifier, deep learning, biological information

**Corresponding author*: adedejikunle2@gmail.com

A. Jaya, K. Kalaiselvi, Dinesh Goyal and Dhiya Al-Jumeily (eds.) Handbook of Intelligent Healthcare Analytics: Knowledge Engineering with Big Data Analytics, (365–386) © 2022 Scrivener Publishing LLC

17.1 Introduction

The biological systems are complex, and as such, there must be adequate information to get accurate positioning and status of some cardiovascular systems in order to detect the diseased parts with absolute precision. The related agonies being encountered by our people living with sickle cell diseases are on the high side in recent times, and the learning curve shows that, if awareness is not put in place to educate those that have not being in this conditions, then the number of cases in the near future would be uncontrollable; consequently, this would worsen the inheritance cases in our communities. In order to assist those already in this condition, due to ignorance, there is need for the invention of an intelligent system or device that can provide medical assistance timely to reduce the diagnosis rigor and time spent by our doctors on these same issues. In view of this, born out this research is to build an advanced biological image classifier to detect early and provide assistance for people living with these cardiovascular problems in our societies.

There are several methods that have been used to image biological organs, such computer tomography, magnetic resonance imaging, and position emission tomography. All produced exciting results, but the method adopted in this book is the classifier algorithm involving deep learning using convolutional neural network (CNN). This learns by pattern of images to know whether a particular body organs or cardiovascular system is in the right position and functioning effectively for the well-being of the patient.

In the segmentation and classification method applied by some researchers using image processing techniques, it was discovered that it is of great advantage to compare the cell morphology between healthy person and sickle cell anemia patients. This enables automatic cells recognition, which is more efficient and accurate diagnosis of sickle cell anemia problems. In the recent time, applications of deep learning using CNNs have been successful and yielded magnificent results in problems involving complex biological system and medical image analysis [2, 18]. In the midst these methods, the present approach employed in this health book has shown a higher performance and reliability as observed in its applications [20].

In a complex biological system, many important organs require adequate flow of blood through them to make them function properly. Any short or inadequate supply of blood through them can result into a serious malfunctioning, such as caused by the effects of sickle cell disorders, which results when there are distortions in the normal shapes of the red blood cells (RBCs) assuming a flat shape and tends to block the veins and arteries. The blockages prevent a constant supply of blood into important parts of the body and make them not to work properly by the damaged sickled cells. An inherited blood disorder, which occurs in patients with SCD, or having irregular-shaped hemoglobin that can affect normal disc-shaped RBCs that leads to heterogeneous shapes of the normal RBCs, is known as the sickle cell disease (SCD), or otherwise called the sickle cell anemia. The well-shaped RBCs can transport with ease within narrow blood vessels to supply the organs that need adequate nutrients and oxygen for proper functioning. The reverse is the case in the irregular-shaped RBCs, which enhances vaso-occlusion phenomena.

The recently study conducted by [2] showed that millions of patients suffered from the sickle cells diseases with greater population around the world having the trait, and about 176,000 deaths recorded in 2013; it was also shown that majority of the cases from 113,000 deaths in 1990, took their origin African countries where the level of awareness is low. The survey raised an alarm that the methods of treating SCD lack some merits and inadequate symptom control measures and should be given active monitoring techniques such as employed in this healthcare book. This will provide adequate prediction of the health status for patients at the different stages of disease development as proposed in this healthcare book [15, 16]. This chapter of the book intends to create a well monitoring device through the classifier algorithm to monitor the health status of the patients, leaving the sickle cell disorders.

The synopsis for this book chapter is highlighted in the next segments: Section 17.2 gave the modeling concepts and deep CNN algorithm, the neural architectures, the related research, different components and the building blocks, and the basic working of the CNN in image classifications. Section 17.3 highlighted methodologies of the conceptual modeling, which includes the dataset analysis, file organization, coding techniques used, the conceptual modeling and programming language used for the algorithm, and the ETL (Extract, Transform, and Load) principles of deep learning and preprocessing. Section 17.4 is purely for results and comparison of other researchers view on the medical imaging, while Section 17.5 gave the findings of this chapter.

17.2 Materials and Methods

17.2.1 Data Acquisition and Pre-Processing

The data used for this research was obtained from various pictures of the normal cells, and abnormal blood cells of people living with sickle cells abnormalities, as obtainable from the electron microscopes. The dataset contains the folder name that represents the class label of all the images in JPG format. Three hundred images of RBCs for the normal patients are named in numerical order, resized, and conform to 224 dpi in resolution to enhance the deep learning algorithm for easier recognition during processing. The algorithm also was designed in a way to assist in cropping and normalization of the pictures. This helped to obtain the grid and the center of the images so as to conform to a configuration of square image of 224 × 224 pixels stored in PIL segment of the program (Figure 17.1).

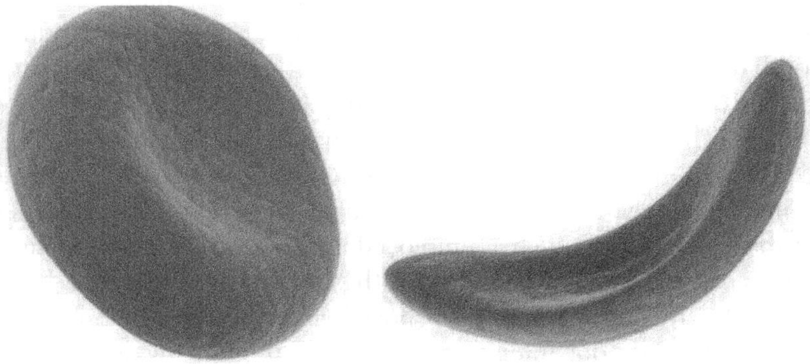

Figure 17.1 Normal cell and Abnormal cell as viewed under microscope. (Courtesy of Adekunle Ajasin University Centre for Bio-Computing and Drugs Development Lab.)

17.2.2 Sickle Cells Normalization Image

In order to achieve at least 95% image classification, it is important to carry out the process of data normalization, which is an aspect of preprocessing step. This will help each input (the pixel value, for the cells images) using the statistical tool of a standard normal distribution. This means that the range of pixel values in one input image is the same as the range in another image. This standardization makes the model to train and reach a minimum error faster.

The data normalization is achieved by continuously subtracting the central value (the average of all pixel values) from individual pixel sets and dividing the approximate values by the standard deviation of all the pixel values.

The distribution of such images as inputs should follow a Gaussian function using the mean value theorems, and a point centered at zero. For image inputs, we assumed the pixel number to be positive, such as

$$X \approx N\,[0,\,1].$$

17.2.3 Gradient Calculation

The mathematical treatment of the gradient of a function can assume the followings:

$$\sigma'(x) = \sigma(x)\,(1 - \sigma(x)) \tag{17.1}$$

The reason for this can be viewed from the following rule of calculating the error using the quotient formula:

$$\begin{aligned}\sigma(x) &= \frac{\delta}{\delta x}\frac{1}{1+e^{-x}}\\ &= \frac{e^{-x}}{(1+e^{-x})^2}\\ &= \frac{1}{1+e^{-x}}\frac{e^{-x}}{1+e^{-x}}\\ &= \sigma(x)(1-\sigma(x))\end{aligned} \tag{17.2}$$

Let us consider set of numbers having m points labeled as $x^{(1)}, x^{(2)}, \ldots,$ $x^{(m)}$ and the error formula was obtained from these set of points as

$$E = \frac{1}{m}\sum_{i=1}^{n}(yi.\ln(\hat{y}i) + (1 - yi)\ln(1 - \hat{y}i) \tag{17.3}$$

where the prediction is given by

$$\hat{y}i = \sigma(Wx + b) \tag{17.4}$$

From these points, it is very easy to calculate the gradient of E, at a point $x = (x_1, \ldots, x_n)$, given by the partial derivatives obtained from the above as

$$\nabla E = \left(\frac{\partial}{\partial w_1}E, \ldots, \frac{\partial}{\partial w_n}E, \frac{\partial}{\partial b}E\right) \tag{17.5}$$

There is a need to simplify these calculations, by looking at the error that each point produces and calculating the derivative of the error accordingly. The sum of the error is the average of the errors at all the points. Therefore, the error produced by each point is obtainable from the expression below:

$$E = y\ \ln(\hat{y}) - (1-y)\ln(1-\hat{y}) \tag{17.6}$$

Having obtained this value, we shall proceed to obtain the derivative of the error of the individual dictators (weights); bearing in mind, first is to calculate rate of change in the estimate against the weights as $\frac{\partial}{\partial w_j}\hat{y}$, and by recalling that $\hat{y} = \sigma(Wx + b)$.

Therefore,

$$\frac{\partial}{\partial w_j}\hat{y} = \frac{\partial}{\partial w_j}\sigma(Wx+b) \tag{17.7}$$

$$= \sigma(Wx+b)(1-\sigma(Wx+b)).\frac{\partial}{\partial w_j}(Wx+b)$$

$$= \hat{y}(1-\hat{y}).\frac{\partial}{\partial w_j}(w_1x_1 + \cdots + w_jx_j + \cdots + w_nx_n + b)$$

$$= \hat{y}(1-\hat{y})\ .\ x_j \tag{17.8}$$

The last expression is measured only in terms of y in the sum and not a constant with respect to w_j; precisely, w_jx_j clearly has derivative x_j.

In lieu on this, the derivative of the error E at a point x, with respect to the weight w_j, can be obtained as

$$\frac{\partial}{\partial w_j}E = \frac{\partial}{\partial w_j}[-y\log(\hat{y}) - (1-y)\log(1-\hat{y})]$$

$$= -y\frac{\partial}{\partial w_j}\log(\hat{y}) - (1-y)\frac{\partial}{\partial w_j}\log(1-\hat{y})$$

$$= -y.\frac{1}{y}.\frac{\partial}{\partial w_j}\hat{y} - (1-y).\frac{1}{1-y}.\frac{\partial}{\partial w_j}(1-\hat{y})$$

$$= -y.\frac{1}{y}.\hat{y}(1-\hat{y})x_j - (1-y).(1-y).\frac{1}{1-y}.(-1)\hat{y}(1-\hat{y})x_j$$

$$= -y(1-\hat{y}).x_j - (1-y)\hat{y}.x_j$$

$$= -(y-\hat{y}).x_j \tag{17.9}$$

A similar calculation shows that

$$\frac{d}{db}E = -(y-\hat{y}) \tag{17.10}$$

The above explained a very important functionality, for a point with coordinates $(x_1, \ldots, x_n)$, label y, and prediction ŷ, on the gradient of the error function at that point is $(-(y - \hat{y})x_1, \ldots, -(y - \hat{y})x_n, -(y - \hat{y}))$. In summary, the gradient is $\nabla E = -(y - \hat{y})(x_1, \ldots, x_n, 1$.

17.2.4 Gradient Descent Step

For the purpose of the problem at hand, the method of gradient descent step was employed by subtracting a multiple of the gradient of the error function at every point, which were obtained in previous section of this chapter, and carrying out an updates on neurons' weights using the following procedures:

$$w'_i \leftarrow w_i - \alpha[-(y-\hat{y})x_i],$$

which is equivalent to:

$$w'_i \leftarrow w_i + \alpha(y-\hat{y})x_i \tag{17.11}$$

Similarly, updates were carried out, on the bias as

$$b' \leftarrow b + \alpha(y-\hat{y}), \tag{17.12}$$

The calculated average value is the term obtained from the expression $\frac{1}{m}$. α, which is of great importance, instead of ordinary α, where α is a constant that yields the calculations of an important parameter, and our adjusted value of the learning rate as $\frac{1}{m}$ α being the aggregate.

17.2.5 Insight to Previous Methods Adopted in Convolutional Neural Networks

The principles of deep learning as one of the artificial intelligence techniques have been used in various fields not limited to science and engineering. Certain features are mimicked and borrowed from the biological neurons to mimic human intelligence [6, 7]. The CNN is an advanced neural network with many layers acting as convolutions as in the work of LeCun *et al.* in 1998, which consisted of a seven-layer neurons purposely applied for forensic purposes on checks in banks. Thereafter, the full knowledge was known in diverse applications with the three configurations employed in the layered neurons, in respect of weights transfer, local receptive fields, and pooling layers [19].

This healthcare book employed the full applications of the CNNs where all the nodes in the mth layer being connected to n^{th} nodes in the $(m - 1)^{th}$ layer, n being the size of the receptive field of the convolutional neural, as compared to the fully connected layers, where all nodes are connected to the initial layer [11, 12]. The advantage is that the total number of parameters within the neurons is minimized, thereby preventing overfitting and invariance computations which can affect the performance of the classifier.

In the research conducted by Alex Krizhevsky *et al.* [3] in 2012, he applied CNNs to image classification and got a very good result that showed the efficacy and wide applications of the CNN in recent times problems involving classifications of images.

17.2.6 Segments of Convolutional Neural Networks

This chapter highlighted the CNNs configurations comprising of various components as they can be employed in healthcare system for biological imaging, and they are described in the following:

17.2.6.1 Convolutional Layer

The basic principle of employing intelligent networks having layers of neurons is premised on the network parameters and the nature of the problems at hand. The problems considered in this book involved the detection of normal sickle cell and learning how to detect the shapes in terms of the edges and corners; this makes the CNN suitable for use as an image classifier for healthcare purpose in this chapter of the book. The initial convolutional layers were meant for recognizing low level features: edges and corner; the outputs of these layers are transferred into the next convolutional layers to learn at higher level of such features as color in RGB order.

According to some previous researches done in CNN, the neurons in the transmitting side are connected to fewer neurons in the receptive layer. Without carrying out filtrations, the obtained features might not be good representations in pattern; therefore, a filter of n × n in dimensions is employed. The filter works on the principles of convolution mathematics; the output from the filter is transformed in the width and height as the dot products between input filters, which is mapped one-to-one as output and the final feature.

Since sickle cells images have multiple features, with higher resolutions, this book has employed multiple filters with associated hyperparameters to suit the problems at hand, as also reported in the research conducted by [4, 5].

17.2.6.2 Pooling Layer

Generally, in neuron-computing researches involving the applications of CNN of this nature, it is important to introduce the pooling layer, which assists in the reduction of spatial size of the features and the number of parameters. This will reduce the computational stresses and number of connections within the convolutional layers. Although, there are three types that are currently in use: max pooling, average pooling, and mixed pooling. Out of these, the average pooling has less industrial applications, but max pooling is more useful methods. This book made use of mixed pooling because of its relative advantages in taken care of overfitting, which also confirmed that the previous research was not misleading the real principles behind it.

Let us assume L_p to be the pooling, which is biologically inspired by a process from complex cells of visual cortex. Mathematically, this can be represented by equation

$$Y_{i,j,k} = \left[\sum_{(m,n) g R_{ij}} (a_{m,n,k})^p \right]^{1/p}$$

It is worth noting that the value P = 1 reduces the above expression to average pooling, and for the value P = L_p makes to be max pooling. From these two expressions, mixed pooling can be obtained by combining the mean and max pooling. This method is adopted in this book and invariably solves problems relating to over fitting more efficiently, as was also employed in the research conducted by [8].

17.2.6.3 *Fully Connected Layer*

They are used to achieve some kinds of linearity in a network; they are placed generally right before the Softmax layers to ensure these characteristics. The major applications are in the versatility of dual transformation characteristics from linearity in the networks to replacement or conversion into convolutional layers, which, in turn, can be converted to its equivalent in the fully connected network.

17.2.6.4 *Softmax Layer*

The usefulness of Softmax function helps to convert and non-linear function to its linear equivalent using log function as if x_1, x_2, x_3,...., x_n represents a non-linear function, its Softmax function at the Softmax can be obtained using $\Sigma \log x_n$. This is a linearly dependent layer with a Softmax classifier which transforms the functions into two bi-stable states and summing up 1. It is used in neuron computing to determine the probability of the final output and maintained at maximum value of 1 as the threshold, the closer to this value the better the performance.

Mathematically, it can be expressed as a probability distribution

$$\sigma(Z)_j = \frac{e^{z_j}}{\sum_{k=1}^{k} e^{z_k}}$$

The notation z is a vector of input units; j is an index for output values, while k is an index to input values with the total number of elements in j being k.

17.2.7 Basic Transformations of Convolutional Neural Networks in Healthcare

In order to understand the basic principles of convolutional neuronal network as it was applied in this book as a sickle cells classifier for healthcare purposes, let us consider a system with mth neurons (the cardiovascular symptoms), we can now view each neuron in a CNN in the mth layer been connected to nth neurons in the (m – 1)th layer, where n is known as the receptive neurons. Let us also assume that it consists of certain steps called the strides; in order to obtain new feature map, a convolution operation must be performed by convolving the input with a weight/kernel according to some set of rules. The convolution operation can thus be viewed mathematically as follows:

$$Z_{i,j,k}^{m} = W_k^m * X_{i,j}^{m-1} + B_k^m$$

To obtain the kth feature map in the mth layer, (i, j) is assumed to be the cache occupying the sickle cells in the mth layer, where the value in the feature map has to be computed.

W_k^m and B_k^m represent the weights and bias for the kth kernel in the mth layer and X_l, j is the input function to be transformed. The transformed output continuously enters the non-linear section of the sigmoid function the maps final feature. This approach correlates the established results from previous researches, which served as precedence to this current research.

In order to obtain the sets of mapped features, the above operations are repeated using the input with trial sets of weights/kernel [9]. However, an occasion can demand when the same weights are shared for the entire spatial locations of the input, and it becomes a special case known as the shared weights concept in CNNs.

In principle of CNN, it is believed that some of the layers are specifically meant to perform certain task. In view of this, the network consists of many convolutional layers which are ordered; the first operational layer detects low level features such as edges and corners of the sickle cells, while the next intelligent layers detect high level features in respect of the sickle cells images. Since all networks must be connected in a way and relieve some of the networks computational stress, the process of pooling came fully into action. It was applied in this book to view the edges of highly colorized RBC images. This involves the conversion from one network into another using pooling and regularization functions to perform the conversion in order to achieve certain features, such processes are convolutional layer, pooling layer, and the Softmax layer, which has an attributes of a fully connected layer [20].

This book also looked at the principle of deconvolutional layers, which was used to perform an inverse of convolutional networks, which helps on how to visualize the exact procedures of how the CNN works. This is achieved when the back propagation is performed by using one-to-one mapping to the input pixels and transforming it to get the output which is matched to the input. It enables the cross-correlation exercises to be carried out in order to see whether the classifier performs the task of recognition or classification according to predetermined objectives [10, 13, 14].

17.2.8 Algorithm Review and Comparison

In the work of Seiichi Uchida *et al.*, they acknowledged that image processing is a large research area, and hence, they tried to employ gray-level transformation and feature extraction techniques, and they used image pattern recognition to classify images to one predefined class. The researchers admitted that some biological systems are microscopic and feature extraction would be suitable for such cases and that it will help biologists to observe and visualize molecules as bio images for analysis. But the algorithm used in this healthcare book properly harnessed a broad classifier that can be used for unlimited number of complex images going through malfunctioning cases as regards those finding it difficult to analyze huge number of bio images [21].

The uniqueness of this algorithm is that it comprises major components required to successfully process images that have complex nature, such as filtration, segmentation, optical flow [17]. These are similar to survey papers and textbook displayed by Gonzales and Woods (2007). The only addition is in the area of image file organization, which made the algorithm used in this healthcare book more accessible.

A wide survey on visual image or object tracking by Yilmaz *et al.* (2006) [23] also showed that the algorithm used in this book was in principle with any standard image classifier in applications of Artificial intelligence to healthcare segment. The survey papers on image registration by McInerney and Terzopoulo (1999), Glasbey and Mardia (1998), and Jain *et al.* (1999) also gave similar results [22], except that the present algorithm gave higher resolutions with higher pixels and gray-level ranging from 0 to 255 that is remarkable.

The present algorithm did not dwell much on Markov random field (MRF) techniques of extractions.

17.2.9 Feedforward

In order to turn the input into an output, the feedforward rule is used as can be seen in the neural networks shown in Figure 17.2.

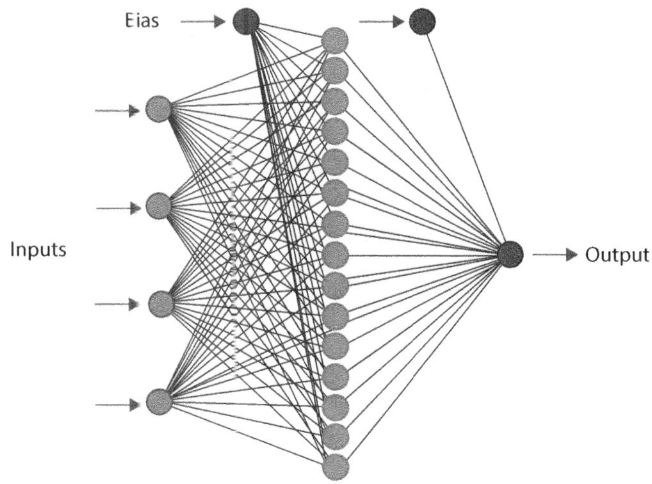

Figure 17.2 Neural network architecture.

17.3 Results and Discussion

17.3.1 Results on Suitability for Applications in Healthcare

This segment discuses the suitability of the classifier as an imaging tool to display abnormalities in patients living with sickle cell problems as discused below:

17.3.2 Class Prediction

Since the images have been stored in the correct format, a callable function has been saved as checkpoint and can be used for predictions using our built model. A check is usually performed to view the first five images for view (known as top-K). From here, the class probabilities were found using the K, which is the largest values. This procedure returns both the highest k possibilities values and the indices equivalence of the classes. Thereafter, these indices are converted to the actual class labels as previously mapped and stored in the JSON file formatting, and this can be added to the model used to load the data.

17.3.3 The Model Sanity Checking

The trained model is now qualified for making predictions; firstly, it is checked to make sure it makes sense in terms of getting the actual images proposed by the research. In respect of high accuracy of validation obtained during the iterations processes, there is need to check for bugs since the

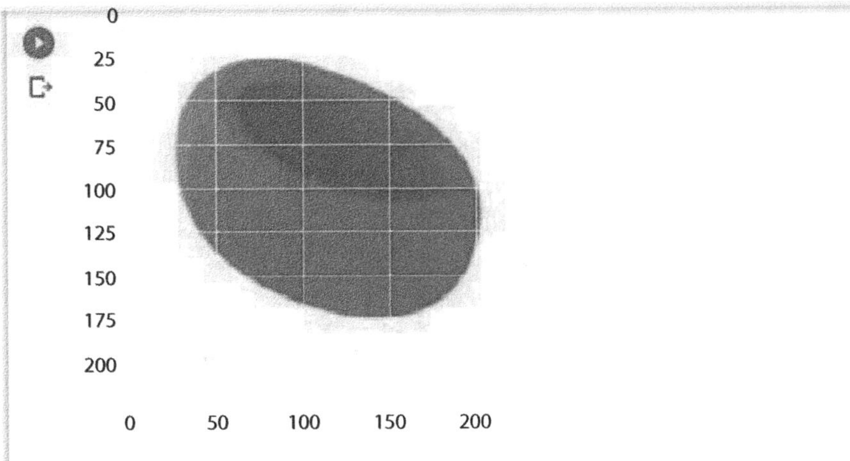

Figure 17.3 The predicted normal red blood cell.

research involves health issues. Thereafter, it is necessary to view few images using the Matplotlib library in the python; this helps to visualize the predicted images as to know if it is working according to the objectives of the research. This gives the image of the predicted normal sickle cell as purposed by the research suitable for use in healthcare imagining problems.

Figure 17.3 depicts the predicted normal cell, that shows that CNN algorithm is efficient in the image processing to detect the normal RBCs as they appear in the blood vessels of human being when it was viewed under the electron microscopes and to check that it is capable of detecting any abnormality in the blood stream using the pre-trained pattern recognition of the neural algorithm. The graph also shows that the system is actually learning and can be used to classify complex biological images supplied to it with 95% accuracy.

The level of accuracy can also be explained using the graph of training losses against the epoch numbers below, that determines the best training loss of the classifier.

17.3.4 Analysis of the Epoch and Training Losses

The graph was used to analyze the epoch number as the datasets went through the CNN algorithm in other to verify the best model of epoch that best describes the classification process. It shows a negative exponential at the beginning, which means that the train loss decreases with the number of epoch and the classifier is actually training to predict the images supplied to the algorithm. Thereafter, it got to the dropout region

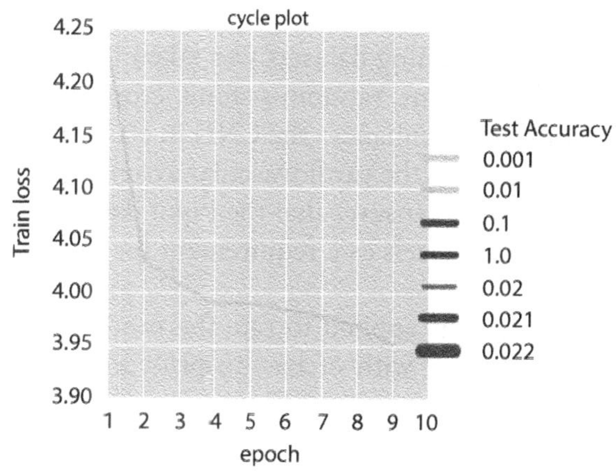

Figure 17.4 The graphs of training losses against epoch numbers.

of the gradient descent where it has acquired knowledge of recognizing the images supplied to the CNN algorithm.

The training loss initially gave higher value meaning that at beginning of classification, the algorithm is not yet efficient in classifications, but as the time progresses, it learnt by experience and adjusted error values. Thereafter, it decreases and got to a dropout where the classifier has finished training and reached the validation stage of recognizing the supplied images to be predicted. In practice, this graph illustrated that the classifier has been built very well (Figure 17.4).

17.3.5 Discussion and Healthcare Interpretations

The discussions are based largely on three important techniques employed in this book: these in respect of loading and preprocessing of the sickle cell image as the dataset and training thealgorithm on the set of data and using the trained algorithm to forecast features in the folders.

17.3.6 Load Data

The dataset actually passed through three processes called the ETL (Extract, Transformed, and Load); the data was extracted from the source zip file, and it was taking into transformation segment that allows the easier interpretation in the classification process. The torchvision was used to load the data into the CNN platform, and this aids the computer vision of recognizing the normal sickle cell as an output.

The image folder containing the dataset was segmented into two: training and validation. The training datasets was taken through the following preprocesses: transformations, random scaling, cropping, and flipping in the python algorithm. This enhances the CNN in classification for a better performance in recognizing the cardiovascular system with normal sickle. A pre-trained network also ensures that the input dataset is conformed to the dimension 224 × 224 pixels as a requirement for the CNN to perform classification better.

The validation dataset is applied to rate the model's efficiency on the unknown data not familiar with or has no prior knowledge of its existence. At this stage, there is no need for the algorithm to perform scaling or rotation transformations, because the process is a subset of the training datasets, which has already passed through transformation. The pre-trained networks from torchvision were trained on the image dataset consisting of three channels (RGB), which were normalized separately on each color. The components of both (train and validation) normalization parameters computed by the algorithm are μ, the means, and σ, standard deviations of the images. These were obtained through iterations to hasten the results expected by the network in terms of the numerical values that are means, at about [0.485, 0.456, 0.406], and for the standard deviations, at around [0.229, 0.224, 0.225], estimated valued as predetermined by the network. Any discrepancies in these values will adjust the color channel to position 0 with the range from −1 to 1, since they have been normalized according to the rule; $Z \approx (0,1)$ in the standard values of means and standard deviations.

17.3.7 Image Pre-Processing

The PIL library was evoked to load the images in jpeg format for each neuron at the input stage of the model. The Softmax function processed the images that have been normalized in standard format described in the previous section for training datasets.

The first thing to do at this stage is to allow the images to pass through another transformation of resizing the images to the configuration 256 pixels, for the shortest side, and maintaining the ratio. The PIL algorithm encoded the images in three colors as integers 0–255, with the expected floats 0–1. These values were then converted using the Numpy array in python library, which is also from a PIL images using np_image = np.array (pil_image). The images already have been normalized with the values [0.485, 0.456, 0.406] for the means and for the standard deviations [0.229,

0.224, 0.225]. The model performance is achieved by using these normalized values obtained during the preprocessing to acquire a more fitted model that is not overfitting.

At the final stage of transformation, PyTorch is employed to carry out reorder of the dimensions, since the color channel is expected to be in the first dimension but has been supplied in the third dimension by the PIL image and Numpy array; therefore, the dimensionality processes can be reordered using the python command [ndarray.transpose]. The same procedures are repeated for the other two remaining dimensions by maintaining the previous and retaining the order of the other two dimensions; this is accomplished by the algorithm in the program segment at the GitHub repository (kunlejelili).

17.3.8 Building and Training the Classifier

After extracting, transforming, and loading the data, these are useful procedures for the building of the CNN classifier algorithm according to the sets of training obtained earlier. The normal process is to employ the pre-trained models that can get the image features from torchvision models. These processes of building and training a new feed-forward classifier in order to achieve the stated objectives of the healthcare and solve the problems at hand are highlighted using the following features:

- Load a (pre-trained network), for the purpose this healthcare research, the VGG networkwas employed with ResNet training model.
- Defined a new, feed-forward algorithm that can be used to predict new image features byapplying ReLU with an appropriate dropout parameter.
- Trained the neuron layers to detect errors in images and corrections using backpropagation and save as checkpoint to get the new features.
- Tracked the valuable parameters to determine the best model and hyperparameters.

During the training, the focus is on updating of the weights of the feed-forward network, in order to achieve the validation accuracy above 70%, to confirm if the classifier is built right; thereafter, different hyperparameters are tried in respect of learning rate and epochs, to determine the best model that can predict the normal sickle cell features and achieve

Table 17.1 Performance estimation for accuracy.

epoch	Train loss	Val loss	Test accuracy
1	4.232	4.042	0.02
2	4.031	3.983	0.02
3	4.008	3.969	0.02
4	3.99	3.984	0.02
5	3.99	3.961	0.02
6	3.984	3.959	0.02
7	3.978	3.949	0.02
8	3.969	3.958	0.02
9	3.951	3.933	0.02
10	3.934	3.925	0.022
Training completed in 39m 58s Best Val Acc: 0.022036			

the healthcare objectives. The output of the training can be viewed in Table 17.1.

The above outputs were obtained from the training and validation datasets of the research in the CNN classifier, to obtain the losses, accuracy, and the best value, and are saved as checkpoint in the next segment of this chapter. The indication from the results showed that, at each epoch run, the training losses are greater than the validation losses, which means that classifier algorithm is actually training very well. The results are reliable and the algorithm can recognize the normal sickle cells and differentiate it from the abnormal ones in the validation sets for the purpose of making predictions in the datasets.

17.3.9 Saving the Checkpoint Suitable for Healthcare

Now that the network has been trained in the previous sections and confirmed as good model for making predictions of the normal sickle cells, this can now be saved for later use in making predictions and hardware design of the classifier for healthcare use. The other important things are the one-to-one mapping of classes to their equivalents, which can be obtained from the JSON files of the images saved as "image_datasets ['train'].class_to_idx"; this has an attribute that makes identifications easier using the command .model.class_to_idx = image_datasets ['train']. class_to_idx. This is achieved through the algorithm segment in the GitHub repository.

17.3.10 Loading the Checkpoint

The callable function is written and saved as checkpoint and ready for use for predictions for the healthcare research; this is achieved without having to retain the network. This callable function can be viewed from a program segment of the classifier at the GitHub repository.

17.4 Conclusion

This chapter of the book has discussed the applications of deep learning algorithm and CNNs effectively, to process any complex biological system and their applications in healthcare services. Besides this, it has provided breakthrough in image recognition to study any abnormality in complex biological system. The chapter concluded that the image classifier is efficient with 96% accuracy to process complex biological images suitable for healthcare systems to obtain information relevant to any abnormality especially patients in sickle cell crisis.

References

1. Abdel-Hamid, O., Mohamed, A., Jiang, H., Penn, G., "Applying convolutional neural networks concepts to hybrid NN-HMM model for speech recognition". *International Conference on Acoustics, Speech, and Signal Processing*, pp. 4277–4280, 2012.
2. Abubakar, I., Tillmann, T., Banerjee, A., Global, regional, and national age-sex specific all-cause and cause-specific mortality for 240 causes of death, a systematic analysis for the Global Burden of Disease Study 2013. *Lancet*, 385, 9963, 2015, 2013.
3. Krizhevsky, A., Sutskever, I., Hinton, G.E., *Image Net Classification with Deep Convolutional Neural Networks*, University of Toronto, Canada, pp. 1–9, 2012.
4. Chegoonian, A.M., Mokhtarzade, M., Zoej, M.J.V., Salehi., M., "Soft Supervised Classification and Improved Method for Coral Reef Classification Using Medium Resolution Satellite Images". *IGARSS*, 2016.

5. Garcia, C. and Tziritas, G., "Face Detection Using Quantized Skin Color Region Merging and Wavelet Packet Analysis". *IEEE Trans. Multimedia*, 1, 264–277, 1998.
6. Ciresan, D.C., Meier, U., Masci, J., Gambardella, L.C., Schmidhuber, J., "Flexible, high performance convolutional neural networks for image classification". *International Joint Conference on Artificial Intelligence*, pp. 1237–1242, 2011.
7. Hubel, D.H. and Wiesel, T.N., Receptive fields and functional architecture of monkey striate cortex. *J. Physiol.*, 195, 215–43, 1998.
8. Yu, D., Wang, H., Chen, P., Wei, Z., Mixed pooling for convolutional neural networks. *Rough Sets and Knowledge Technology*, pp. 364–375, 2014.
9. Osuna, E., Freund, R., Girosi, F., Training Support Vector Machines: an application to face detection. *Conference on Computer Vision and Pattern Recognition*, pp. 130–136, 1997.
10. Al-Ahmadi, F.S. and d. Hames, A.S., Comparison of Four Classification Methods to Extract Land Use and Land Cover from Raw Satellite Images for Some Remote Arid Areas, Kingdom of Saudi Arabia. *JKAU; Earth Sci.*, 20, 1, 167–191, 2009.
11. Simonyan, K. and Zisserman, A., Two-Stream Convolutional Networks for Action Recognition in Videos. *Neural Inf. Process. Syst.*, 1, 568–576, 20142014.
12. Kurian, J. and Karunakaran, V., A Survey on Image Classification Methods. *Int. J. Adv. Res. Electron. Commun. Eng. (IJARECE)*, 1, 4, 2012.
13. Zeiler, M.D. and Fergus, R., Stochastic pooling for regularization of deep convolutional neural networks. *International Conference on Learning Representations*, 2013.
14. Sampat, M.P., Bovik, A.C., Aggarwal, J.K., Castleman, K.R., "Supervised Parametric and Non-Parametric Classification of Chromosome Images." *Pattern Recognit.*, 38, 8, 1209–1223, 2005.
15. Jain, M., Singh, P., Bhopal, T.L., Review of Image Classification Methods and Techniques. *Int. J. Eng. Res. Technol.*, 2, 8, 2278–0181, 2013.
16. Milton, J.N., Gordeuk, V.R., Taylor, J.G., Gladwin, M.T., Steinberg, M.H., Sebastiani, P., Prediction of fetal hemoglobin in sickle cell anemia using an ensemble of genetic risk prediction models. *Circ. Cardiovasc. Genet.*, vol. 7, 2, 98–113, 2014.
17. Girshick, R., Donahue, J., Darrell, T., Malik, J., Rich feature hierarchies for accurate object detection and semantic segmentation. *Conference on Computer Vision and Pattern Recognition*, pp. 580–587, 2013.
18. Ashenayi, K., Veltri, R., O'Dowd, G., Miller, G., Hurst, R., d. Bonner, R., A Comparison of Neural Network and Fuzzy c-Means Methods in Bladder Cancer Cell Classification. *IEEE World Congress on Computational Intelligence*, 1994.
19. Lecun, Y., Bottou, L., Bengio, Y., Haffner, P., Gradient based learning applied to document recognition. *Proc. IEEE*, 86, 2278–2324, 1998.

20. Tang, Y., Huang, J., Zhang, F., Gong, W., *Deep Residual Network With a Fully Connected Reconstruction Layer for Single Image Super-Resolution*, vol. 1000089, pp. 2–32, An Innovation Co. Ltd, Beijing, China, 2018.
21. Uchida., S., *Image Processing and recognition for Biological Images, Department of Advanced Information Technology*, vol. 744, pp. 819–0395, Kyushu University, Mootoka, Nishi, Fukuoka, Japan, 2013.
22. Jain, A.K., Murty, M.N., Flynn, P.J., Data clustering: a review. *ACM Comput. Surv.*, vol. 31, 3, 264–323, 1999.
23. Yilmaz, A., Javed, O., Shah, M., Object tracking : a survey. *ACM Comput. Surv.*, vol. 38, 4, pp. 13–21, 2006.

18

New Trends and Applications of Big Data Analytics for Medical Science and Healthcare

Niha K.[1*] **and Aisha Banu W.**[2]

[1]*Department of Computer Science and Engineering, Saveetha Engineering College, Chennai, India*
[2]*Department of Computer Science and Engineering, Crescent Institute of Science and Technology, Chennai, India*

Abstract

In this paper, the deep learning–based algorithms have been used to detect spike and wave patterns using single-channel EEG signals. The database comprises data collected from three different subjects as normal, abnormal, and absent. Here, the algorithm performed well in detecting spike and wave patterns, which is a key indicator to find an absence seizure subject. This work based on a new deep learning algorithm comprising four steps: pre-processing strategy selection, extraction of a spike and wave pattern using convolution layer, classification of a spike and wave pattern from the normal and abnormal pattern using gated recurrent unit, and, at the last, layer support vector machine (SVM) that has been used to get a result, which is categorized under the applications of a clinical domain. As an observation, the proposed algorithm enhances the feature extraction by feeding the raw input into the convolutional layer that improves the detection, solves the vanishing gradient problem with gated recurrent unit, and gives a stable result with less training time SVM. Our proposed algorithm achieves an accuracy of 0.95 on spike and wave detection with 1.37-s run time when compared with traditional algorithms. Also, it can be used to detect other bio-signals such as electroencephalography, electromyography, and electrooculogram.

**Corresponding author*: niha.k.cse@gmail.com

A. Jaya, K. Kalaiselvi, Dinesh Goyal and Dhiya Al-Jumeily (eds.) Handbook of Intelligent Healthcare Analytics: Knowledge Engineering with Big Data Analytics, (387–412) © 2022 Scrivener Publishing LLC

***Keywords*:** Convolutional layer, deep learning algorithm, electroencephalography, gated recurrent unit, normal and abnormal pattern, pattern recognition, spike and wave pattern, support vector machine

18.1 Introduction

Nowadays, most common applications of bio-signals in clinical research [1] are used for monitoring or caretaking of patients. Different algorithms under the umbrella of AI have been used in this domain. They are artificial neural network (ANN), support vector machine (SVM), convolutional neural network (CNN), recurrent neural network (RNN), and so on. Among these algorithms, deep learning algorithms have an ability to lean by itself and adapt themselves to the assigned task. These algorithms automatically predict the accurate results based on the previous observation/results. The basic goal of these algorithms is to automatically perform their operation with respect to time based on the specific task or adapt according to the circumstance. These algorithms are important for online classification of a linear or non-linear tasks [2]. In terms of the processing EEG signal, different methods were available such as time domain, frequency domain, and time-frequency domain. These methods are used for classification and also used to extract patterns from bio-signals. With the advancement of deep learning algorithms along with the above methods, feature extraction and classification made easy.

EEG (electroencephalography) signal processing comes under bio-signal processing, which refers to a bio-electric signals in the living being, which can be monitored and measured continuously. Some of them are EEG, ECG (electrocardiogram), EMG (electromyogram), and EOG (electrooculography). EEG determines the brain activity, ECG measures hear activity, EMG determines specific muscle activity, and EOG measures eye movement. These signals have few important properties: they are non-stationary or stationary, nonlinear or linear, and random (stochastic) or deterministic. All these signals are low and very low amplitude. Because of this instrument, amplifiers are used to obtain the low noise and consistent signal. These signals are contaminated with unwanted signal components. The processing of these signals is tedious for extracting informative data.

In this work, spike and wave patterns have been detected and classified based on three classes, which is a key indicator of absence seizure. An absence seizure is a type of epilepsy that occurs among 4 to 14 years of children (one in 20 cases of epilepsy) but possible to occur at any age. The human brain plays a vital role in our body which controls the entire system. The intense and abnormal electrical activity in the brain leads to an absence seizure. The availability of absence seizure data is less when compared to the other seizure data for research. Also, with technical aspects, researchers have not used the deep learning techniques for automatic feature extraction and classification of absence seizure. This tedious task is made easy by the evolution of deep learning algorithms.

18.2 Related Work

In this research, deep learning play a very important role by solving major problems discussed for extracting informative features that have been used by Martin *et al.* [3]. Deep learning enables direct learning on input without prior domain knowledge for feature extraction. It is a part of machine learning, which includes multiple levels of different abstractions to learn data representation. The current situation of research let researchers select the most frequently used deep learning structure for feature extraction as a CNN. It includes many convolutional layers and pooling layers for feature extraction and reductions.

18.3 Convolutional Layer

Convolutional layers have been used by Xiaoyan *et al.* [4], which works well on time-series data for analysis with a fixed length. It comprises several convolution kernels (which are nothing but a filter) in the same layer with the same size. This layer performs an operation on a signal input to generate a 1D feature vector by extracting features from the different kernels. Weight sharing has been fulfilled by selecting a specific feature on whole feature vector by detecting features using convolution kernel. This weight sharing and local connectivity reduce the network complexity and parameters of training. The mathematical formulation of the convolutional layer is given as

$$b_k = \sum_{n=0}^{N-1} a_n h_{k-n} \tag{18.1}$$

where a denotes input vector, b denotes output vector, N denotes number of elements in a, h denotes activation function (filter), and k and n denote elements of the vector. In this research work, the convolution layer has been taken for feature extraction. The feature vectors extracted have been used as input for classification. It [5] comes under the roof of the neural network and has three important layers to process input. They are 1) convolution layer, 2) pooling layer, and 3) fully connected layer. In that, each layer count (i.e., number of layers) may vary based on the requirements of the researcher, which are used for image or signal processing (feature extraction), classification, segmentation, and for autocorrelated data.

In this paper, the time-series data have been considered as input for the three-dimensional structures. Each set of neurons analyzes a specific region of input to obtain an output from the predictions of intermediate layer. CNN consists of different kinds of layers.

- (i) Convolutional layer: This layer composed of several convolutional kernels with the same size in the sale layer, which have been explained above and perform feature extraction.
- (ii) Pooling layer: This layer holds the method for averaging the filtered input.
- (iii) Fully connected layer: This layer has been used for classifying the processed input with several computations to attain the results.

18.4 Pooling Layer

The output from the convolution layer was not suitable for calculation because of the increase in the number of feature maps lead to dimensionality reduction. This has been done by average or max pooling on each feature map for the further process. The max-pooling is nothing but finding the maximum parameter from the pre-determined range and the average pooling has been calculated based on the size of the pre-determined range as output.

18.5 Fully Connected Layer

All nodes in this fully connected layer have been connected to all nodes in the previous layer. The most used activation function soft-max has an ability to predict which feature vector belongs to which class. If it is the final layer, then it get an output of the pooling layer and is given as

$$h(x) = f(w^{l+1} \, . \, x^{l+1} + b^{l+1}) \tag{18.2}$$

where w is the weights and b is the bias.

Nowadays, many researchers have started to use CNN. This classifier successfully handles large-scale video and image input. Some researchers like Acharya *et al.* [6] have applied CNN for EEG signals for seizure detection. Because of its extensive use, it becomes familiar in the research community.

18.6 Recurrent Neural Network

RNN [7] is a type of neural network, which has been used to deal with the numerical sequence of data. In this network, the input has been processed through hidden layers, where the output from the previous layer fed into the next layer as input. Also, RNN [8] consists of "memory" which remembers information of operations performed by each node for each input through time to produce the output in the hidden layer. As same to other neural networks, weight and bias are the two parameters to all nodes in the hidden layer. Because of these parameters, the activation has converted from independent to the dependent for all nodes, which reduces the complexity for memorizing the operations of current and previous nodes of the hidden layer. In RNN classifiers, time step is denoted as t, input as X, and hidden state and h, where the length of X represents the size or dimension of the input and h as length represents the hidden units. The mathematical formulation of RNN is given as

$$h_t = f(h_{t-1}, X_t) \tag{18.3}$$

where above formula is for calculating the current state. In this denotes current, previous, and input state. The mathematical formulation for calculating activation function is given as

$$h_t = tanh(W_{hh}h_{t-1} + W_{xh}X_t) \tag{18.4}$$

Here, W_{hh} and W_{xh} are weight at the recurrent neuron and input neuron. To calculate the output, the mathematical formulation is given as

$$Y_t = W_{hy}h_t \tag{18.5}$$

Here, W_{hy} and Y_t denote weights at output and output. This classifier is used with the convolution layer to get an effective classification. It is difficult to process a long sequence and training is difficult because of the vanishing gradient problem. To overcome this, LSTM and gated recurrent unit (GRU) are designed.

18.7 LSTM and GRU

LSTM (long short-term memory) [9] model has a hidden state and cell state at each time step t. Cell states are controlled by gates, which are composed of multiplication operation and sigmoid neural net layer. Here, the output is always between 0 and 1 (0 means nothing fired, and 1 means everything fired). This model [79 10] consists of three important gates. They are the input gate, forget gate, and output gate. In this, forget gate in LSTM decides how much data to store and how much data to ignore with the help of the neural net layer. The cell states dimension is the same as the hidden state. The mathematical formulation to calculate forget gate is given as

$$f_t = \sigma(W_f.[h_{t-1}, x_t]+b_f) \tag{18.6}$$

The input gate decides which data to store in the cell state with sigmoid and tanh activation function, where the sigmoid function decides which value to update and tanh function creates the new vector values [80]. The formula for the input gate with sigmoid and tanh activation function is given as

$$i_t = \sigma(W_i.[h_{t-1}, x_t]+b_i) \tag{18.7}$$

$$\tilde{C}_t = tanh(W_C.[h_{t-1}, x_t]+b_C) \tag{18.8}$$

Here, the ReLu function ranges between 0 and ∞, and tanh function ranges between −1 and 1. At last, the output has obtained by combining the vector using the sigmoid and tanh function, i.e., how much data to send to the next layer and how much data to obtain as output. The mathematical formulation of output is as

$$o_t = \sigma(w_o[h_{t-1}, x_t] + b_o) \quad (18.9)$$

$$h_t = o_t * tanh\,(o_t) \quad (18.10)$$

$$C_t = f_t * C_{t-1} + (1 - f_t).\tilde{c}_t) \quad (18.11)$$

The output implies that it take care of the values of input and forgets the layer that should not be same at any time step. With this, the GRU has a different structure by combining both forget and input gate into a single gate as an update gate and another gate is reset gate. The reset gate determines how to combine the new input with the previous data, and the update gate defines how much of the previous data to keep around [11]. The mathematical formulations are given as

$$\text{update gate, } z_t = \sigma(w^{(z)}x_t + U^{(z)}h_{t-1}) \quad (18.12)$$

$$\text{reset gate, } r_t = \sigma(w^{(r)}x_t + U^{(r)}h_{t-1}) \quad (18.13)$$

$$\text{current memory content, } h_t' = \tanh(Wx_t + r_t \circledcirc U\, h_{t-1}) \quad (18.14)$$

$$\text{final memory at current time step, } h_t = z_t \circledcirc h_{t-1} + (1 - z_t) \circledcirc h_t' \quad (18.15)$$

Here, z_t and r_t are update gate and reset gate vector, x_t and h_t are input and output vector, W and U are parameters matrices.

Because of the practical scenarios, GRU has been in use when compared to LSTM model. With respect to the seizure detection application, researchers like Subasi *et al.* [12] are quite interested in exploring deep learning models. The most used models are CNN and RNN deep learning models. Even though Sachin *et al.* [13] figure out the frequently used classifier, it is difficult to directly compare the results of different studies because of network structure and input modalities. Some papers on seizure detection based on preprocessing, input formulation, feature extraction, and classification have been tabulated in Table 18.1.

From the above observation, the use of deep learning algorithms for EEG signal is in the boom and the CGRUSVM algorithm has not been

Table 18.1 Seizure detection based pre-processing, input formulation, feature extraction, and classification methods.

Author	Year	Approach/Objective	Algorithm/Method	Limitations
J. E. Kline *et al.* [14]	2015	Manually removed the movement artifact from the recorded EEG signal.	Moving average, wavelets.	Use of manual artifact removal method.
Kilicarslan *et al.* [15]	2016	Removal of eye blinks, eye motions, amplitude drifts, and recording biases.	Robust adaptive noise cancelling (ANC) scheme.	Use of automatic artifact removal method.
Z. Lasefr *et al.* [16]	2017	Signal noise is removed to achieve maximum spectrum efficiency and signal values are used as input for epilepsy detection.	Chebyshev filter and Infinite Impulse Response (IIR) filter. Extracted feature: energy.	Use of manual artifact removal method and signal to frequency domain using DWT.
L. Wang *et al.* [17]	2017	Signal noise are reduced by wavelet threshold de-noising. Calculated features (statistical) and signal values are used to formulate input for epilepsy detection.	Daubechies wavelet. Extracted feature: statistical and relative power, and entropy.	Use of automatic artifact removal method. Signal to frequency domain using DWT.
X. Wei *et al.* [18]	2018	Image as input have been considered for automatic epilepsy detection without removal of artifacts.	Convolutional neural network (CNN)	Does not apply any artifact removal strategy. Does not apply for absence seizure data.

(Continued)

Table 18.1 Seizure detection based pre-processing, input formulation, feature extraction, and classification methods. (*Continued*)

Author	Year	Approach/Objective	Algorithm/Method	Limitations
Yıldız *et al.* [19]	2015	26 features (calculated features) (statistical) are used as input to analysis epilepsy.	Some of them are mean, standard deviation, maximum, minimum, and median.	Does not used raw EEG signal. Lags in automatic feature extraction.
R. Hussein *et al.* [20]	2019	Image as input have been considered for automatic epilepsy detection.	Long short-term memory (LSTM)	Does not utilize absence seizure data.
T. Kjaer *et al.* [21]	2017	Absence seizure detection at non-clinical setup based on patient specific model.	SVM classifier.	Non clinical data. Lags in automatic feature extraction. Does not applied deep learning algorithm.
M. Zihlmann *et al.* [22]	2017	To classify atrial fibrillation (AF) using DL.	LSTM, CNN	Signal to frequency domain using logarithmic transform. Does not apply for absence seizure data.

(*Continued*)

Table 18.1 Seizure detection based pre-processing, input formulation, feature extraction, and classification methods. (*Continued*)

Author	Year	Approach/Objective	Algorithm/Method	Limitations
E. Alickovica *et al.* [23]	2018	Automatic epileptic seizure detection and prediction	DWT, WPC KNN, RF, and SVM	Small dataset Does not apply deep learning algorithm. Signal to frequency domain Does not utilize absence seizure data.
I. Ullah *et al.* [24]	2018	Addressed ternary classification and small dataset problem.	An ensemble of pyramidal one-dimensional CNN. Augmentation based on window size and stride	Does not apply for absence seizure data.
S. Huang *et al.* [25]	2019	The input have been taken as vibration signal (simulated signal) in the mechanical operation process for feature extraction.	one-dimensional CNN (1D-CNN)	Does not utilize absence seizure data.

used for spike and wave detection (absence seizure detection). Based on this, the CGRUSVM algorithm has been applied on a single channel EEG database.

18.8 Materials and Methods

The data in the single-channel EEG database has been obtained from the clinical setup. The procedures have been given as EEG recording session usually takes 20–30 min including preparation time. In this paper, 16-channel montage with 21 electrodes is placed based on the international 10–20 system on to the scalp. Once the recording starts, the electrodes send an impulse from brain to the recording machine, and then, the machine converts impulse into visual patterns that have been saved on the computer. Further, saved patterns are analyzed for diagnosis and have been recorded using RMS acquisition head box with RMS acquisition module and RMS analysis modules for visualization. The basic configurations have been set up as analog to the digital with frequency (sampling) of 256 Hz and filtered with a filter [band pass of 1.0 Hz (LF) to 15 Hz (HF)] and a notch filter (50 Hz) to display recorded patterns. Based on this, the recorded signals of 51 subjects have been taken with the recording time of 30 min. The subject's age group differs from 4 to 15 years. The details of gender and category are tabulated in Table 18.2.

The recorded signal has been further processed using the MATLAB platform and different python and deep learning packages to find the category of the subject. The saved signal patterns have been exported to excel which acts as a database. Then, those data have been imported to the MATLAB platform for channel selection, input formulation, augmentation, and feature extraction and classification. By using this database, CNN, GRU, and CGRUSVM algorithms have been evaluated with respect to accuracy based on pre-processing strategy selection, feature extraction, and classification.

18.8.1 Pre-Processing Strategy Selection

In general, EEG signal is noisy with unwanted signals like artifacts of electrode displacement, cable movement, EMG from eye blink, and neck muscles, which are removed by Urigüen *et al.* [26] and Teplan [27], as pre-processing techniques. There are three strategies for removing artifacts; they are manual removal, automatic removal, and non-removal strategies. Some of them are independent component analysis, which have been used by Delorme *et al.* [28] for decomposition, which has been used

Table 18.2 Details of normal, abnormal, and absence subject.

Subject	Age	Gender	Category
Subject 1	7	Male	Abnormal
Subject 2	9	Female	Normal
Subject 3	12	Female	Abnormal
Subject 4	7	Male	Normal
Subject 5	15	Female	Abnormal
Subject 6	6	Female	Absence
Subject 7	8	Female	Normal
Subject 8	7	Male	Abnormal
Subject 9	5	Male	Normal
Subject 10	9	Male	Absence
Subject 11	12	Male	Abnormal
Subject 12	6	Male	Normal
Subject 13	11	Male	Abnormal
Subject 14	8	Female	Absence
Subject 15	8	Male	Normal
Subject 16	6	Male	Normal
Subject 17	14	Female	Abnormal
Subject 18	9	Male	Absence
Subject 19	8	Female	Normal
Subject 20	5	Female	Abnormal
Subject 21	12	Male	Normal
Subject 22	13	Male	Absence
Subject 23	10	Male	Abnormal
Subject 24	7	Male	Normal
Subject 25	9	Female	Absence

(Continued)

Table 18.2 Details of normal, abnormal, and absence subject. (*Continued*)

Subject	Age	Gender	Category
Subject 26	11	Male	Abnormal
Subject 27	6	Female	Normal
Subject 28	8	Female	Normal
Subject 29	13	Female	Normal
Subject 30	11	Female	Abnormal
Subject 31	5	Male	Normal
Subject 32	6	Male	Abnormal
Subject 33	12	Male	Normal
Subject 34	7	Female	Absence
Subject 35	15	Male	Abnormal
Subject 36	6	Female	Abnormal
Subject 37	8	Male	Absence
Subject 38	7	Female	Normal
Subject 39	5	Male	Abnormal
Subject 40	9	Female	Normal
Subject 41	12	Male	Abnormal
Subject 42	6	Female	Absence
Subject 43	11	Male	Normal
Subject 44	8	Female	Abnormal
Subject 45	8	Male	Normal
Subject 46	6	Female	Absence
Subject 47	14	Male	Abnormal
Subject 48	4	Male	Normal
Subject 49	14	Male	Abnormal
Subject 50	6	Male	Normal
Subject 51	7	Female	Abnormal

from early days; the spectral power fluctuations, which have been used by Kline *et al.* [29]; artifact subspace reconstruction (ASR), which have been used by Nathan *et al.* [30] as automatic artifact removal; adaptive noise canceling (ANC) scheme have been used by Kilicarslan *et al.* [31]; regression method has been used by Flumeri *et al.* [32]; wavelet transform has been used by Mahajan *et al.* [33]; empirical mode decomposition (EMD) has been used by Chavez *et al.* [34]; filtering methods have been used by Somers *et al.* [35]; and hybrid methods have been used by Chen *et al.* [36] and Wang *et al.* [37], which remove the artifacts from input data.

The most frequently used artifact removal methods have been discussed with respect to the pre-processing strategy and artifact removal method on single-channel input. The manual and automatic removal strategies are most used in general and also in seizure detection scenarios in the existing works when compared to non-removal strategy. Also, there was no comparative to analyze the non-removal or artifact removal process. Nowadays, to handle EEG signal with no-removal strategy, deep learning–based algorithms are more capable. So, in this paper, the single-channel EEG signal input has been taken based on the non-removal strategy.

18.8.2 Feature Extraction and Classification

As discussed in related works, lots of traditional feature extraction techniques exist in seizure detection applications. From that, a deep learning–based convolution layer have been chosen in this paper for absence seizure (spike and wave pattern) feature extraction and has not been used for absence seizure pattern detection in existing work using single-channel EEG input signal. In general, based on the frequency bands, EEG signals are divided into the low, medium, and high frequency with respect to peeks of the spectrum by Wen *et al.* [38]. The extracted features based on the frequency ranges as given above have been classified using GRU classifier into three classes as normal, abnormal, and absence seizure. In the end, SVM acts as a decision function to decide which segment belongs to which class. A deep learning algorithm based on spike and wave pattern detection workflow has been shown in Figure 18.1.

Before performing the feature extraction and classification, the inputs have been augmented. Augmentation is a strategy that helps researchers to populate the amount of data instead of collecting new data for training. In terms of signal processing, augmenting input has been carried out based on sliding window technique. In this work, incorporating stride concept (moving the window forward) along with window size (WS) in terms of deep learning has been used. Here, before augmenting input, there are 90

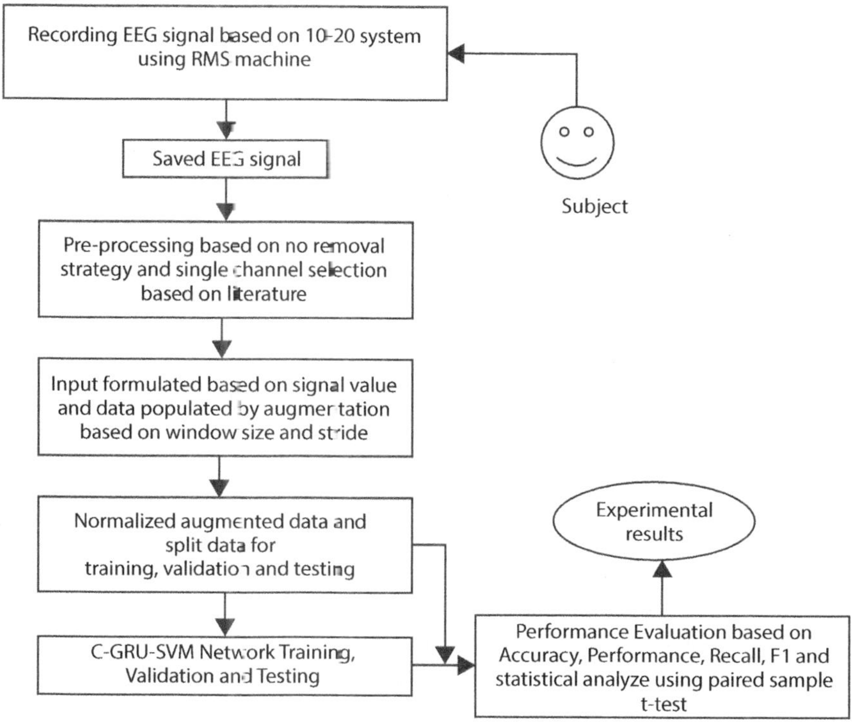

Figure 18.1 Deep learning–based absence seizure detection work flow.

instances that have been denoted as S_i that is not sufficient for training deep learning model. So, before augmenting, the data have been divided into training set and test set as 90% and 10% of total signal, respectively. The training set has been used for augmentation, which has been further used for training deep learning model. Assume the time series as with time and the cropping period (sampling period) remains constant ΔT. Here, WS is calculated in terms of seconds and overlapping (OL) duration of two consecutive window remains zero because it falls under non-overlapping (NOL) category. The number of instances is calculated as

$$S_i = \sum_{i=0}^{N} N * n_{seg} \tag{18.16}$$

where S_i is the augmented output
n_{seg} is the number of segments
N is the number of instances

The above proposed method has been used to populate the input data by consider the WS with 72 samples each and stride as 57 (i.e., the forward movement of the window) for augmentation. These have been applied on available input data to get populated input for training deep learning model. Before augmentation, 90 instances of 2,700 s are chosen from the EEG recording with the help of clinicians containing 368,730 sample points each, which means single instance consisting of 30 s with 4,097 sample points.

The quantity of instance is sufficient to train deep learning modal. So, the above method has been applied to augment these instances by cropping the data into 0.5-ms segment, which means WS of 72 sample and stride (moving window forward) of 57, which means cropped segments from each instance. In total, 5,130 instances are obtained after augmentation. These augmented inputs have been used for feature extraction and classification using deep learning model. The segmented first eight single instance as augmented output have been shown in Figure 18.2.

The convolution layer has been taken for feature extraction. Here, the convolution kernel plays a major role in extracting frequency components. In general, based on the frequency bands, EEG signals are divided into low, medium, and high frequency with respect to the peaks of the spectrum. They are as follows: low, 1.0 to 5.5 Hz; medium, 9.0 to 14.0 Hz; and high, 18.0 to 21.0 Hz. Also, the frequencies of normal, abnormal, and absence

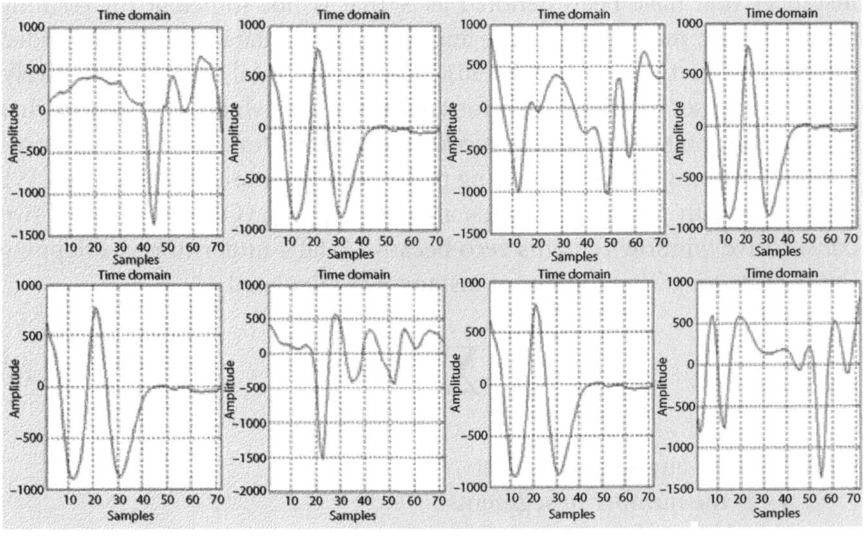

Figure 18.2 First eight segments of single instances after augmentation.

seizure EEG signal of three classes are 8 Hz and above, 7 Hz and less, and 3 Hz (spike and wave complexes), respectively. Based on these frequency components, the number of convolution layers used in this research is con_lay = 2 with con_fill = 10, 5 (Kernel) as the number of filters each. There are 5,130 instances with 368,730 sample points of three types of input signal being considered for training, and the leftover data have been considered for testing. The size of each kernel in two convolution layers is 1 × 5 and 1 × 3, which are applied to 1 × 72 input. In terms of deep learning, input is referred to as feature vector, output referred to as feature map, and kernels referred to as filters, and these are depicted in Figure 18.3.

The convolution layer output has been calculated by considering the single frame with sample size of 70.

$$\text{Convolution layer output} = \text{input size} - (\text{filter size} - 1)$$

$$\text{Input size} = 70$$

$$\text{Filter size} = 3$$

$$\text{Convolution layer output} = 70 - (3 - 1) = 68.$$

The learnable parameters of the convolution layer are weight and bias. The weight has been learned by network itself and bias = number of filters. Here, the size of input is 70 × 3 (height and width). Initially, the learnable parameter of the input layer is zero.

The convolution layers learnable parameters are calculated as follows:

$$\text{Learnable parameters of convolution layer} = \text{inputs} \times \text{outputs} + \text{biases}$$

$$\text{Conv\_param} = 3\times\ (3 \times 3 \times 10)$$

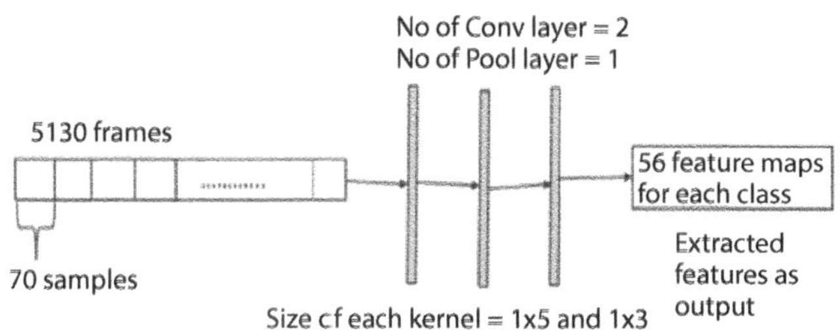

Figure 18.3 Feature extraction process with its parameters.

= 270 parameters

i.e., 3 (input) × (3 × 3) (size of filter) × 10 (no. of filter) + 10 (no. of filters).

If the last layer is convolutional layer, then input for the next layer is number of filters. Based on the above process, the features have been extracted using the different convolution kernels and a few outputs of convolution layer have been shown in frequency and time domain in Figure 18.4. The output of the convolution layers based on the above process is not suitable for calculation because of increase in the number of feature maps that lead to dimensionality reduction. This has been done by max pooling on each feature map for further process. The max pooling is the process of finding the maximum parameter from the pre-determined range as output. Based

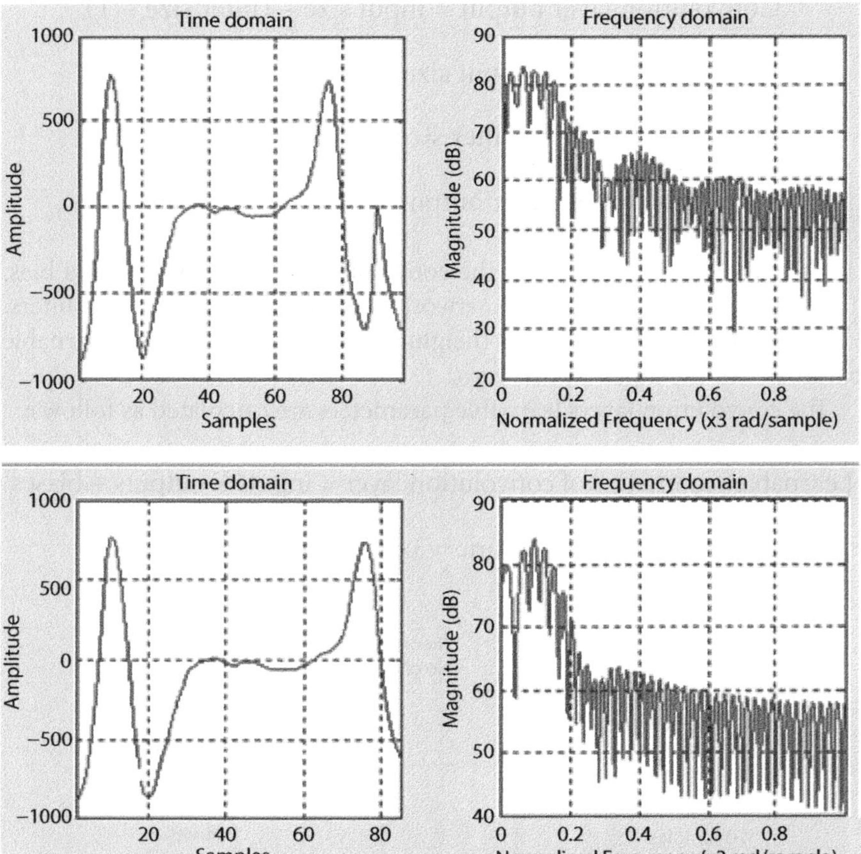

Figure 18.4 Convolution layer output of absence seizure pattern in time and frequency domain.

Table 18.3 Schematics of convolution layer.

Layer (type)	Output	Parameter
Input	(None, 70, 3)	0
Con_0	(None, 68, 3)	270
Con_1	(None, 66, 3)	910
Max_pool	(None, 56, 3)	0

on this, the obtained output is 56 feature maps, which have been extracted and used as input for classification. The schematics of the convolution layer have been shown in Table 18.3.

The feature classification categories extracted features into different classes. From the existing research work, it is observed that feature extraction and classification are two different processes. But in this work, with the help of deep learning technique, both the processes have been carried out using the single model. Here, the GRU has been used with different structure by combining both forget and input gates into a single gate as update gate and another one gate is reset gate. The reset gate and the update gate determine how to combine the new input with the previous data and how much of the previous data to keep around.

At each input of the sequence, the model considers not just the current input, but what it remembers about the preceding input. This model comprises hidden layer and fully connected (fc) layer with 100 GRU units and 100 hidden nodes. The fully connected layer is applied across the entire length of the input sequence in a time distributed fashion. This layer's output is fed into another hidden layer which consists of 100 GRU units. The output of the hidden layer is fed into the penultimate layer of SVM to find the class probability for the input data sequence with decision function of SVM. This process has been shown in Figure 18.5. The models loss has been calculated and the loss minimization done using Adam optimization. This repeated process repeats until it reaches the desired accuracy. The learnable parameters of GRU have been calculated as follows:

$$\text{Parameters of GRU layer} = 3(n^2+nm+n)$$

Here, n = Size of hidden units
m = Size of input

$$\text{With } m = 56, n = 100$$

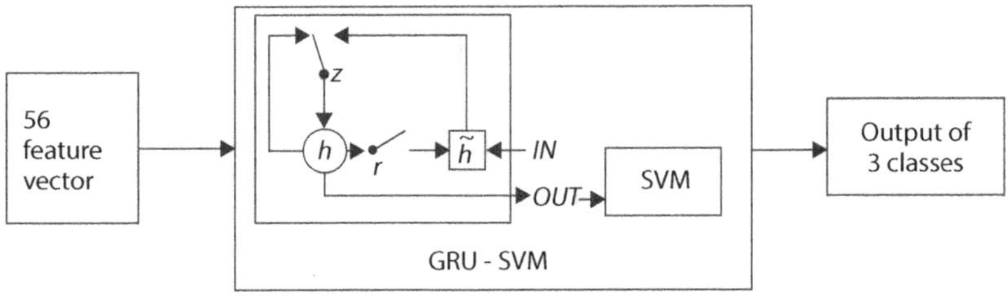

Figure 18.5 Working of GRU-SVM.

Table 18.4 Schematics of GRU-SVM layers.

Layer (type)	Output	Parameters
GRU_0	(56,100,1)	47,100
Fc	(56,100,1)	10,100
GRU_1	(56,100)	60,300
SVM_C	(56,3)	303

$$\text{Number of parameters} = 3(\,(100)^2+(100)(56)+100)$$
$$= 47{,}100$$

Here, the hyperparameters and learnable parameters play a major role. Hyperparameter are 1) input features, 2) no of hidden nodes and layers, 3) learning rate, and 4) activation function. Learnable parameter would be learned by the algorithm based on biases and weights. The hyperparameters and their values for C-GRU-SVM are as follows: convolution layer, 2 layers with 10 and 5 kernels; GRU, 2 hidden layer with 100 nodes; activation function, sigmoid and tanh; epoch, 200; optimizer, Adam; batch size, 70; and learning rate, 0.01 (rescale by 0.1 at each 100 epoch). The schematics of GRU-SVM have been depicted in Table 18.4.

18.9 Results and Discussions

The result of the CGRUSVM has been compared with the results of the CNN and GRU deep learning algorithms, which have been discussed in detail with the performance metrics. The metrics are precision, recall, F1,

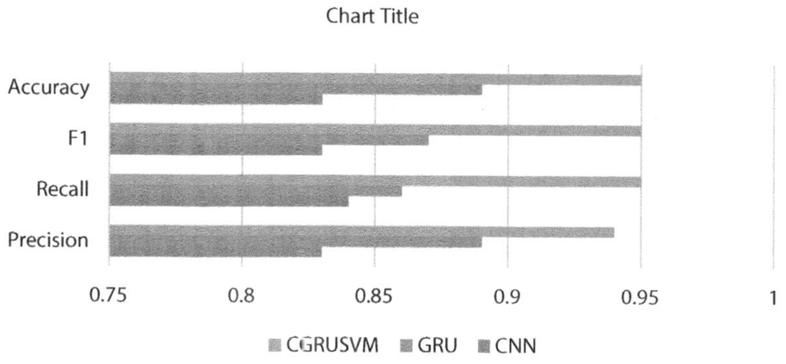

Figure 18.6 Performance of the classifiers.

and accuracy. These metrics are computed using the confusion matrix, which has been used for CGRUSVM algorithm as well as GRU and CNN algorithms, and the values are depicted in Figure 18.6.

Apart from these performance analyses, a statistical analysis technique has been used in this study to test whether a proposed model is different from other models with the help of a pared sample t-test. A paired sample t-test is used to compare the performance of two different machine learning or deep learning model. It consist of four main assumptions. These are as follows: (i) dependent variable must be continuous, (ii) observations are independent of one another, (iii) dependent variable should be approximately normally distributed, and (iv) dependent variable should not contain any outliers. Procedure for paired sample t-test consist of four steps by calculating:

1. The sample mean.
2. The sample standard deviation.
3. The test statistic.
4. Under the null hypothesis observe the probability of test statistics value by comparing t and t-distribution with (n – 1) degrees of freedom. The value can be found from a table in a statistical book or with statistical software.

The researchers decide a statistical significance cutoff value based on the calculated p-value. The usual value considered so far is 0.05 or less than 0.05. That means the variability of both the model is not the same if

Table 18.5 p-value of a classification model.

Model	p-value
C-RNN-SVM vs. CNN	0.021
C-RNN-SVM vs. GRU	0.042

the p-value is less than 0.05. The above statistical analysis technique has been widely used in the seizure detection domain for the comparison of a different models. Here, the proposed C-GRU-SVM model performed with an accuracy of 0.95; hence, we consider this model as a baseline. In this work, based on the above procedure, the p-value has been calculated to determine the difference between the two models using the formulation given in the following, and the computed values are tabulated in Table 18.5.

$$\text{Paired sample t-test} = m \,/\, (s/\sqrt{n}) \tag{18.17}$$

where t = t-statistics
m = mean of the sample
s = mean of the sample
n = sample size

18.10 Conclusion

The main goal of deep learning is almost the same as the other traditional learning methods, which learn seizure-related features from data and classify it to detect seizure subject. The major difference is that, in terms of processing signal input, the traditional learning method should have prior knowledge to build a feature set from the feature extraction process. Whereas in a deep learning approach, it enables training on the data without prior knowledge and the model is trained using to differentiate the seizure from the input. A traditional pre-processing method of the input signal may have a chance, which leads to an information loss. However, in the deep learning approach, this loss has been controlled by accepting the least processed input for automatic feature extraction and classification of seizure data to detect seizure subject without prior knowledge. In the future, multiple combinations of algorithms (deep learning) such as CNN, deep belief network, and LSTM (RNN) can be used along with SVM for absence seizure (spike and wave pattern) detection.

18.11 Acknowledgement

The authors thank Dr. S. Velusamy, DM - Neurology, MD - Pediatrics, MBBS - Neurology, and General Physician, who has 22 years of experience, for his continuous support throughout this work.

References

1. Siuly, S., Li., Y., Wen, P., Clustering technique-based least square support vector machine for EEG signal classification, in: *Computer Methods and Programs in Biomedicine in Elsevier Science*, vol. 104, pp. 358–372, 2008.
2. Walter., C., Cierniak., G., Gerjets., P., Rosenstiel., W., Bogdan, M., Classifying mental states with machine learning algorithms using alpha activty decline. *European Symposium on Artificial Neural Networks, Computational Intelligence and Machine Learning*, pp. 405–410, 2011.
3. Zihlmann., M. and Perekrestenko, D., Convolutional Recurrent Neural Networks for Electrocardiogram Classification, in: *IEEE CinC*, pp. 1–4, 2018.
4. Wei, X., Zhou, L. *et al.*, Automatic seizure detection using three dimensional CNN based on multi-channel EEG. *BMC Med. Inform. Decis. Mak.*, 2, 1–26, 2018.
5. Cun, Y.L., Convolutional networks for images, speech, and time series, in: *Handbook of Brain Theory & Neural Networks*, pp. 1–8, MIT Press, Cambridge, 1995.
6. Acharya, U.R., Oh, S.L., Hagiwara, Y., Tan, J.H., Adeli, H., Deep convolutional neural network for the automated detection and diagnosis of seizure using EEG signals. *Comput. Biol. Med.*, 100, 270–278, 2017.
7. Graves, A., *Generating sequences with recurrent neural networks*, vol. 13, ArXiv, New York, 2013.
8. Li, X., Song, D., Zhang, P., Yu, G., Hou, Y., Hu, B., Emotion recognition from multi-channel EEG data through convolutional recurrent neural network. *IEEE Int. Conf. Bioinformatics Biomedicine*, pp. 352–359, 2016.
9. Li, Y., Huang, J., Zhou, H., Zhong, N., Human emotion recognition with electroencephalographic multidimensional features by hybrid deep neural networks. *Appl. Sci.*, 7, 1060, 2017.
10. Tsiouris, K.M., Pezoulas, V.C., Zervakis, M., Konitsiotis, S., Koutsouris, D.D., Fotiadis, D., II, A long short-term memory deep learning network for the

prediction of epileptic seizures using EEG signals. *Comput. Biol. Med.*, 99, 24–37, 2017.
11. Talathi, S., *Deep Recurrent Neural Networks for seizure detection and early seizure detection systems*, arxiv, New York, 2017.
12. Kline, J.E. and Helen, Isolating gait-related movement artifacts in electroencephalography during human walking. *J. Neural Eng.*, 15, 26–44, 2015.
13. Kilicarslan, G., A robust adaptive de-noising framework for real-time artifact removal in scalp EEG measurements. *J. Neural Eng.*, 6, 14–21, 2016.
14. Wang, L. and Xue, W., Automatic Epileptic Seizure Detection in EEG Signals using Multi-Domain Feature Extraction and Nonlinear Analysis. *Entropy*, 19, 2–22, 2017.
15. Lasefr, Z., Shiva, S., Ayyalasomayajula, V.N.R., Elleithy, K., An Efficient Automated Technique for Epilepsy Seizure Detection Using EEG signals. *IEEE*, vol. 142, 2017.
16. Yıldız, Bergil, The Investigation of Channel Selection Effects on Epileptic Analysis of EEG Signals. *Balkan J. Electr. Comput. Eng.*, 15, 236–241, 2015.
17. Hussein, R. and Palangiet, H., Optimized deep neural network architecture for robust detection of epileptic seizures using EEG signals. *Clin. Neurophysiol.*, 19, 25–37, 2019.
18. Wei, X., Zhou, L. *et al.*, Automatic seizure detection using three dimensional CNN based on multi-channel EEG. *BMC Med. Inform. Decis. Mak.*, 18, 1–26, 2018.
19. Hussein, R. and Palangiet, H., Optimized deep neural network architecture for robust detection of epileptic seizures using EEG signals. *Clin. Neurophysiol.*, 19, 25–37, 2019.
20. Kjaer, T. and Sorensen, H., Detection of Paroxysms in Long-Term, Single-Channel EEG-Monitoring of Patients with Typical Absence Seizures. *Neurovascular devices Syst.*, 17, 26–49, 2017.
21. Alickovica, E. and Kevricb, J., Performance evaluation of empirical mode decomposition, discrete wavelet transform, and wavelet packed decomposition for automated epileptic seizure detection and prediction. *Biomed. Signal Process. Control*, 39, 94–102, 2018.
22. Ullah, I. and Hussain, M., An Automated System for Epilepsy Detection using EEG Brain Signals based on Deep Learning Approach, in: *Expert Systems with applications*, vol. 107, pp. 61–71, Elsevier, New York, 2018.
23. Huang, S. and Tang, J., Signal Status Recognition Based on 1DCNN and Its Feature Extraction Mechanism Analysis. *Sensors*, 19, 9, 1–19, 2019.
24. Segan, S., MD Director of SBH Stroke Center and Attending Neurologist, St Barnabas Hospital, in: *Absence Seizures: Overview, Etiology, Epidemiology, Medscape*, 2018.
25. Albuja, A.C. and Murphy, P.B., *Absence Seizure*, Treasure Island (FL): StatPearls Publishing, Florida, 2020.
26. Urigüen, J.A. and Garciazapirain, B., EEG artifact removal - State-of-the-art and guidelines. *J. Neural Eng.*, 12, 3, 3–12, 2015.

27. Albuja, A.C. and Murphy, P.B., *Absence Seizure*, Treasure Island (FL): StatPearls Publishing, Florida, 2021.
28. Delorme, A., Sejnowski, T., Makeig, S., Enhanced detection of artifacts in EEG data using higher-order statistics and independent component analysis. *J. Neural Eng.*, 9, 16, 14–43, 2019.
29. Kline, J.E., Huang, H.J., Snyder, K.L., Ferris, D.P., Isolating gait-related movement artifacts in electroencephalography during human walking. *J. Neural Eng.*, 12, 4–22, 2015.
30. Nathan, K. and Contreras-Vidal, J.L., Negligible motion artifacts in scalp electroencephalography (EEG) during treadmill walking. *Front. Hum. Neurosci.*, 9, 1–12, 2016.
31. Kilicarslan, A., Grossman, R.G., Contreras-Vidal, J.L., A robust adaptive denoising framework for real-time artifact removal in scalp EEG measurements. *J. Neural Eng.*, 3, 13, 2–13, 2016.
32. Flumeri, G.D., Aricó, P., Borghini, G., Colosimo, A., Babiloni, F., A New Regression-based Method for the Eye Blinks Artifacts Correction in the EEG Signal, without Using Any EOG Channel, in: *Proceedings of the 2016 38th Annual International Conference of the IEEE Engineering in Medicine and Biology Society (EMBC)*, pp. 16–20, 2016.
33. Morshed, B.I., Unsupervised eye blink artifact denoising of EEG data with modified multiscale sample entropy, kurtosis, and wavelet-ICA. *IEEE J. Biomed. Health Inform*, 19, 158–165, 2015.
34. Chavez, M., Grosselin, F., Bussalb, A., Fallani, F.D.V., Navarro-Sune, X., Surrogate-based artifact removal from single channel EEG. *IEEE Trans. Neural Syst. Rehabil. Eng.*, 26, 540–550, 2018.
35. Somers, B., Francart, T., Bertrand, A., A generic EEG artifact removal algorithm based on the multi-channel Wiener filter. *J. Neural Eng.*, 15, 25–52, 2018.
36. Chen, X., He, C., Peng, H., Removal of muscle artifacts from single-channel EEG based on ensemble empirical mode decomposition and multiset canonical correlation analysis. *J. Appl. Math.*, 14, 1–10, 2014.
37. Wang, G., Teng, C., Li, K., Zhang, Z., Yan, X., The removal of EOG artifacts from EEG signals using independent component analysis and multivariate empirical mode decomposition. *IEEE J. Biomed. Health Inf.*, 20, 1301–1308, 2016.
38. Lee, S., McKeown, M.J., Wang, Z.J., Chen, X., Removal of high-voltage brain stimulation artifacts from simultaneous EEG recordings. *IEEE Trans. Biomed. Eng.*, 66, 50–60, 2019.

Index

Printed and bound by CPI Group (UK) Ltd, Croydon, CR0 4YY
17/09/2026
14964247-0007